# Earth Surface Science: Properties, Processes and Interactions

# Earth Surface Science: Properties, Processes and Interactions

Edited by **Matt Weilberg**

⬚ SYRAWOOD
PUBLISHING HOUSE

New York

Published by Syrawood Publishing House,
750 Third Avenue, 9th Floor,
New York, NY 10017, USA
www.syrawoodpublishinghouse.com

**Earth Surface Science: Properties, Processes and Interactions**
Edited by Matt Weilberg

International Standard Book Number: 978-1-68286-019-9 (Hardback)

Printed in the United States of America.

# Contents

# Preface

This book has been a concerted effort by a group of academicians, researchers and scientists, who have contributed their research works for the realization of the book. This book has materialized in the wake of emerging advancements and innovations in this field. Therefore, the need of the hour was to compile all the required researches and disseminate the knowledge to a broad spectrum of people comprising of students, researchers and specialists of the field.

Earth surface science helps in understanding the structure and nature of the various layers that form the earth's surface. This subject area integrates chemistry, physics, mathematics, and biology to develop a comprehensive understanding about the earth. The topics included in this book on earth surface science, such as remote sensing, interactions between various layers such as lithosphere, biosphere, hydrosphere, etc. are of utmost significance and bound to provide incredible insights to readers. This book is a complete source of knowledge on the present status of this important field and is highly recommended for students and researchers pursuing earth sciences.

At the end of the preface, I would like to thank the authors for their brilliant chapters and the publisher for guiding us all-through the making of the book till its final stage. Also, I would like to thank my family for providing the support and encouragement throughout my academic career and research projects.

**Editor**

# The mass distribution of coarse particulate organic matter exported from an Alpine headwater stream

J. M. Turowski[1,2], A. Badoux[1], K. Bunte[3], C. Rickli[1], N. Federspiel[1,4], and M. Jochner[1,5]

[1]Swiss Federal Research Institute WSL, Zürcherstrasse 111, 8903 Birmensdorf, Switzerland
[2]Helmholtz Centre Potsdam, GFZ German Research Centre for Geosciences,
Telegrafenberg, 14473 Potsdam, Germany
[3]Engineering Research Center, Colorado State University, Fort Collins, CO 80523, USA
[4]CSD Engineers SA, Hessstrasse 27d, 3097 Liebefeld (Berne), Switzerland
[5]Institute of Geography of the University of Berne (GIUB), Hallerstrasse 12, 3012 Berne, Switzerland

*Correspondence to:* J. M. Turowski (turowski@gfz-potsdam.de)

**Abstract.** Coarse particulate organic matter (CPOM) particles span sizes from 1 mm, with a dry mass less than 1 mg, to large logs and entire trees, which can have a dry mass of several hundred kilograms. Pieces of different size and mass play different roles in stream environments, from being the prime source of energy in stream ecosystems to macroscopically determining channel morphology and local hydraulics. We show that a single scaling exponent can describe the mass distribution of CPOM heavier than 0.1 g transported in the Erlenbach, a steep mountain stream in the Swiss pre-Alps. This exponent takes an average value of $-1.8$, is independent of discharge and valid for particle masses spanning almost seven orders of magnitude. Similarly, the mass distribution of in-stream large woody debris (LWD) in several Swiss streams can be described by power law scaling distributions, with exponents varying between $-1.8$ and $-2.0$, if all in-stream LWD is considered, and between $-1.3$ and $-1.8$ for material locked in log jams. We found similar values for in-stream and transported material in the literature. We had expected that scaling exponents are determined by stream type, vegetation, climate, substrate properties, and the connectivity between channels and hillslopes. However, none of the descriptor variables tested here, including drainage area, channel bed slope and the percentage of forested area, show a strong control on exponent value. Together with a rating curve of CPOM transport rates with discharge, the scaling exponents can be used in the design of measuring strategies and in natural hazard mitigation.

## 1 Introduction

Coarse particulate organic matter (CPOM) plays multiple roles in stream systems. Usually defined as pieces of organic matter with a diameter larger than 1 mm (e.g., Fisher and Likens, 1972, 1973; Cummins, 1974; Webster et al., 1999), it spans the range from leaves, needles and wood fragments to twigs and branches to logs and entire trees. Large woody debris (LWD), at the top of this range, is often defined as having a minimum diameter of 0.1 m and a minimum length of 1 m (e.g., Abbe and Montgomery, 2003; Faustini and Jones, 2003; Mao et al., 2008; Wohl and Jaeger, 2009), but others have used minimum lengths of 0.5, 1.5 or 3 m, and various di-

ameters (e.g., Hassan et al., 2005; Jackson and Sturm, 2002; Martin and Benda, 2001). Each size group within the wide range of CPOM sizes fulfills different geomorphic and ecological roles (e.g., Harmon et al., 1986). Leaves and twigs and their decay products are consumed by in-stream shredders and suspension feeders. This so-called allochthonous organic matter, which is produced outside of the stream system, enters the channel, for example, as litter fall from adjoining forest. Especially in headwater catchments it is the prime source of energy in stream ecosystems (e.g., Fisher and Likens, 1973; Minshall, 1967). Larger pieces of woody debris change flow hydraulics, and thus have an effect on flow velocity and sediment transport rates, which affects habitat and breeding grounds for fish and other aquatic animals (e.g., Abbe and Montgomery, 1996; Bilby and Ward, 1991; Brooks et al., 2004; Keller and Swanson, 1979; MacFarlane and Wohl, 2003; Montgomery and Piégay, 2003). Log jams are often a major element of stream morphology, and floating logs may pose a natural hazard (e.g., Comiti et al., 2006; Curran, 2010; Kraft and Warren, 2003; Manga and Kirchner, 2000; Mao et al., 2008; Mazzorana et al., 2011; Rickenmann, 1997).

Wood budgets are a common tool in geomorphology to assess availability and transport of woody material, but often focus on LWD (e.g., Benda and Sias, 2003; Keller and Swanson, 1979; Martin and Benda, 2001; Reeves et al., 2003). Budgets of smaller organic material are necessary in ecological studies to assess the availability of food in a stream (e.g., Fisher and Likens, 1973; Webster and Meyer, 1997), but typically focus on small streams where LWD rarely moves. Studies investigating the entire size range of transported material, from leaves to logs, are rare. It is currently unclear how the different CPOM size groups, especially LWD and leaf-size fractions, relate to each other. Size distributions have been reported for LWD only, either based on piece diameter or volume of material stored in the stream (Harmon et al., 1986; Hogan, 1987; Jackson and Sturm, 2002; Rickli and Bucher, 2006), or piece length of transported material (MacVicar and Piégay, 2012). Here, we hypothesize that the use of dry mass as a descriptor variable leads to a scaling relation consistently connecting all CPOM size groups transported by a stream, from leaves to large logs. We have measured transport rates and dry masses of CPOM pieces heavier than 0.1 g moving in the Erlenbach, a headwater stream in the Swiss pre-Alps, using several sampling methods over a large range of discharges. In addition, we collected data on in-stream LWD size distributions for the Erlenbach, and compared it to data from ten other mountain streams in Switzerland (Rickli and Bucher, 2006), and from the literature. We discuss the use of scaling relations in the design of sampling campaigns, data analysis and natural hazard mitigation.

## 2 Field site

The Erlenbach is a small headwater stream located near Einsiedeln in the Swiss pre-Alps (Fig. 1), where scientific observations have been conducted since the late 1960s (Hegg et al., 2006). The channel bed has a mean gradient of around 18 % and drains an area of 0.7 km$^2$ at the main observation site. About 40 % of the total catchment area is forested, with the remaining 60 % consisting of wetland and Alpine meadows. The forest dominantly comprises Norway spruce (*Picea abies*) and European Silver Fir (*Abies alba*), intermingled with some Alder (*Alnus sp.*), and a wide variety of shrubs and ground plants (Schleppi et al., 1999). The Erlenbach is a step-pool channel with high sediment load, which is mainly

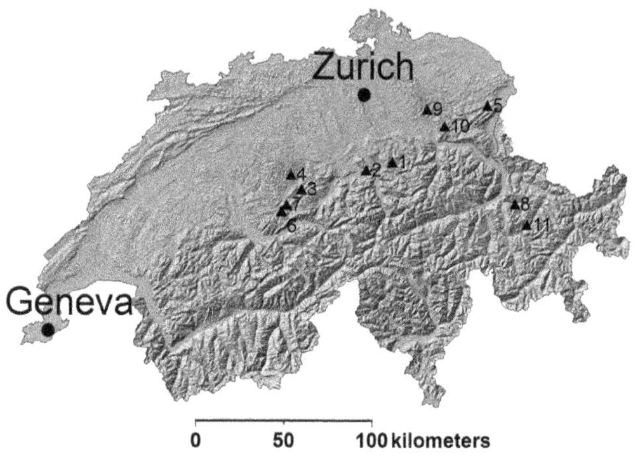

**Figure 1.** Location map of the Erlenbach (1) and the ten streams studied by Rickli and Bucher (2006) in Switzerland (see also Table 1).

supplied by a series of slow-moving landslides along the channel banks (Schuerch et al., 2006; Turowski et al., 2009; Molnar et al., 2010). There are two discharge gauges located immediately upstream and downstream of a sediment retention basin where automatic basket samplers and indirect bedload sensors for sediment transport measurements are available (Rickenmann et al., 2012; Turowski et al., 2011). Discharge is continuously recorded at 10 min intervals, and at 1 min intervals during bedload transport events. Unless otherwise stated, we used the 10 min data of the upper gauge throughout this paper. The mean discharge at the Erlenbach is 39 L s$^{-1}$, and during dry weather it is typically below 10 L s$^{-1}$. Floods, driven mainly by convective summer storms, are common, and stream flow quickly responds to heavy rainfall. The annual return discharge is approximately 2000 L s$^{-1}$. The highest discharge recorded in the last 30 yr occurred on 20 June 2007. At ∼ 11 100 L s$^{-1}$ (10 min data), its return frequency is estimated at around 50 yr (Turowski et al., 2009, 2013).

## 3 Methods

Transported CPOM was sampled using three different methods, ensuring the possibility of sampling a wide range of particle sizes and discharges (Fig. 2). At low to intermediate discharges (from 1 to ∼ 1000 L s$^{-1}$, with most samples below 250 L s$^{-1}$), when the stream was wadeable, we used bedload traps (Bunte et al., 2004, 2007). Automatic basket samplers were used at higher discharges (from 200 to 1500 L s$^{-1}$, with most samples above 400 L s$^{-1}$) (Rickenmann et al., 2012). Two samples were obtained from measuring woody material accumulated in the retention basin after two large floods that occurred in 1995 and 2010 (cf. Turowski et al., 2009, 2013).

Bedload traps are portable samplers installed on metal plates placed on the channel bed. They were specifically

**Figure 2.** (left panel) Typical CPOM pieces caught in the basket samplers, including twigs, chips of woody material, bark fragments, scales of conifer cones, and alder catkins. The typical dry mass of the pieces shown ranges from a tenth of a gram up to a few grams. (right panel) LWD material caught in the outlet grid of the retention basin in the extreme event from 1 August 2010 (picture taken on the 2 August 2010). The retention basin and the basket samplers can be seen in the background.

designed for obtaining samples of gravel bedload (Bunte et al., 2004, 2007), but are also suitable for sampling CPOM. Bedload traps consist of an aluminium frame, 0.3 m wide and 0.2 m high, to which a net 1–2 m long is attached. Bedload trap samples were obtained from three channel locations across a step-pool unit located ~ 20 m upstream of the discharge gauge: at the crown and the foot of a step and at the end of the associated pool. Bedload traps sample the whole flow depth and sampling efficiency is mainly determined by the minimum and maximum size of particles that can be sampled. In the flows the traps were used for, it was possible to sample the largest transported CPOM particles. A single trap was sufficient to cover the entire flow width at the crown and the foot of the step. At the pool exit, two traps were used to sample the flow width of approximately 2.6 m, one located near the thalweg, the other on a submerged gravel bank. The transport rates measured with these two traps were assumed to be representative for unmeasured parts of the channel cross section, in the main flow channel and on the gravel bank, respectively. To obtain transport rates over the whole flow width, we divided the transported mass by trap width and multiplied by the width of the relevant section (i.e., the gravel bank or the main channel), and added the values for both sections. Interpolated transport rates for the whole channel width measured by the two traps at the pool exit were assumed to be representative for the entire channel, as they were similar to transport rates measured at the crown and the foot of the step at similar discharges. The traps used at the Erlenbach have a net opening size of 6 mm. At most discharges, we sampled for 10–20 min. However, the sampling time was adjusted to match the supply rate of material (both bedload and CPOM). At the highest discharges of around $1000 \, \text{L s}^{-1}$, sampling times were reduced to a few minutes to avoid overfilling of the nets. At the lowest discharges ($< 10 \, \text{L s}^{-1}$) sampling times of up to nearly seven hours were needed to collect a sufficient amount of material. Samples were taken in spring 2012 during snowmelt events, and in summer 2012 at higher

discharges in rainfall-driven flow events and at baseflow discharges.

The automatic basket samplers consist of metal cubes with 1 m edges (Rickenmann et al., 2012, see also Fig. 2b). The baskets' walls and floors are made of a metal mesh of square holes with an edge length of 10 mm. Sampling is triggered when thresholds of stage level and indirect bedload sensor activity are exceeded. The entire flow width is sampled at discharges below around $1500 \, \text{L s}^{-1}$, and the opening size of the baskets is large enough to capture the largest CPOM particle sizes transported in these flows. Sampling efficiency is determined mainly by the lower limit of sampled sizes (10 mm). Basket samples were taken from May to November of 2009 to 2012 during rainfall-driven floods.

Organic matter from basket and trap samples was separated from clastic material and weighed wet in the field. Subsequently, the samples were oven-dried over 24 h at 80 °C in the laboratory, and the dry mass was obtained. For a total of 18 out of 42 basket samples, all pieces of organic matter heavier than about 1 g were individually weighed and measured. For ten samples taken at lower background discharges with the bedload traps, we individually weighed all pieces heavier than 0.1 g, or all pieces heavier than 0.01 g, if the total number of pieces in the sample was small.

After two large floods, in July 1995 and August 2010 with peak discharges of $\sim 10\,100 \, \text{L s}^{-1}$ and $\sim 9300 \, \text{L s}^{-1}$, respectively (10 min data) (Turowski et al., 2009, 2013, see also Fig. 2b), pieces of woody debris with diameters larger than 0.05 m were collected from the sediment retention basin, and their lengths and diameters were measured in the field. The measurements were converted to mass assuming a cylindrical shape and a dry wood density of $410 \, \text{kg m}^{-3}$, which is typical for the Norway spruce (*Picea abies*) that is common in the catchment. Substantial volumes of LWD have been observed in the retention basin after events that exceeded this discharge, whereas only a few LWD pieces have been found in the retention basin after events with lower peaks. Thus,

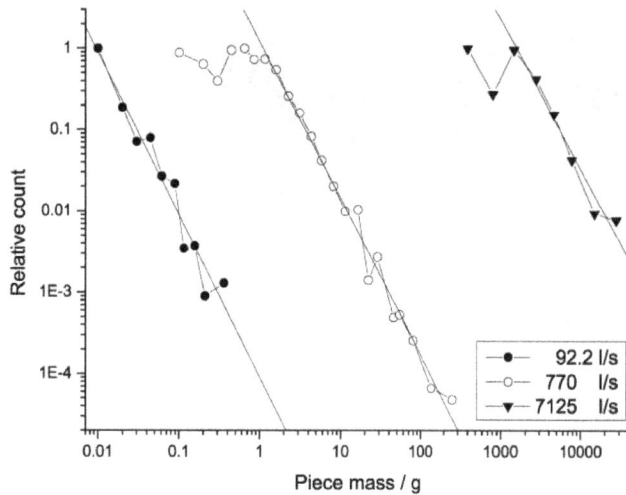

**Figure 3.** Examples of histograms of CPOM particle masses at three different discharges, spanning two orders of magnitude in discharge and more than six orders of magnitude in particle mass. To reduce the extent of the axes, and to demonstrate the general similarity of the CPOM piece count vs. mass relations, each of the histograms was normalized such that the most common fraction plots at a relative count of one. The sample collected at $92.2\,\mathrm{L\,s^{-1}}$ is typical of those taken with bedload traps, the one at $770\,\mathrm{L\,s^{-1}}$ of those with basket samplers, and the one at $7125\,\mathrm{L\,s^{-1}}$ was collected from the debris basin.

to obtain a representative discharge for the retention basin data points, we assumed that large pieces of wood are dominantly transported at discharges higher than $5000\,\mathrm{L\,s^{-1}}$, using measurements at 1 min resolution. The $5000\,\mathrm{L\,s^{-1}}$ threshold was exceeded for 20 and 18 min, respectively, during the two events, and accordingly we averaged all discharge measurements exceeding it.

In addition to sampling material transported by the stream, the length and diameter of all woody debris pieces longer than 1 m (LWD) stored in log jams were recorded in the field. LWD masses were calculated assuming a cylindrical shape and a dry wood density of $410\,\mathrm{kg\,m^{-3}}$.

## 4   Results and analysis

The frequency of occurrence of sampled CPOM particles is tightly related to their masses; the number of CPOM pieces in transport strongly decreased with increasing particle mass. At particle masses greater than a threshold value of $\sim 1\,\mathrm{g}$ for the basket samples and $\sim 3\,\mathrm{kg}$ for the retention basin samples, the relative frequencies of occurrence of masses in the samples plot on a straight line in log–log space (Fig. 3). No such threshold was observed for the bedload trap data because the smaller net size trapped finer particles. The relationships can be described by the equation

$$C = kM^{-\alpha}. \tag{1}$$

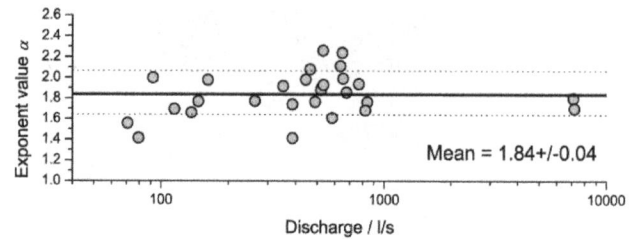

**Figure 4.** Scaling exponents of the mass distribution as function of discharge for the Erlenbach samples. No trend is visible. The solid line gives the mean value, dotted lines depict one standard deviation around the mean.

Here, $C$ is the relative fraction of CPOM of the respective particle mass $M$, $k$ is a constant and $\alpha$ is the scaling exponent. The relative fraction is proportional to the measured fraction of CPOM in each mass class, but may be normalized for example by bin width. For each CPOM sample, scaling exponents were obtained by fitting a linear relation to the log-transformed data in the long falling branch. Minimum and maximum values of $\alpha$ were 1.41 and 2.26, respectively, and the mean of 28 samples was $1.84 \pm 0.04$ (range gives standard error of the mean). For all discharges where samples contained enough individual pieces of material to derive a relation between piece numbers and mass, scaling exponents were independent of discharge (Fig. 4).

By dividing total CPOM mass by sampling time, we obtained CPOM transport rates. These show a clear relation with discharge, defined by a straight line in log–log space (Fig. 5, grey symbols). However, it needs to be borne in mind that we used different sampling devices with different net sizes (bedload traps: 6 mm; basket samplers: 10 mm; LWD from the retention basin). Below we will show how the scaling relation of Eq. (1) can be used to make the samples comparable.

As long as the scaling exponent $\alpha > 1$, the total mass $M_{\mathrm{tot}}$ of the sample above a minimum mass $M_{\mathrm{min}}$ can be obtained by using the integral

$$M_{\mathrm{tot}} = \int_{M_{\mathrm{min}}}^{\infty} kM^{-\alpha}\,\mathrm{d}M = \frac{kM_{\mathrm{min}}^{1-\alpha}}{\alpha - 1}. \tag{2}$$

The ratio between the total mass $M_{1,\mathrm{tot}}$ larger than a threshold mass $M_{1,\mathrm{min}}$, and the total mass $M_{2,\mathrm{tot}}$ larger than a threshold mass $M_{2,\mathrm{min}}$ can thus be written as

$$\frac{M_{1,\mathrm{tot}}}{M_{2,\mathrm{tot}}} = \left(\frac{M_{1,\mathrm{min}}}{M_{2,\mathrm{min}}}\right)^{1-\alpha}. \tag{3}$$

To be able to compare the samples taken with bedload traps and basket samplers, we estimated the minimal particle mass that is sampled representatively by the baskets at 1 g. The lack of a further increase in relative counts for smaller particles indicates that not all material of smaller masses was

**Table 1.** Characteristics of the Swiss streams and scaling exponents (see also Fig. 1).

| | Stream | Community, Canton[1] | Drainage area (km²) | % forested | Mean elevation (m a.s.l.) | Channel bed slope | Mean channel width (m) | Scaling exponent (transport) | Scaling exponent (in stream)[2] | Scaling exponent (log jams)[2] |
|---|---|---|---|---|---|---|---|---|---|---|
| 1 | Erlenbach | Brunni, SZ | 0.7 | 40 | 1347 | 0.105 | 4.0 | 1.84 | N.A.[3] | 1.26 (79) |
| 2 | Brüggenwaldbach | Gersau, SZ | 0.81 | 33 | 804 | 0.341 | 8.7 | | 2.04 (472) | 0.93 (14) |
| 3 | Steinibach | Flühli, LU | 1.49 | 17 | 1160 | 0.164 | 8.8 | | 1.89 (679) | 1.75 (106) |
| 4 | Seeblibach | Romoos, LU | 1.16 | 47 | 950 | 0.104 | 6.3 | | 1.85 (781) | 1.68 (173) |
| 5 | Ibach | Weissbad, AI | 1.64 | 26 | 880 | 0.070 | 8.3 | | 1.90 (485) | 1.64 (112) |
| 6 | Büetschligraben | Schangnau, BE | 2.24 | 18 | 1040 | 0.155 | 10.9 | | 1.86 (567) | 1.64 (33) |
| 7 | Steiglebach | Marbach, LU | 3.02 | 40 | 1180 | 0.074 | 10.2 | | 1.85 (572) | 1.39 (177) |
| 8 | Grossbach | Molinis, GR | 2.40 | 38 | 1190 | 0.278 | 11.2 | | 1.88 (687) | 0.97 (8) |
| 9 | Chreuelbach | Goldingen, SG | 0.88 | 65 | 920 | 0.132 | 6.9 | | 1.78 (857) | 1.69 (98) |
| 10 | Geissbach | Ebnat-Kappel, SG | 1.63 | 45 | 1080 | 0.095 | 9.5 | | 1.84 (749) | 1.52 (104) |
| 11 | Ursprung | Wiesen, GR | 1.33 | 50 | 1610 | 0.089 | 8.9 | | 1.87 (879) | 1.71 (252) |
| | Literature data | Source | | | | | | | | |
| | Ain River, France | MacVicar and Piégay (2012) | 3630 | | | 0.0015 | 65 | 1.8 | | |
| | small, British Columbia | Hogan (1987) | 3.9 | | | 0.024 | 18.8 | | 1.72 | |
| | medium, British Columbia | Hogan (1987) | 6.9 | | | 0.013 | 20.2 | | 1.61 | |
| | large, British Columbia | Hogan (1987) | 20.2 | | | 0.007 | 29.3 | | 1.90 | |

[1] See Fig. 1; [2] the number of data points used in the fit is given in brackets; [3] in the Erlenbach, only material in log jams was surveyed.

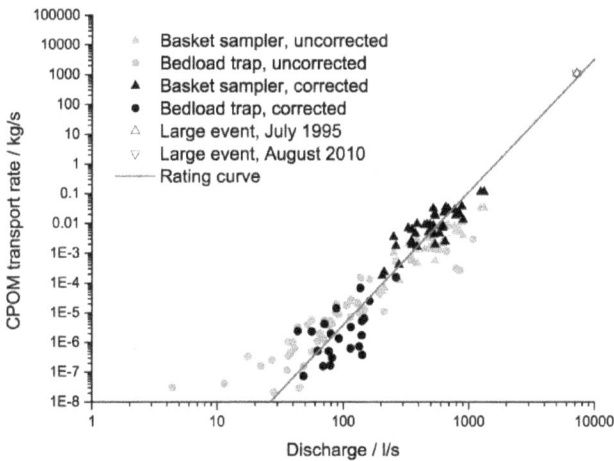

**Figure 5.** CPOM transport rates as a functions of discharge. Both the untreated data (light grey) and the values corrected for transport rates of all particles heavier than 0.1 g (black) are shown. The rating curve is of the form $Q_{CPOM} = aQ^b$, with $a = 4.42 \times 10^{-15}$ and $b = 4.47 \pm 0.21$, with an $R$ value of 0.94. The two data points from the large events (open triangles) plot nearly at the same location. They were not used in the regression to obtain the rating curve.

The corrected samples were plotted against discharge to obtain a rating curve (Fig. 5, black symbols). The rating exponent of 4.47 was obtained by fitting a linear relationship to the log-transformed data. The data points from the two extreme events were not included in the fit. They are the least reliable data points, because only particles heavier than about 3 kg (dry mass) were sampled representatively, and the measured masses of 237 (1995 event) and 219 kg (2010 event) were subsequently extrapolated down to piece masses of 0.1 g using the average scaling exponent of 1.84 (Fig. 4). The resulting points plot close to the rating curve derived from the trap and basket measurements and confirm the validity of the CPOM transport rating relation at high discharges (Fig. 5).

In nine log jams along the Erlenbach channel, we measured a total of 79 pieces of LWD. The size distribution of this material likewise decreased with particle mass (Fig. 6), although with a value of 1.26 the scaling exponent was lower than observed for transported material. Scaling exponents similar to those from the Erlenbach were also obtained from 6700 wood pieces measured in ten other Swiss mountain streams (Rickli and Bucher, 2006; see Table 1). Power function scaling exponents ranged within 1.78 and 2.04 for all wood pieces in the channel and within 1.39 and 1.75 for those locked in log jams (Fig. 7) (the two extraordinary low values at the Brüggenwaldbach and the Grossbach were derived from small sample sizes and are therefore unreliable). Scaling exponents appeared unrelated to basin area size, stream width and gradient, and the percent forested area (Fig. 8).

sampled (see Fig. 3). Particles lighter than 1 g were omitted from the samples, and the remainder was used for further analysis. For basket samples where the mass distribution was not measured, we used the average fraction of mass contributed by 1 g particles to the total mass (= 0.48) from the basket samples where it was measured. This value was multiplied with the total measured CPOM mass. For each basket sample, the total mass above 0.1 g was then calculated using Eq. (3). For the bedload trap samples, we only used samples for which the mass distribution was measured and which included individual pieces heavier than 0.1 g.

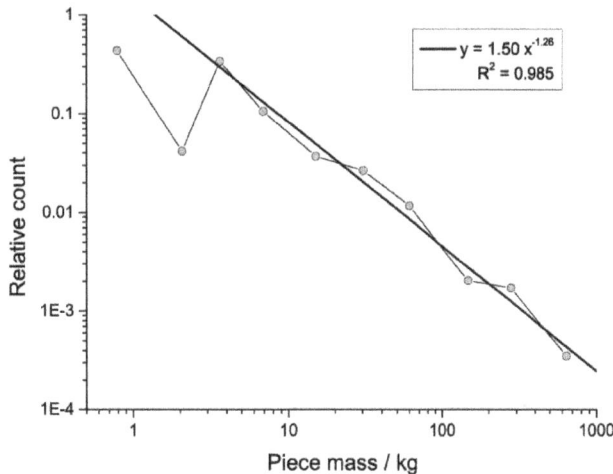

**Figure 6.** Scaling distributions of piece mass for LWD locked in log jams in the Erlenbach channel.

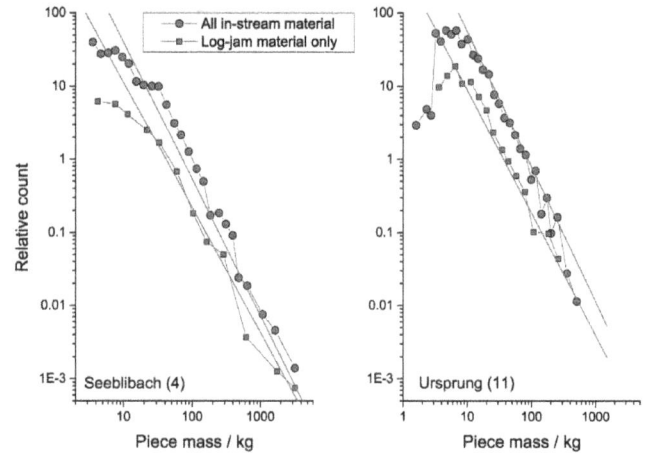

**Figure 7.** Scaling distributions of piece mass for LWD in the Seeblibach (4) and Ursprung (11) (see Table 1), as examples for distributions observed in the Swiss mountain streams (Rickli and Bucher, 2006). Distributions both for material locked in log jams (square symbols) and for all material (circles) stored in the channel are shown.

## 5  Discussion

### 5.1  Masses of CPOM input and piece break down

Various processes in the stream work together to produce the observed mass distribution of CPOM particles from the original mass distribution of organic material supplied to the stream. Coarse particulate organic matter enters the stream either as litter fall directly from the trees or blown in by wind, or via the stream banks either as material advected into the channel by landslides and snow creep, or flushed into it by overland flow (cf. Benda and Sias, 2003; Fisher and Likens, 1973; Hassan et al., 2005; Harmon et al., 1986). Litter typically comprises leaves with dry masses below 0.1 g and twigs with masses smaller than about 5 g. Larger material, often entire trees, can enter the channel due to wind throw or by riding on top of landslides. A number of slow-moving landslide complexes have been identified along the banks of the Erlenbach, causing efficient channel–hillslope connectivity (Schuerch et al., 2006). Trees, tree trunks and large branches are abundant in the Erlenbach channel and along the banks. Thus, based on preliminary data from litter traps installed above the stream channel and from field observations, we can assume that the input distribution of CPOM masses is strongly bimodal, with one peak lying within 0.1 to 1 g (leaves and twigs), and one at > 100 kg (whole trees). The mass distribution of the material flushed out of the stream therefore does not correspond to the input distribution, and in-stream processes must break down larger material while it resides in the channel. These processes can be broadly divided into physical, chemical and biological processes (although these may interact) (Harmon et al., 1986; Merten et al., 2013; Webster et al., 1999). The physical processes include breakage and detachment of pieces by fluid stressing, grinding of woody debris by gravel bedload, wood–wood interaction (for example a floating log impacting sta-

tionary material), or the break-up of log jams (Harmon et al., 1986). In addition, wetting/drying cycles and possibly freezing/thawing may lead to swelling and cracking of wood. Parts of woody material can be dissolved in water and carried out in solution (e.g., Yoshimura et al., 2010). This loss of material may destabilize the structure of the debris and make it more prone to physical erosion. Finally, bacteria and fungi colonize dead wood and decompose it, while animals, for example fresh-water invertebrates or certain types of insects such as Plecoptera larvae, may attack CPOM pieces to obtain food or to create shelter (e.g., Webster and Benfield, 1986; Montemarano et al., 2007). As pointed out by Hassan et al. (2005), little is known about the relative importance of different wood depletion processes. Merten et al. (2013) found that physical breakage was responsible for a mass loss of around 7.3 % in LWD, while decay processes contributed around 1.9 % in streams in northern Minnesota, USA. However, it is currently unclear how these results transfer to other regions. In addition, it is not known what piece sizes are produced by the various deterioration processes, how fluvial transport affects the CPOM/LWD size distribution, and how exactly the different processes play together to produce the observed output mass distribution. Finally, it would be interesting to compare the mass distribution of in-stream CPOM to the distributions of branch mass in living trees. However, although the necessary data must have been recorded for studies of tree allometry (e.g., Bertram, 1989: Dassot et al., 2012), we were unable to obtain such data, or to find relevant accounts in the literature.

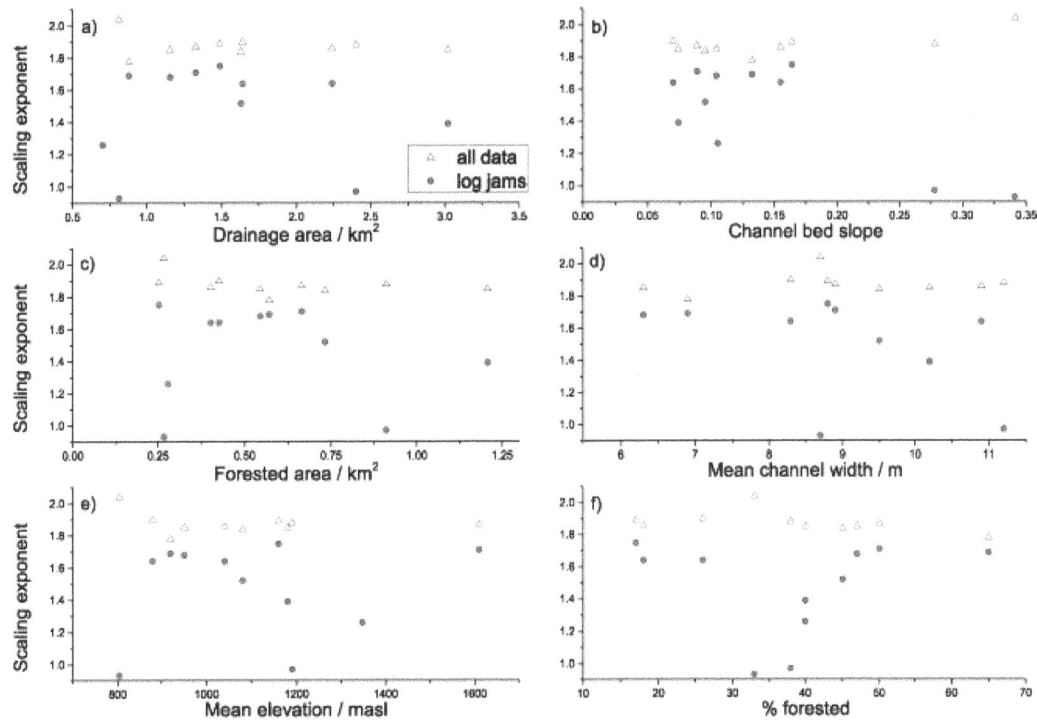

**Figure 8.** Scaling exponents of eleven Swiss mountain streams (Table 1) as functions of (**a**) drainage area, (**b**) channel bed slope, (**c**) forested area, (**d**) mean channel width, (**e**) mean elevation above sea level, and (**f**) percent fraction of the catchment covered by forest. Note that the two lowest scaling exponents for log jam material at 0.93 and 0.97, corresponding to the Brüggenwaldbach and Grossbach (Table 1), are based on a small number of measurement and are probably spurious. No strong correlations or trends are obvious.

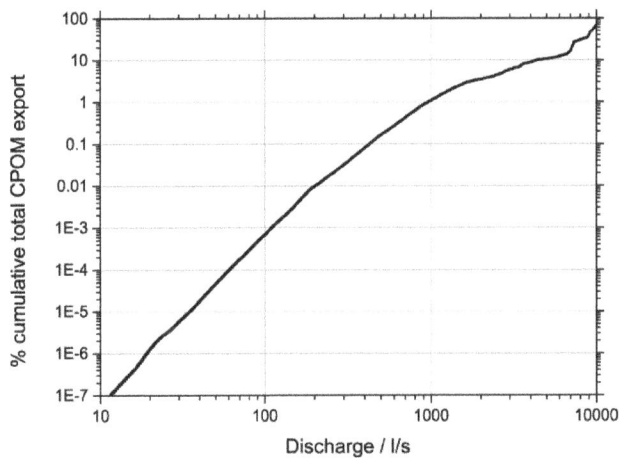

**Figure 9.** Percentage of the total cumulative CPOM export from the catchment as a function of discharge, obtained by integrating the rating curves with the 10 min discharge data 1983–2012. Only about 10 % of the total export is delivered by discharges lower than $\sim 4000\,\mathrm{L\,s^{-1}}$, which is on average exceeded for less than 6 min $\mathrm{yr^{-1}}$. Thus, large events dominate the signal.

## 5.2 Implications for sampling strategies

Using the rating curve and measured discharge, estimations of CPOM export during individual events and over longer timescales can be calculated. This assumes that the rating curve derived from the data has not changed over the considered period and is representative for the average transport throughout the year (rather than for a specific season). At the Erlenbach, due to the strong dependence of CPOM on discharge with a rating curve exponent of 4.47, CPOM transport is dominated by large discharge events (Fig. 9). For instance, applying the rating curve to the measured discharge in the extreme event in August 2010, this event alone exported nearly $2 \times 10^6$ kg of CPOM, comprising 94 % of the total export for the year 2010. Without deriving a detailed rating curve, similarly strong dependencies on discharge have been reported previously (e.g., Bormann et al., 1969; Fisher and Likens, 1973; Wallace et al., 1995). The strong dependence of CPOM transport on discharge suggests that measurements conducted at low flows give an incomplete picture of overall CPOM transport. However, if the scaling exponent of the mass distribution is known, and if for a given event a certain size fraction has been sampled representatively, the complete CPOM export can be calculated. Our study results suggest that CPOM export for all size fractions can be estimated from the volumes of LWD exported in a large event, for example by measuring piece sizes trapped in a reservoir or by video monitoring the passage of wood pieces (e.g., MacVicar and Piégay, 2012; MacVicar et al., 2009; Moulin and Piégay, 2004; Seo et al., 2008; West et al., 2011). In addition to scaling down from LWD deposits with the aim to estimate total CPOM export, the scaling relations also permit scaling up from measurements of small CPOM to estimate LWD transport. The combination of a rating curve of CPOM transport rates with discharge and the mass distribution obtained from samples collected at low and intermediate flows can be used to discuss the mobility of LWD.

### 5.3  Mobility of large woody debris and hazard assessment

Floating wood is a source of hazard in many mountain streams (e.g., Comiti et al., 2006; Mazzorana et al., 2011; Rickenmann, 1997). For example, woody debris can jam in channel constrictions or at bridges, which may trap sediment and force water to overflow and leave the channel, ultimately causing overbank sedimentation. In this case the arrival frequency of pieces larger than a potentially hazardous size is an important parameter to consider. In addition, logs floating at the surface of the stream can move at high speeds, and can cause considerable damage upon impact on infrastructure elements such as bridge piers. We will here give an example of how the scaling relation and the knowledge of the rating curve can be used to estimate the frequency of arrival of LWD of a minimum weight. Using the definition that LWD pieces have a diameter larger than 0.1 m and a length larger than 1 m (Wohl and Jaeger, 2009), and assuming a dry wood density of $410 \, kg \, m^{-3}$ and a cylindrical shape, LWD has a minimum dry mass of $\sim 3.2 \, kg$. As an example, we will use a mass of 3 kg as the minimum value for the calculation. The values for a different mass limit can easily be calculated using Eq. (3). According to Eq. (3), the mass of all transported pieces heavier than 3 kg is $1.73 \times 10^{-4}$ times the total load of pieces heavier than 0.1 g. Using the rating curve, the transport rate of LWD and the frequency of piece arrivals can thus be obtained. According to our results, at the Erlenbach a potentially hazardous piece of wood with a mass of 3 kg or more arrives on average every ten minutes at a discharge of $\sim 3400 \, L \, s^{-1}$, every minute at a discharge of $\sim 5800 \, L \, s^{-1}$, every ten seconds at a discharge of $\sim 8600 \, L \, s^{-1}$ and every second at a discharge of $14\,400 \, L \, s^{-1}$. The uncertainty of these estimates is dependent on the quality of the original sampling data, the observed range of discharge and mass of pieces, and of the statistical methods used in the analysis. An independent confirmation of our estimates would require additional field efforts and is beyond the scope of the present paper. Nevertheless, calculations similar to the ones made above may help to assess the potential hazard of LWD in floods of certain sizes. Mobility of wood of all sizes may of course be limited by the ready availability of the material in or near the stream.

### 5.4  Transferability of the results to other catchments

We have presented results on the mass distribution of transported material for a single small headwater catchment, the Erlenbach. We hypothesize that a similarly consistent, i.e., discharge-independent scaling relation of transported CPOM masses can be found in other streams, with a scaling exponent that is characteristic for the catchment. Clearly, forest type, vegetation growth rates, and the distribution of forest throughout the catchment should be important, as they determine the type, amount, and sizes of woody debris available for delivery to the channel. All of these parameters are dependent on the climatic regime, which also determines wind speeds, wind direction and the rate of occurrence of extreme precipitation events. These, in turn, determine how often twigs and branches fall into or close to the channel, if trees can be uprooted, and when landslides are triggered. It seems reasonable to expect CPOM supply to vary in time. More material should be available after severe storms, particularly when they occur in the growth season, or in autumn, when broadleaf trees lose their leaves. Once CPOM has entered the channel, hydraulics, and the relative importance of different decay and transport processes in the stream play a role in modifying CPOM particle mass distributions. Finally, we expect the scaling exponent to be a function of the distance of the sampling locations from source regions, and of the drainage area, because channel processes and channel-hillslope coupling can change along a stream, and because material will break down in transport.

Transported CPOM is recruited from material stored in or near the stream. The distributions should therefore be related. The mass distribution of transported material could differ from in-stream material because of size-dependent entrainment and transport behavior of different CPOM pieces. In addition, CPOM can have a wide range of different shapes, which can affect particle mobility. Small CPOM pieces can be elongated (twigs, conifer needles), rounded, or platy (wood chips, bark fragments, scales of conifer cones, leaves, see Fig. 2a), while LWD is typically elongated (logs and branches). In addition, twigs and logs may have branches or root wads, which can lower mobility and enhance the likelihood of trapping (e.g., Abbe and Montgomery, 1996; Braudrick and Grant, 2000).

The data collected by Rickli and Bucher (2006) on distributions of in-stream LWD in ten mountainous catchments of Switzerland provides some evidence for the transferability of the Erlenbach observations. We found consistent power-law distributions of frequency per mass class vs. particle mass for all of these streams, indicating that there is some generality in this type of distribution (cf. Fig. 7 and Table 1). Albeit, the scaling exponent for LWD in log jams is smaller than the scaling exponent for all in-stream wood (Table 1). Similarly, for the Erlenbach, the log jam scaling exponent is smaller than for the transported material, implying that coarser material is moved less frequently than smaller material. This reflects the selective transport of large pieces of wood and the fact that jamming makes coarse material less mobile. While scaling exponents differ between mobile and stored LWD, the scaling exponents for in-stream wood do not show significant correlation (tested using Kendall's tau rank correlation coefficient and $R^2$ from a linear regression) with any of the tested predictor variables: mean elevation above sea level, drainage area, channel bed slope, channel width, and percent forested area (Fig. 8). However, the range of conditions in the investigated streams is small and a final assessment would need a larger database. For example, the channel bed slope of all studied streams is at least 7 %, while the maximum drainage area is about 3 km$^2$. In contrast, the percentage of forested area varies between 17 and 65 %, covering a fairly large range. The lack of correlation to the scaling exponent may imply that the distribution of forest along the stream is more important in determining the scaling exponent than the percentage of forested area. Typically, forest is denser near the banks of a stream than far from them, due to the ready availability of water, or due to land use management.

Not many reports of size distributions of CPOM can be found in the literature, and the majority of the available studies used piece diameter as descriptor variable (e.g., Harmon et al., 1986; Jackson and Sturm, 2002). We were able to find a single study using volume as a descriptor variable (Hogan, 1987). We digitized the data for the unlogged reaches, and found power law scaling with exponent values of 1.72 for the small watershed (3.9 km$^2$), 1.61 for the medium watershed (6.9 km$^2$) and 1.90 for the large watershed (20.2 km$^2$) (here, the depiction of "small", "medium" and "large" is after Hogan's (1987) own terminology). These values are larger than the scaling exponent for log jam material in the Erlenbach, but are comparable to the exponents for in-stream material of the other Swiss streams. MacVicar and Piégay (2012) reported the distribution of LWD piece length in transport, observed using a video camera during floods of the Ain River, France. The Ain River is a much larger stream than the Erlenbach, with a drainage area of 3630 km$^2$ (compared to 0.7 km$^2$ at the Erlenbach) and a width of ~65 m (compared to ~4 m at the Erlenbach). In general, LWD pieces longer than the channel width are rarely transported (Bilby and Ward, 1989; Nakamura and Swanson, 1993), and it has been shown in field studies that LWD moves further and more frequently in larger streams (e.g., Lienkamper and Swanson, 1987). Thus, a slightly greater abundance of long pieces in the Ain River in comparison to the Erlenbach may be expected. The observed mass distribution yields a well-defined power-law scaling with a scaling exponent of 1.8 (B. MacVicar, personal communication, 2013), which is similar to the one observed at the Erlenbach (1.84) (see also the discussion version of the paper and the reviewers' comments for more details on the Ain River data).

In summary, we have found similar scaling exponents for in-stream LWD of eleven Swiss mountain streams, including the Erlenbach, and small forested catchments in British Columbia, Canada (Hogan, 1987). In addition, we found similar scaling exponents for transported LWD in the Erlenbach headwater stream, Switzerland (this study), and the Ain River, France (MacVicar and Piégay, 2012). This suggests that, at least for LWD, piece mass scales as a declining power law with scaling exponents between about 1.4 and 2.0 (cf. Table 1).

## 6 Conclusions

We have demonstrated that the mass of coarse particulate organic matter (CPOM) transported in the Erlenbach, a steep mountain headwater stream in the pre-Alps of Switzerland, displays a well-defined scaling behavior. The scaling is consistent over almost seven orders of magnitude of particle mass and independent of discharge. Such scaling information can be used to make comparable CPOM transport rates collected with different sampling methods and to estimate the total masses of exported material from small samples comprising only a few size classes of CPOM. Currently, our results have been demonstrated to hold fully for the Erlenbach only. However, the comparison of the Erlenbach data with the scaling distributions of large woody debris transported in the Ain River, France, and of in-channel material in ten small Swiss mountain streams and forested catchments in British Columbia, Canada, suggests that a similarly consistent scaling behavior between CPOM masses and the number of pieces exist for other streams. We found that the watershed/channel parameters examined in the eleven Swiss data sets did not determine LWD scaling; however, the number of streams and the ranges of the observed values are too small to make final conclusions. Thus, it remains to be determined in how far the scaling exponent depends on stream type, vegetation, climate, substrate properties, and on the connectivity between channels and hillslopes. High quality data on piece mass of CPOM in streams spanning the full range of masses are rare, and a full evaluation of our findings is not possible at this stage. If a scaling similar to the one observed at the Erlenbach, which consistently holds over nearly seven orders of magnitude in particle mass and three orders of magnitude in discharge, is correct, measurements of scaling exponents on a small fraction of transported or stored material can be used to extrapolate from LWD to total CPOM or vice versa. This may be useful to assess the hazard potential of large woody debris during floods, it may simplify the planning and logistics of measurement campaigns, and it may help in obtaining woody debris budgets for stream reaches.

**Acknowledgements.** We thank the current and former members of the mountain hydrology team, especially A. Klaiber, A. Ludwig, A. Pöhlmann, and C. Wyss for help in the field. B. Fritschi and D. Rickenmann designed the basket samplers and contributed through discussions. Further discussions with R. Hilton, N. Hovius, J. Kirchner, K. Krause, and P. Schleppi are acknowledged. We are grateful to D. Hunziker, A. Klaiber, A. Pöhlmann, C. Schär, and M. Sieber, who spend many hours counting, weighing and measuring pieces of woody debris. A. Beer helped to make Fig. 1. B. MacVicar, H. Piégay, J. Warburton, an anonymous reviewer, and AE M. Krautblatter provided constructive comments on the discussion paper. This study was supported by SNF grants 200021_124634/1 and 200021_137681/1, and WSL.

Edited by: M. Krautblatter

## References

Abbe, T. B. and Montgomery, D. R.: Large woody debris jams, channel hydraulics and habitat formation in large rivers, Regul. River., 12, 201–221, 1996.

Abbe, T. B. and Montgomery, D. R.: Patterns and processes of wood debris accumulation in the Queets river basin, Washington, Geomorphology, 51, 81–107, 2003.

Benda, L. E. and Sias, J. C.: A quantitative framework for evaluating the mass balance of in-stream organic debris, Forest Ecol. Manag., 172, 1–16, 2003.

Bertram, J. E. A.: Size-dependent differential scaling in branches: the mechanical design of trees revisited, Trees, 4, 241–253, 1989.

Bilby, R. E. and Ward, J. W.: Changes in characteristics and function of woody debris with increasing size of streams in western Washington, Trans. Am. Fish. Soc., 118, 368–378, 1989.

Bilby, R. E. and Ward, J. W.: Characteristics and function of large woody debris in streams draining old-growth, clear-cut, and second-growth forests in southwestern Washington, Can. J. Fish. Aquat. Sci., 48, 2499–2508, 1991.

Bormann, F. H., Likens, G. E., and Eaton, J. S.: Biotic regulation of particulate and solution losses from a forest ecosystem, Bioscience, 19, 600–611, 1969.

Braudrick, C. A. and Grant, G. E.: When do logs move in rivers?, Water Resour. Res., 36, 571–583, 2000.

Brooks, A. P., Gehrke, P. C., Jansen, J. D., and Abbe, T. B.: Experimental reintroduction of woody debris on the Williams River, NSW: Geomorphic and ecological responses, River Res. Appl., 20, 513–536, 2004.

Bunte, K., Abt, S. R., Potyondy, J. P., and Ryan, S. E.: Measurement of coarse gravel and cobble transport using portable bedload traps, J. Hydraul. Eng., 130, 879–893, doi:10.1061/(ASCE)0733-9429(2004)130:9(879), 2004.

Bunte, K., Swingle, K. W., and Abt, S. R.: Guidelines for using bedload traps in coarse-bedded mountain streams: Construction, installation, operation and sample processing, 91 pp., General Technical Report RMRS-GTR-191, U.S. Forest Service, Rocky Mountain Research Station, Fort Collins, Colorado, http://www.fs.fed.us/rm/pubs/rmrs_gtr191.pdf, 2007.

Comiti, F., Andreoli, A., Lenzi, M. A., and Mao, L.: Spatial density and characteristics of woody debris in five mountain rivers of the Dolomites (Italian Alps), Geomorphology, 78, 44–63, doi:10.1016/j.geomorph.2006.01.021, 2006.

Cummins, K. W.: Structure and function of stream ecosystems, BioScience, 24, 631–641, 1974.

Curran, J. C.: Mobility of large woody debris (LWD) jams in a low gradient channel, Geomorphology, 116, 320–329, doi:10.1016/j.geomorph.2009.11.027, 2010.

Dassot, M., Colin, A., Santenoise, P., Fournier, M., and Constant, T.: Terrestrial laser scanning for measuring the solid wood volume, including branches, of adult standing trees in the forest environment, Comput. Electron. Agr., 89, 86–93, doi:10.1016/j.compag.2012.08.005, 2012.

Faustini, J. M. and Jones, J. A.: Influence of large woody debris on channel morphology and dynamics in steep, boulder-rich mountain sreams, western Cascades, Oregon, Geomorphology, 51, 187–205, 2003.

Fisher, S. G. and Likens, G. E.: Stream ecosystem: Organic energy budget, BioScience, 22, 33–35, 1972.

Fisher, S. G. and Likens, G. E.: Energy flow in Bear Brook, New Hampshire: An integrative approach to stream ecosystem metabolism, Ecol. Monogr., 43, 421–439, 1973.

Harmon, M. E., Franklin, J. F., Swanson, F. J., Sollins, P., Gregory, S. V., Lattin, J. D., Anderson, N. H., Cline, S. P., Aumen, N. G., Sedell, J. R., Lienkaemper, G. W., Cromack, K., and Cummins, K. W.: Ecology of coarse woody debris in temperate ecosystems, Adv. Ecol. Res., 15, 132–302, 1986.

Hassan, M. A., Hogan, D. L., Bird, S. A., May, C. L., Gomi, T., and Campbell, D.: Spatial and temporal dynamics of wood in headwater streams of the Pacific Northwest, J. Am. Water Resour. As., 41, 899–919, 2005.

Hegg, C., McArdell, B. W., and Badoux, A.: One hundred years of mountain hydrology in Switzerland by the WSL, Hydrol. Process., 20, 371–376, doi:10.1002/hyp.6055, 2006.

Hogan, D. L.: The influence of large organic debris on channel recovery in the Queen Charlotte Islands, British Columbia, Canada, in: Erosion and Sedimentation in the Pacific Rim (Proceedings of the Corvallis Symposium, August, 1987), IAHS Publ. no. 165, 1987.

Jackson, C. R. and Sturm, C. A.: Woody debris and channel morphology in first- and second-order forested channels in Washington's coast ranges, Water Resour. Res., 38, 1177, doi:10.1029/2001WR001138, 2002.

Keller, E. A. and Swanson, F. J.: Effects of large organic material on channel form and fluvial process, Earth Surf. Process., 4, 361–380, 1979.

Kraft, C. E. and Warren, D. R.: Development of spatial pattern in large woody debris and debris dams in streams, Geomorphology, 51, 127–139, 2003.

Lienkaemper, G. W. and Swanson, F. J.: Dynamics of large woody debris in streams in old-growth Douglas-fir forests, Can. J. Forest Res., 17, 150–156, 1987.

MacFarlane, W. A. and Wohl, E.: Influence of step composition on step geometry and flow resistance in step-pool streams of the Washington Cascades, Water Resour. Res., 39, 1037, doi:10.1029/2001WR001238, 2003.

MacVicar, B. and Piégay, H.: Implementation and validation of video monitoring for wood budgeting in a wandering piedmont river, the Ain River (France), Earth Surf. Proc. Land., 37, 1272–1289, doi:10.1002/esp.3240, 2012.

MacVicar, B., Piégay, H., Henderson, A., Comiti, F., Oberlin, C., and Pecorari, E.: Quantifying the temporal dynamics of wood in large rivers: field trials of wood surveying, dating, tracking, and monitoring techniques, Earth Surf. Proc. Land., 34, 2031–2046, doi:10.1002/esp.1888, 2009.

Manga, M. and Kirchner, J. W.: Stress partitioning in streams by large woody debris, Water Resour. Res., 36, 2373–2379, 2000.

Mao, L., Andreoli, A., Comiti, F., and Lenzi, M. A.: Geomorphic effects of large wood jams on a sub-antarctic mountain stream, River Res. Appl., 24, 249–266, doi:10.1002/rra.1062, 2008.

Martin, D. J. and Benda, L. E.: Patterns of instream wood recruitment and transport at the watershed scale, Trans. Am. Fish. Soc., 130, 940–958, 2001.

Mazzorana, B., Hübl, J., Zischg, A., and Largiader, A.: Modelling woody material transport and deposition in alpine rivers, Nat. Hazards, 56, 425–449, doi:10.1007/s11069-009-9492-y, 2011.

Merten, E. C., Vaz, P. G., Decker-Fritz, J. A., Finlay, J. C., and Stefan, H. G.: Relative importance of breakage and decay as processes depleting large wood from streams, Geomorphology, 190, 40–47, doi:10.1016/j.geomorph.2013.02.006, 2013.

Minshall, G. W.: Role of allochthonous detritus in the trophic structure of a woodland springbrook community, Ecology, 48, 139–149, 1967.

Molnar, P., Densmore, A. L., McArdell, B. W., Turowski, J. M., and Burlando, P.: Analysis of changes in the step-pool morphology and channel profile of a steep mountain stream following a large flood, Geomorphology, 124, 85–94, doi:10.1016/j.geomorph.2010.08.014, 2010.

Montemarano, J. J., Kershner, M. W., and Leff, L. G.: Crayfish effects on fine particulate organic matter quality and quantity, Fund. Appl. Limnol. – Archiv für Hydrobiologie, 168, 223–229, doi:10.1127/1863-9135/2007/0169-0223, 2007.

Montgomery, D. R. and Piégay, H.: Wood in rivers: interactions with channel morphology and processes, Geomorphology, 51, 1–5, 2003.

Moulin, B. and Piégay, H.: Characteristics and temporal variability of large woody debris trapped in a reservoir on the river Rhone (Rhone): Implications for river basin management, River Res. Appl., 20, 79–97, doi:10.1002/rra.724, 2004.

Nakamura, F. and Swanson, F. J.: Effects of coarse woody debris on morphology and sediment storage of a mountain stream system in western Oregon, Earth Surf. Proc. Land., 18, 43–61, 1993.

Reeves, G. H., Burnett, K. M., and McGarry, E. V.: Sources of large wood in the main stem of a fourth-order watershed in coastal Oregon, Can. J. Forest Res., 33, 1363–1370, doi:10.1139/x03-095, 2003.

Rickenmann, D.: Schwemmholz und Hochwasser, Wasser, Energie, Luft, 89, 115–119, 1997.

Rickenmann, D., Turowski, J. M., Fritschi, B., Klaiber, A., and Ludwig, A.: Bedload transport measurements at the Erlenbach stream with geophones and automated basket samplers, Earth Surf. Proc. Land., 37, 1000–1011, doi:10.1002/esp.3225, 2012.

Rickli, C. and Bucher, H.: Einfluss ufernaher Bestockung auf das Schwemmholzvorkommen in Wildbächen, technical report, Swiss Federal Research Institute WSL, 94 pp., www.wsl.ch/fe/gebirgshydrologie/wildbaeche/projekte/schwemmholzvorkommen/rickli_shber_261007.pdf, 2006.

Schleppi, P., Muller, N., Edwards, P. J., und Bucher, J. B.: Three years of increased nitrogen deposition do not affect the vegetation of a montane forest ecosystem, Phyton, 39, 197–204, 1999.

Schuerch, P., Densmore, A. L., McArdell, B. W., and Molnar, P.: The influence of landsliding on sediment supply and channel change in a steep mountain catchment, Geomorphology, 78, 222–235, doi:10.1016/j.geomorph.2006.01.025, 2006.

Seo, J. I., Nakamura, F., Nakano, D., Ichiyanagi, H., and Chun, K. W.: Factors controlling the fluvial export of large woody debris, and its contribution to organic carbon at watershed scales, Water Resour. Res., 44, W04428, doi:10.1029/2007WR006453, 2008.

Turowski, J. M., Yager, E. M., Badoux, A., Rickenmann, D., and Molnar, P.: The impact of exceptional events on erosion, bedload transport and channel stability in a step-pool channel, Earth Surf. Proc. Land., 34, 1661–1673, doi:10.1002/esp.1855, 2009.

Turowski, J. M., Badoux, A., and Rickenmann, D.: Start and end of bedload transport in gravel-bed streams, Geophys. Res. Lett., 38, L04401, doi:10.1029/2010GL046558, 2011.

Turowski, J. M., Badoux, A., Leuzinger, J., and Hegglin, R.: Large floods, alluvial overprint and bedrock erosion, Earth Surf. Proc. Land., 38, 947–958, doi:10.1002/esp.3341, 2013.

Wallace, J. B., Whiles, M. R., Eggert, S., Cuffney, T. F., Lugthart, G. J., and Chung, K.: Long-term dynamics of coarse particulate organic matter in three Appalachian mountain streams, J. N. Am. Bent. Soc., 14, 217–232, 1995.

Webster, J. R. and Benfield, E. F.: Vascular plant breakdown in freshwater ecosystems, Annu. Rev. Ecol. Evol. Syst., 17, 567–594, 1986.

Webster, J. R. and Meyer, J. L.: Organic matter budgets for streams: A synthesis, J. N. Am. Benthol. Soc., 16, 141–161, 1997.

Webster, J. R., Benfield, E. F., Ehrman, T. P., Schaeffer, M. A., Tank, J. L., Hutchens, J. J., and D'Angelo, D. J.: What happens to allochthonous material that falls into streams? A synthesis of new and published information from Coweeta, Freshwater Biol., 41, 687–705, 1999.

West, A. J., Lin, C.-W., Lin, T.-C., Hilton, R. G., Liu, S.-H., Chang, C.-T., Lin, K.-C., Galy, A., Sparkes, R. B., and Hovius, N.: Mobilization and transport of coarse woody debris to the oceans triggered by an extreme tropical storm, Limnol. Oceanogr., 56, 77–85, doi:10.4319/lo.2011.56.1.0077, 2011.

Wohl, E. and Jaeger, K.: A conceptual model for the longitudinal distribution of wood in mountain streams, Earth Surf. Proc. Land., 34, 329–344, doi:10.1002/esp.1722, 2009.

Yoshimura, C., Fujii, M., Omura, T., and Tockner, K.: Instream release of dissolved organic matter from coarse and fine particulate organic matter of different origins, Biochemistry, 100, 151–165, doi:10.1007/s10533-010-9412-y, 2010.

# 2

# Erosional response of an actively uplifting mountain belt to cyclic rainfall variations

**J. Braun, C. Voisin, A. T. Gourlan, and C. Chauvel**

ISTerre, Université Grenoble Alpes and CNRS BP 53, 38041 Grenoble CEDEX 9, France

*Correspondence to:* J. Braun (jean.braun@ujf-grenoble.fr)

**Abstract.** We present an approximate analytical solution to the stream power equation describing the erosion of bedrock in an actively uplifting mountain range subject to periodic variations in precipitation rate. It predicts a time lag between the climate forcing and the erosional response of the system that increases with the forcing period. The predicted variations in the sedimentary flux coming out of the mountain are also scaled with respect to the imposed rainfall variations in a direct proportion to the discharge exponent, $m$, in the stream power law expression. These findings are confirmed by 1-D and 2-D numerical solutions. We also show that the response of a river channel is independent of its length and thus the size of its catchment area, implying that all actively eroding streams in a mountain belt will constructively contribute to the integrated signal in the sedimentary record. We show that rainfall variability at Milankovitch periods should affect the erosional response of fast uplifting mountain belts such as the Himalayas, Taiwan or the South Island, New Zealand, and predict 1000 to 10 000-year offsets between forcing and response. We suggest that this theoretical prediction could be used to independently constrain the value of the poorly defined stream power law exponents, and provide an example of how this could be done, using geochemical proxy signals from an ODP borehole in the Bengal Fan.

## 1 Introduction

Much work has been devoted to studying the potential links that might exist between climate and surface processes, and in particular the erosion of high-relief, tectonically active mountain belts (Tucker and Slingerland, 1997; Whipple et al., 1999; Zhang et al., 2001; Boggart et al., 2003; Whipple and Meade, 2006; Willett et al., 2006; Strecker et al., 2007; Whipple, 2009; Wobus et al., 2010; DiBiase and Whipple, 2011; Champagnac et al., 2012; Herman et al., 2014). This is in turn important to understand whether climate can affect tectonics as suggested by several now highly quoted studies (Molnar and England, 1990; Willett et al., 1993; Montgomery, 1994; Montgomery et al., 2001). Because glacial erosion is more efficient than fluvial processes, it is now reasonably well established that Cenozoic climate cooling culminating in the onset of periodic glaciations in the Quaternary has led to enhanced erosion rates in many high-latitude or high-elevation mountain belts (Egholm et al., 2009; Herman et al., 2014), as long as ice is not frozen to the bedrock

(Thomson et al., 2010). In non-glaciated environments, rainfall intensity plays an important role in controlling erosion (Roe et al., 2005), but there is growing evidence that variability in rainfall may be as important as mean precipitation in limiting or enhancing the rate of surface erosion by mountain streams and rivers (Lague and Hovius, 2005; DiBiase and Whipple, 2011). Variations in rainfall also affect the rate of chemical erosion as water availability is, with temperature, one of the main controls on silicate weathering (Dixon et al., 2009; Maher and Chamberlain, 2014). The thawing and freezing of soil-mantled slopes is also known to highly amplify the rate of soil creep (Anderson, 2002), implying that variability in temperature at or near freezing point must affect the rate of transport of chemically weathered rocks. Vegetation type and cover are also a function of climate (and elevation) and there is now mounting evidence that vegetation and erosion are linked (Hupp and Osterkamp, 1996; Dosseto et al., 2010), providing another potential, and so far poorly studied, link between erosional efficiency and climate.

The climate of the late Cenozoic and of the Quaternary in particular is dominated by large variations in continental ice cover both near the poles and in regions of elevated topography, which are controlled by variations in the Earth's orbital parameters, the so-called Milankovitch cycles (Milankovitch, 1941). Other aspects of the Earth's climate are also known to vary between glacial and non-glacial periods, such as the strength of the monsoon, or the latitudinal distribution of precipitation (Rossignolstrick, 1983; Prell and Kutzbach, 1987; Demenocal and Rind, 1993; Clemens et al., 1991, 1996). There is clear evidence that since the onset of large amplitude, 100 ka period glacial cycles approximately 1 Ma ago, glacial erosion is strongly enhanced during periods of extended ice cover (Valla et al., 2011; Herman et al., 2014), but other potential effects of climatic variations at Milankovitch periods on the erosional response of mountain belts, such as variations in rainfall intensity and/or variability, have not been extensively studied, potentially because of the large difference between the Milankovitch orbital periods and the typical tectonics/erosion timescales ( $\geq 1$ Ma).

The response of geomorphic systems to cyclic climate variations at either longer or shorter periods, has, however, been the subject of several studies. Paola et al. (1992) and Willgoose (1994, 2005) showed that, under the assumption that sediment transport rate is related to slope and runoff, the response of an equilibrium system (i.e., having reached a steady-state form by balancing uplift and erosion) to a sinusoidal variation in rainfall depends on the period of forcing in comparison to the response time of the system (the time it takes to reach equilibrium). If the forcing is much faster than the response time, the system is constantly out of equilibrium; when the forcing is slow, the system is able to remain at equilibrium; for intermediary forcing timescales, the geomorphic system's response lags the rainfall variations. Humphrey and Heller (1995) computed the response of a geomorphic system composed of an eroding mountain (obeying a linearized version of the stream power law) and a depositional foreland to demonstrate that an oscillating forcing (in precipitation for example) causes cyclic sedimentation patterns that show maximum amplitude at the contact point between the erosional and depositional systems. They also note that the time delay between forcing and response is mostly a function of the system size, not the period of forcing, and that the response is always damped in comparison with the forcing.

Using a more complex model combining the effects of soil formation and transport on hillslopes to the transport and erosion by river channels, Tucker and Slingerland (1997) demonstrated that cycles in runoff intensity cause a nonlinear response of geomorphic system, which strongly depends on the period of the cycles in comparison with the response time of the system. They also confirmed the results of Rinaldo et al. (1995), who showed that drainage density varies during climate cycles as the balance between fluvial and hillslope transport and erosion evolves through time.

Using a diffusive model for fluvial sediment transport by rivers, Castelltort and Van Den Driessche (2003) showed that Milankovitch-period variations in sedimentary supply from a high-relief, fast-eroding source (the mountain) are strongly buffered and therefore unlikely to be preserved in the depositional record. Using a more sophisticated model that includes the effect of grain size on transport capacity, Armitage et al. (2011) demonstrated that variability in rainfall is mostly imprinted in the sedimentary record as variations in grain size and its distribution with distance to the source. In a more recent study, Simpson and Castelltort (2012) argue that high-frequency rainfall cycles can be propagated and amplified to sedimentary basins if one assumes a potentially strong feedback between discharge and channel gradient. Similarly, but focusing on Milankovitch cycles timescales, Godard et al. (2013) demonstrated, using a surface processes model that combines fluvial erosion and hillslope processes, that geomorphic systems behave as "forced oscillators" where climate forcing is amplified in the sedimentary response at a relatively specific range of frequencies, although their response to the relatively short-period Quaternary climate variability is strongly damped when diffusive processes become dominant over river incision.

Here we present an analytical solution to the stream power law equation forced by climate-driven cycles in precipitation. We show that it is a natural behavior of this equation to predict an amplification and introduce a time lag between forcing and response. We then use a simple numerical solution of the same equation to fully appreciate the response of a fast-eroding tectonic system to periodic perturbations in precipitation, with a particular focus on forcing at the Milankovitch periods. We then interpret these solutions in terms of the consequences they have on the behavior of natural systems and their response to cyclic rainfall variability at a range of forcing periods.

## 2 The stream power law

We will assume that bedrock incision is the dominant mode of erosion in an active mountain belt. Under the assumption that fluvial erosion can be represented by the stream power law (Howard, 1994) and that drainage area increases as a power of the distance to the water divide (Hack, 1957), the evolution of bedrock height, $h$, as a function of time is given by the following partial differential equation, or PDE (see Appendix for a detailed derivation):

$$\frac{\partial h}{\partial t} = U - K v^m (L-x)^{mp} \left(\frac{\partial h}{\partial x}\right)^n, \qquad (1)$$

where $L$ is the length of the river channel, $U$ is rock uplift, $v$ is precipitation rate and $K, n, m$ and $p$ are constants. The distance $x$ is measured from base level ($h(x=0)=0$). Let us note that, as drainage area tends towards zero at the divide, erosion rate is arbitrarily nil at $x=L$. This is commonly handled by defining a critical slope, $S_c$, beyond which the stream power law is no more valid and colluvial and hillslope processes become dominant (Whipple and Tucker, 1999). The slope ($n$) and area ($m$) exponents in the stream power law are not well constrained. Their commonly accepted ranges are $0.2 < m < 0.8$ and $0.5 < n < 2$. The ratio $m/n$ controls the concavity of stream profiles that have reached an equilibrium between uplift and erosion (see Eq. A11 in the Appendix) and, where estimated, it ranges between 0.4 and 0.6. Independently, $p$, the exponent in Hack's law relating drainage area to distance to the divide can be extracted from digital elevation model analysis; its commonly accepted value is close to 2.

## 3  Response to precipitation change

In the Appendix, we show that this equation can be written in dimensionless form as follows:

$$\frac{\partial h'}{\partial t'} = 1 - (1 - x')^{mp}\left(\frac{\partial h'}{\partial x'}\right)^n, \tag{2}$$

where $x' = x/L$, $h' = h/H$, $t' = t/\tau$, and $H = \left(\frac{U}{K\,v^m}\right)^{1/n}\frac{L^{1-mp/n}}{1-mp/n}$ and $\tau = H/U$. We also show that the variation in normalized height, $\delta h'$, resulting from a small perturbation in precipitation, $\delta v'$, obeys the following PDE:

$$\frac{\partial \delta h'}{\partial t'} \approx -m\delta v' - \frac{n}{1-mp/n}(1-x')^{mp/n}\frac{\partial \delta h'}{\partial x'}. \tag{3}$$

The first term of the right-hand side of the equation corresponds to the direct response of the system to the perturbation (it is directly and linearly proportional to $\delta v'$), whereas the second term expresses how the resulting change in slope, $\frac{\partial h'}{\partial x'}$, modifies the response of the system to future perturbation.

In the Appendix, we show that there exists an approximate solution to this equation, which yields the response of the system to a small periodic perturbation in precipitation rate, $v = v_0 + \delta v_0 \sin\left(\frac{2\pi t}{P}\right)$, in the form of the corresponding variation in the sedimentary flux, $Q = Q_0 + \delta Q_0$, i.e., the erosion rate integrated over the channel length. The solution can be expressed as

$$\frac{\delta Q_0/Q_0}{\delta v_0/v_0} = \frac{\delta Q'}{\delta v'} = G\sin\left(\frac{2\pi(t+\theta)}{P}\right), \tag{4}$$

where $G$ is a dimensionless gain and is given by

$$G = \frac{2\pi\tau m/P}{\sqrt{(2\pi\tau/P)^2 + \left(\frac{mp}{1-mp/n}\right)^2}} \tag{5}$$

and $\theta$ is a phase shift, or time lag, between the forcing (the perturbation in precipitation rate) and the response (the resulting perturbation in sedimentary flux), given by

$$\theta = \frac{P}{2\pi}\tan^{-1}\left(\frac{P}{2\pi\tau}\frac{mp}{1-mp/n}\right). \tag{6}$$

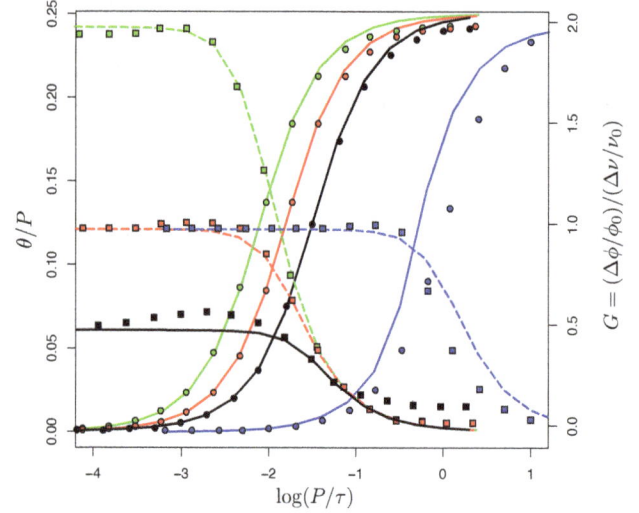

**Figure 1.** Predictions of the quasi-analytical solution (solid and dashed lines) and the numerical simulation (squares and circles). The solid lines and circles represent the time lag (divided by the forcing period, $P$) and the dashed lines and squares represent the gain, both as a function of the forcing period, $P$ (normalized by the characteristic time of the system, $\tau$). The green, red and blue lines/symbols correspond to values of $(m, n)$ equal to (0.5, 1) and (1, 2) and (2, 4), respectively, such that $m/n = 1/2$ in all three cases. The blue lines/symbols correspond to the case where $(m, n) = (1, 3)$, i.e., $m/n = 1/3 \neq 1/2$, which explains why, in that case, the quasi-analytical solution is less accurate.

$P$ is the period of forcing and $\tau$ is the characteristic time of the system, i.e., the time it takes for erosion to balance uplift ($\tau = H/U$). Note that this solution is only valid for values of the ratio $mp/n$ close to unity (see Appendix).

The predicted time lag and gain are shown in Fig. 1 as functions of the period of the perturbation, for various values of the exponent $m$ and $\frac{mp}{1-mp/n} \approx 1$. We see that for very short periods, i.e., in comparison with the characteristic timescale of the system ($P \ll \tau = H/U$), the sedimentary flux is in phase with the perturbation. This is because the channel geometry does not have the time to respond to the perturbation in precipitation rate. As the forcing period increases, the time lag grows until it reaches a quarter of the forcing period and the two signals are completely out of phase. This can be derived from Eq. (6), where

$$\lim_{P \ll \tau} \theta = 0 \tag{7}$$
$$\lim_{P \gg \tau} \theta = P/4. \tag{8}$$

For intermediary periods ($\tau/100 < P < \tau$), the system is out of phase, with the sedimentary response lagging behind the climate forcing. These results are in agreement with the qualitative analysis of Willgoose (1994) based on the results of a landscape evolution model where sediment transport is assumed proportional to slope and surface runoff (proportional to precipitation and drainage area). Our results also show that the time lag increases with $m$ (compare the red curve corresponding to $m = 1$ with the black curve corresponding to $m = 2$ and the green curve corresponding to $m = 0.5$ in Fig. 1).

We note also that the gain decreases with the forcing period. The gain is maximum and equal to $m$ when the forcing period is small compared to $\tau$ and it tends towards zero when the forcing period is much larger than $\tau$. This can be derived from Eq. (5):

$$\lim_{P \ll \tau} G = m \tag{9}$$

$$\lim_{P \gg \tau} G = 0. \tag{10}$$

In this latter situation ($P \gg \tau$), the channel has the time to adjust to the variable precipitation and remains at steady state such that the sedimentary flux perfectly balances the uplift rate, $U$, and there is no perturbation (the gain is nil).

It is worth noting that this behavior is similar to the response of a general linear system with finite memory as described by Howard (1982, see his Fig. 1) while considering the general response of geomorphological systems to step or periodic forcing. Although clearly derived from a non-linear equation, our solution could be regraded as the application of this general principle to bedrock channel incision parameterized by the stream power law.

## 4 Numerical solution

To test the accuracy of this approximate solution, we have solved the stream power law using the following implicit finite difference scheme (Braun and Willett, 2013):

$$h_i(t + \Delta t) = h_i(t) + U \Delta t - K \Delta t v^m (L - x)^{mp}$$
$$\left( \frac{h_i(t + \delta t) - h_{i-1}(t + \delta t)}{\delta x} \right)^n. \tag{11}$$

The non-linear dependence of erosion rate on slope (when $n \neq 1$) was dealt with by using a Newton–Raphson iterative scheme (Braun and Willett, 2013). We imposed the uplift rate and the mean precipitation rate. We adjusted the constant $K$ such that the resulting steady-state topography reaches a set value, $H_{max}$. Note that the value of the mean precipitation rate is not important, as it appears as a multiplier of the constant $K$ in the stream power law. The model was run to steady state and a sinusoidal variation in precipitation rate was then imposed of amplitude equal to one-tenth of the mean precipitation rate. The amplitude of this perturbation is of no importance to the solutions we present, as long as it remains small in comparison with the mean precipitation rate. Increasing the amplitude of the perturbation towards values close to the assumed mean precipitation causes the time lag to increase and the amplification to decrease, but the characteristics of the solution remain unchanged. The channel length was set to 200 km but its value does not influence the solution either, implying that the response of an incising river to a perturbation in precipitation rate does not depend on its size or the

size of its catchment. This is a consequence of using a value of $mp/n \approx 1$.

We computed the gain as the ratio of the amplitude of the variations in sedimentary flux normalized by the mean sedimentary flux (at steady state) by the amplitude of the imposed variations in precipitation rate normalized by the mean precipitation rate. We also computed the phase shift between the imposed precipitations and the computed sedimentary flux by performing a cross-correlation between the two signals. The results are also shown in Fig. 1 where they are compared to the semi-analytical solution for different combinations of the parameters $m$, $n$ and $p$.

The first interesting result is that the numerical solution is almost identical to the analytical solution when $mp/n \approx 1$, implying that the analytical solution is a good approximation of the general behavior of the system. For values of the exponents such that $mp/n \neq 1$, the analytical solution is not so accurate as shown in the case where $m = 1$, $n = 3$ and $p = 2$ (blue curve in Fig. 1), but the shape of the solution is very similar with the analytical solution clearly overestimating the time lag. In all cases, the numerical solution predicts a time lag between climate and the resulting erosional response that is a function of the forcing period, reaching a maximum of $P/4$ when the forcing period is similar to the characteristic time, $\tau$. Similar to the analytical solution, the gain, $G$, or ratio of the relative variation in sedimentary flux (normalized by its steady-state value) to relative variation in precipitation (normalized by its steady-state value), scales with $m$ for periods that are small compared to the characteristic timescale of the system and tends towards zero for longer periods.

To better understand the solution and thus the behavior of the system, we show in Fig. 2 snapshots of the evolution of the models in terms of the departure in erosion rate from the mean (steady-state) value. We selected the model run in which $n = 4$, $m = 2$, $p = 2$, $P = 100$ ka and $U = 6$ mm a$^{-1}$. We recognize that the values of $m$ and $n$ we selected are large and likely to be outside the range of acceptable values, but this choice produces a clear time lag and amplification of the erosional signal that we can more easily use to illustrate the behavior of the system. The solution shows the propagation of damped waves in erosion rate (Fig. 2b) with the erosion rate varying locally by as much as 50 % of the imposed mean value (here 6 mm a$^{-1}$). As rainfall increases at the beginning of a cycle (black curve in Fig. 2b), the incision rate increases near the base level. The maximum in anomalous erosion rate propagates up the profile (blue curve in Fig. 2b) until the precipitation rate starts to decrease. The erosion rate decreases then drastically near the base level (red curve in Fig. 2b), and a similar but opposite-sign wave of reduced erosion rate propagates towards the head of the channel (purple curve in Fig. 2b). The cycle then repeats itself. Note that the height of the river profile is also affected by similar topographic waves but, interestingly, whereas the predicted erosion rate can vary by as much as 50 % locally, the topographic "waves" never exceed 100 m in amplitude (i.e., < 2 % of the maximum to-

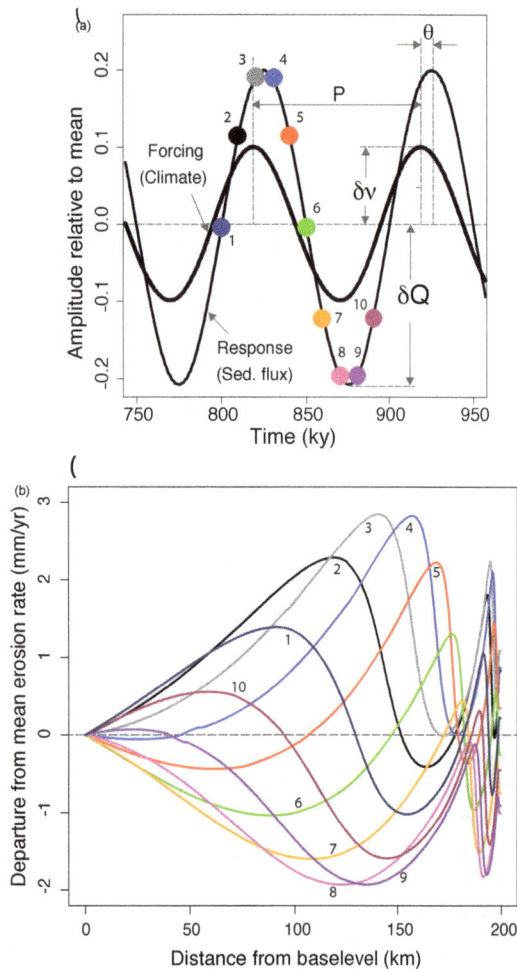

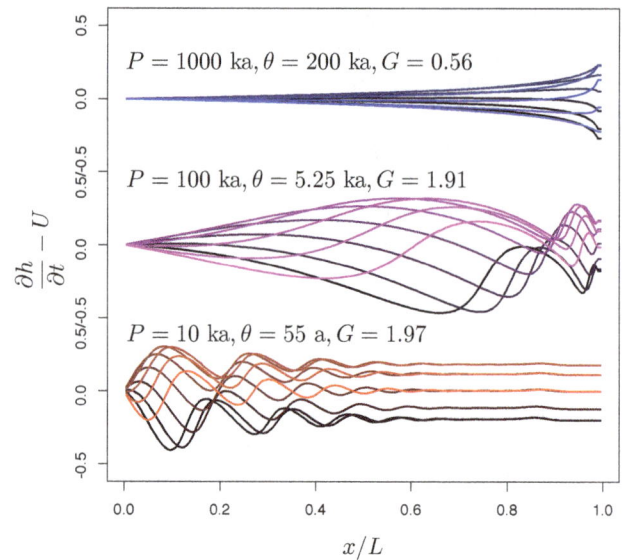

**Figure 2.** (**a**) Evolution of precipitation (climate) and the resulting sedimentary flux through a couple of imposed cycles (here at 100 ka) derived from the numerical solution of the stream power law for exponent values of $n = 4$, $m = 2$ and $p = 2$. The phase lag $\theta$ and the relation between the variations in precipitation $\delta v$ and the resulting variations in sedimentary flux ($\delta Q$), i.e., the gain, are also shown. (**b**) Erosion rate anomaly along the stream profile at various times during a rainfall cycle; each curve corresponds to a time that is indicated by a circle of the same color in (**a**).

**Figure 3.** Computed normalized erosion rate ($\frac{\partial h}{\partial t}/U$) as a function of the normalized distance ($x/L$) along the river profile for different values of the forcing period, $P$. Each curve in a given set corresponds to a selected snapshot during a rainfall cycle. Top (blue) curves, middle (purple) curves and bottom (red) curves correspond to $P = 1000$, 100 and 10 ka, respectively. $n = 4$, $m = 2$ and $p = 2$ in all three model runs.

pography), anywhere along the profile. The wavelength of these waves is such that the topographic perturbation they cause is likely to be almost undetectable.

The formation of these waves corresponds to the response of a system at equilibrium to a small perturbation. If the rate of change of the perturbation is rapid (in comparison to the characteristic time of the system), the system does not have the time to adjust and the perturbation in rainfall results in an instantaneous and proportional response in erosion rate. Consequently, the erosional response is in phase and in proportion to the perturbation (here amplified by the power $m$ to which the discharge, and thus rainfall, is raised in the stream power law). On the contrary, when the perturbation rate is

very slow (in comparison to the characteristic time of the system), the system is able to adjust, the system remains at or close to steady state and the integrated rate of erosion remains constant and equal to the imposed uplift rate. These two end-member behaviors are shown in Fig. 3 (blue and red curves), as well as the case where the forcing period is similar to the characteristic time of the system (purple curves). We see that, for long forcing periods, the response is strongly damped as the system is able to change its shape (slope) and adjust to the variation in precipitation rate such that erosion rate remains almost equal to uplift rate along the entire length of the river profile; the only remaining anomaly in erosion rate is near the headwaters of the stream and is therefore strongly out of phase. For short forcing periods, the perturbation propagates rapidly over the entire length of the profile such that it is able to offset the erosion rate over its entire length, therefore leading to a large amplitude response that is, however, in phase with the forcing.

In Fig. 4, we show the geometry of these waves/perturbations at the start of a precipitation cycle (corresponding to point 1 on Fig. 2a) for a range of values of the exponents $m$ and $n$. We kept all other parameters at constant values, i.e., $U = 6$ mm yr$^{-1}$ and $P = 100$ ka. As usual the $K$ parameter is adjusted such that the resulting steady-state river profile maximum height is 6000 m. We note that, to first order, the wavelength of these waves determines the time lag – i.e., the shorter the wavelength, the

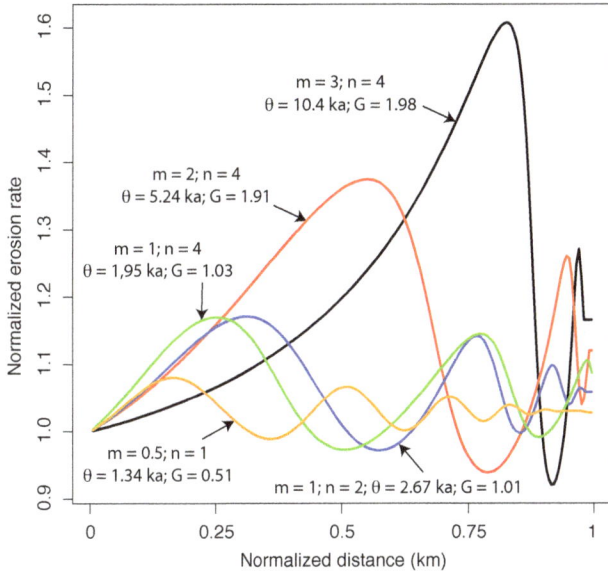

**Figure 4.** Computed normalized erosion rate ($\frac{\partial h}{\partial t}/U$) as a function of the normalized distance ($x/L$) along the river profile for a forcing at $P = 100$ ka and for various values of the exponents $m$ and $n$; for each run we indicate the value of the computed time lag, $\theta$, and gain, $G$.

shorter the time lag – whereas their amplitude determines the gain.

Finally, we performed a large number of simulations, keeping the uplift rate constant at $6\,\mathrm{mm\,a^{-1}}$, but varying $m$, $n$ and $P$. The results are shown in Fig. 5. For small forcing periods (Fig. 5a; $P = 1$ ka) compared to the characteristic time, the offset is nil and the gain is directly proportional to $m$ (Fig. 5a). For intermediate values of the forcing period (Fig. 5b and c), which we arbitrarily selected to correspond to the 41 and 100 ka Milankovitch periods, the gain remains proportional to $m$, especially for values of $m$ and $n$ that are close to the limit $m/n = 0.5$ (or $mp/n = 1$). If we assume that the ratio between $m$ and $n$ is well constrained, this implies that the gain is also proportional to $n$. This simple relationship breaks down for large forcing periods (Fig. 5d) as the gain tends towards zero, independently of $m$ or $n$.

The time lag, $\theta$, is nil for small values of the forcing period. For intermediate values of the forcing period (Fig. 5a), it increases with $m$ as well as with $n$, such that the increase in time lag along the $m/n = 0.5$ line is relatively small: it varies between 250 and 1000 a for $P = 41$ ka and between 1000 and 5500 a for $P = 100$ ka (Fig. 5b and c). These contour plots also show that the time lag increases mostly as the ratio $m/n$ tends towards 1 (the thick black diagonal line).

## 5  Discussion

Our solutions demonstrate that time lags are a natural response of erosional systems to climate (rainfall) variability

if they obey and are controlled by the stream power law. The sedimentary flux responds to an external climate forcing – variable precipitation – in a way that depends on how the forcing period compares to the characteristic timescale of the system, $\tau$, which is itself proportional to mountain height and inversely proportional to mean uplift rate. When the forcing period, $P$, is within the range $\tau/100 < P < \tau$, a substantial time lag is predicted in the erosional response to a cyclic precipitation pattern (see Fig. 1). Time lags associated with forcing at Milankovitch periods should therefore be measurable in most orogenic systems that have a characteristic timescale of a few million years (Whipple and Meade, 2006) and, particularly, in fast uplifting/eroding mountain belts, such as the Southern Alps in New Zealand or the Taiwan orogen, which both experience uplift and erosion rates of the order of $10\,\mathrm{mm\,a^{-1}}$.

We also predict that the erosional response is multiplied by a factor $m$, the area exponent in the stream power law, for forcing at periods smaller than $\tau$. Although $m$ is likely to be smaller than unity, it is possible that, if $m > 1$, the sedimentary signal be enhanced, which may explain the strong imprint that Milankovitch cycles have on the sedimentary record (De Boer and Smith, 1994) despite the relatively small changes in both solar insulation and temperature that are associated with the corresponding variations in the Earth's orbital parameters. At long forcing periods (compared to $\tau$), the gain tends towards zero, inhibiting detection of the time lag.

We have also shown that the erosive response of a river to a change in precipitation rate does not depend on its length. This ensures that all streams and catchments in a given mountain belt respond in a synchronous manner. It is a direct consequence of the stream power law combined with Hack's law. To test whether this still holds when taking into account the complex geometry and varied topology of river networks, we used the plan-form two-dimensional landscape evolution model FastScape (Braun and Willett, 2013) to perform a simulation similar to the 1-D models presented here above. We used the following model parameters: $n = 4$, $m = 2$, $K = 3 \times 10^{-14}\,\mathrm{m^{1-3m}\,a^{-1-m}}$ and $U = 6\,\mathrm{mm\,a^{-1}}$ to allow for a direct comparison with the results shown in Fig. 2. The solution is shown in Fig. 6 and is also provided as a small animation (see Supplement).

On the one hand, and as in the 1-D model, the computed topography (Fig. 6d–f) remains relatively unchanged throughout the precipitation cycle with local variations of the order of a few tens of meters only. On the other hand, the erosion rate (Fig. 6a–c) changes dramatically from step to step. The model predicts a wave of erosion rate in each of the model catchments. The wave propagates at the same rate in all catchments, regardless of their size or geometry (panels a to c) demonstrating that Hack's law, used in the 1-D analytical solution (Eq. 1) and in the 1-D numerical model (Fig. 2), is a good approximation to the topology of catchments and the plan-form relationship between drainage area

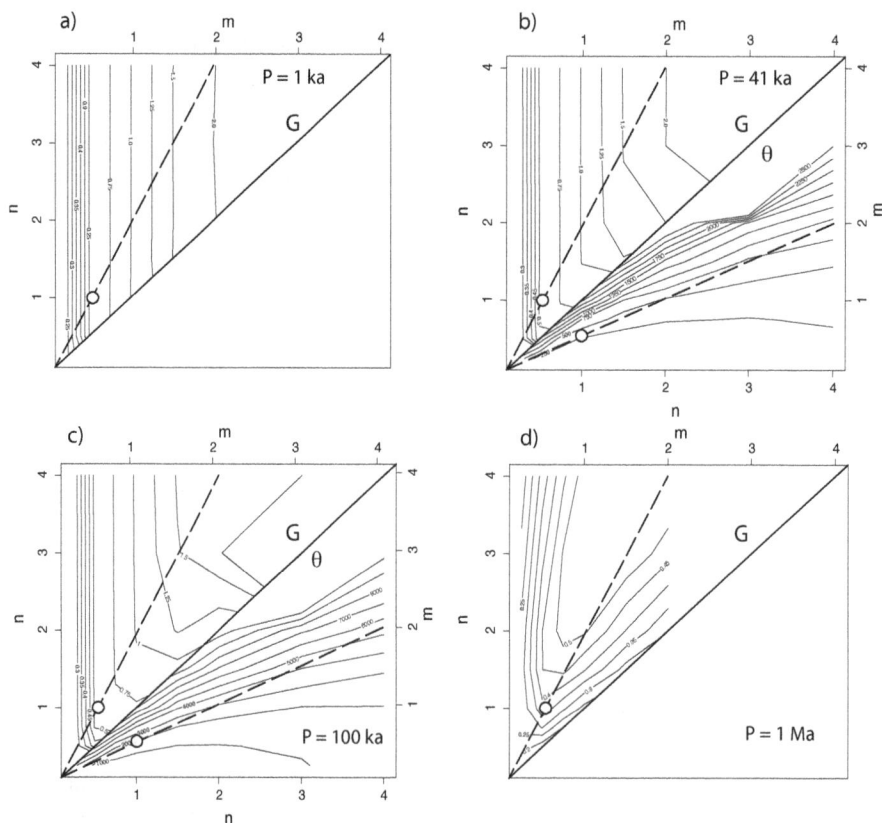

**Figure 5.** Contour plots of computed gain $G$ (upper left half of each panel), and time lag $\theta$ (lower right half of **b** and **c**), as a function of the stream power law exponents $m$ and $n$ for different forcing periods: **(a)** 1 ka, **(b)** 41 ka, **(c)** 100 ka and **(d)** 1 Ma. $U$ is 6 mm a$^{-1}$ and $p = 2$ in all model runs. Contour labels for $\theta$ are in years; $G$ is a dimensionless quantity. The dashed lines correspond to $m/n = 0.5$, a commonly accepted value derived from river profile concavity measurements. The circles correspond to the preferred values for $n$ ($= 1$) and $m$ ($= 0.5$). Gain and time lag were only computed for values of $n \geq m$.

and distance to the divide. We also note that drainage geometry and, a fortiori, drainage density does not change during a climate cycle. This is contrary to the results of simulations performed by Rinaldo et al. (1995) and Tucker and Slingerland (1997) implying that hillslope processes (not included in the FastScape model runs presented here) must control drainage density in a varying climate.

Another important outcome of our study is that, although the response of the stream power law to small cyclic variations in precipitation produces nearly undetectable changes in river longitudinal profiles, the erosional waves they trigger are measurable and, potentially, amplified (depending on the value of $m$). These waves could cause in situ measurements of erosion rate by cosmogenic isotope methods, for example, to be strongly variable both in space and time, rendering estimates of local or catchment-averaged exhumation rate rather difficult.

A direct comparison of our results (Fig. 1) with those of Godard et al. (2013) (see their Fig. 2) shows that the amplification of the climate cycles in the sedimentary record near the "forced oscillator" periods they evidenced is reproduced

by our model; it corresponds to the slight increase in the gain (or amplitude response) that is seen on all curves presented in Fig. 1 ahead of the transition to low gain values. Note that this slight increase in gain is relatively subtle compared to the main one we evidence here, which scales with $m$.

Unlike Godard et al. (2013), however, we did not include in our computations the effect of the hillslope response to variations in stream incision rate caused by rainfall cycles. Here, we focused our attention on the stream power law representing bedrock incision which we considered as the main controlling agent on the rate of landscape evolution in active mountain belts. In a mountain that has reached steady state between fast uplift and erosion, it is likely that hillslopes are close to or at a critical state (slope) and should therefore respond almost instantaneously to variations in stream incision rate, at least for forcing periods of the order of a few tens of thousands of years (the Milankovitch periods, for example). For slowly uplifting areas, this might not be the case and further work should concentrate on including a reasonable representation of hillslope process but also of sediment transport capacity by rivers in the calculations presented here.

Erosion Rate (m/yr)

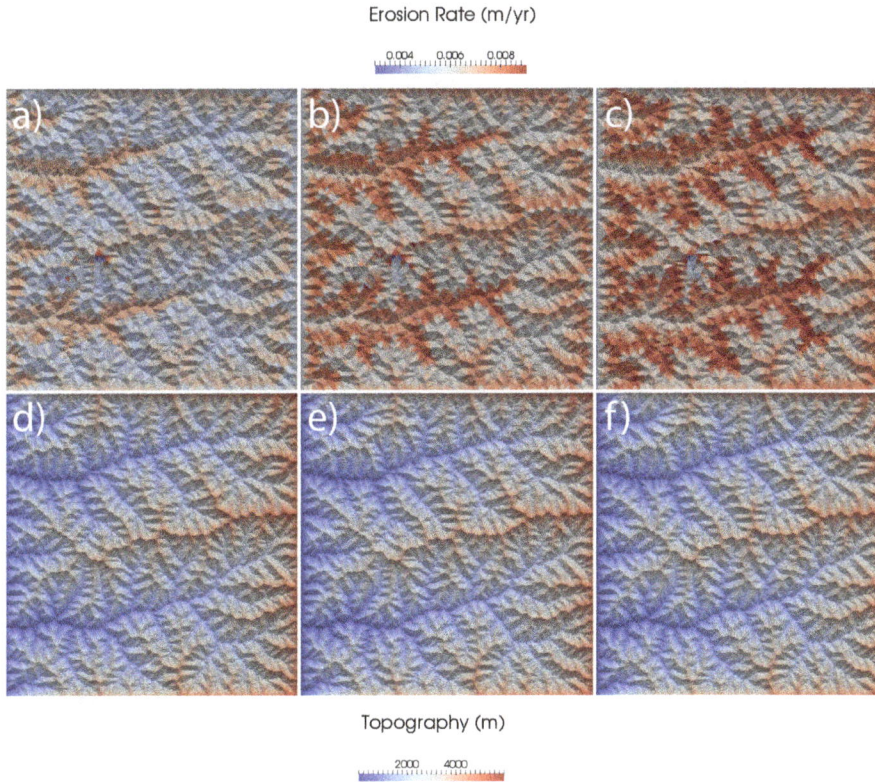

Topography (m)

**Figure 6.** Evolution of **(a)**–**(c)** the erosion rate and **(d)**–**(f)** the topography predicted by the FastScape landscape evolution model showing the propagation of waves of erosion during an imposed precipitation cycle (**a** and **d** correspond to the time when precipitation rate is at its mean value; **b** and **e** correspond to the time when precipitation rate has increased to half of its maximum amplitude, i.e., one-eighth into the precipitation cycle; **c** and **f** correspond to the time when precipitation rate has increased to its maximum amplitude, i.e., one-quarter into the precipitation cycle). Note the synchronism between all drainage basins, regardless of their drainage area and/or geometry.

Observations of a potential time lag between climate forcing and the erosional response of an active tectonic area are rare. In a recent paper, Gourlan et al. (2014) argue that they observe a time lag between $\delta^{18}O$ and $\epsilon_{Nd}$ records derived from a well-studied ODP site (ODP 758) located in the southern part of the Bengal Fan (Gourlan et al., 2010). This is, potentially, an appropriate site to observe changes in continental riverine input related to changes in the erosional flux from the Himalayas. The data sets they use are rather unique for they provide records of both climate and $\epsilon_{Nd}$ (a proxy for the intensity of the riverine sedimentary input from Himalayan rivers) at high resolution and on the same samples. This allows for a direct time correlation between the two data sets, even if the exact age of each sample is only constrained by correlating the local $\delta^{18}O$ signal with globally averaged sea surface temperature data (Gourlan et al., 2010). A careful spectral analysis of the two signals shows the existence of a well-defined time lag between $\delta^{18}O$ and $\epsilon_{Nd}$ at Milankovitch periods, which increases with the forcing period. This time lag is 1000, 2000 and 7000 ($\pm$500) years at the 23, 41, and 100 ka Milankovitch periods, respectively. Gourlan et al. (2014) argue that the delay between temperature changes recorded by $\delta^{18}O$ and the erosion flux out of the

Himalayas recorded by $\epsilon_{Nd}$ must be a consequence of how the variability in summer monsoonal rainfall affects erosion in the Himalayas. Estimates from global circulation models suggest that Indian monsoonal rainfall intensity varies in phase with temperature at orbital cycle periods (Braconnot, 2004) with an amplitude of a 1–2 mm day$^{-1}$, which is approximately 10 % of the present-day precipitations.

Using the numerical model described above, we searched through parameter space to find the best fitting model parameters that would provide a close fit to the observed time lags. We varied $m$, $n$, $p$ and $U$ and, for each run, adjusted $K$ so that the steady-state maximum mountain height is 6000 m. There is no single solution to this search. In Fig. 7, we show the fit of three model runs corresponding to various values of the model parameters.

We combined the solutions of many model runs performed at the three Milankovitch periods (23, 41 and 100 ka) (Fig. 8) but assuming a constant value of $U = 6 \, \text{mm a}^{-1}$ and $p = 2$, to show that the range of acceptable $m$ and $n$ values defines a region in $[m, n]$ space (dark grey shaded area in Fig. 8) that is sub-parallel to the commonly accepted range for $m$ and $n$ defined by $m/n = 0.5$. The corresponding gain factors range from 1 to 2, depending on the value chosen for $m$. For

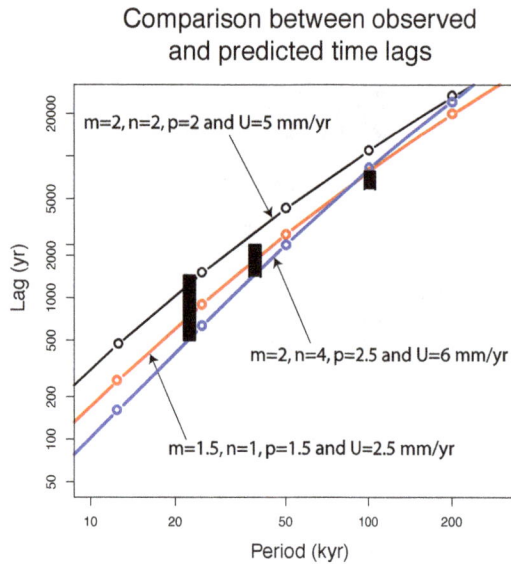

Figure 7. Comparison between observed time lags (i.e., derived from the spectral analysis of the geochemical proxies from Gourlan et al. (2014)) and the model predictions for different values of the $m$, $n$ and $p$ exponents and the assumed mean (steady-state) imposed uplift rate.

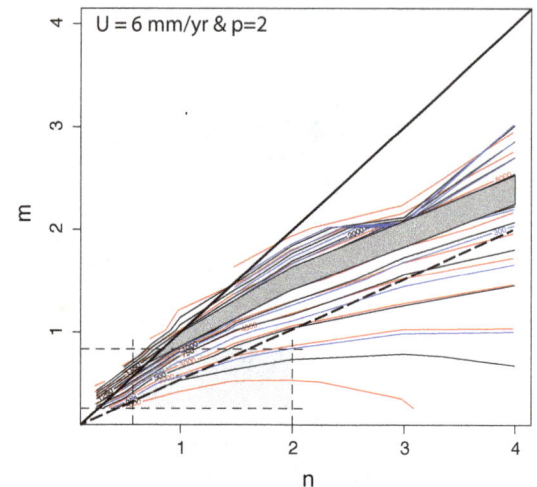

Figure 8. Contour plots of predicted time lags at the three Milankovitch periods (blue at 23 ka, black at 42 ka and red at 100 ka) as a function of $m$ and $n$. The dark grey shaded area corresponds to the values of $m$ and $n$ that satisfy the three time lags derived from the spectral analysis of geochemical data from the Bengal Fan (Gourlan et al., 2014); the light grey shaded area corresponds to the range of commonly accepted values for $m$ and $n$.

large $n$ values, the optimum $m/n$ ratio tends towards its more commonly accepted value of 0.5.

We note, however, that only a small sub-ensemble of the best fitting values of the $m$ and $n$ model parameters (dark grey shaded area) are within the most commonly accepted ranges ($0.2 < m < 0.8$ and $0.5 < n < 2$, light grey-shaded area between light dashed lines). This could imply that the time lags observed between the geochemical data sets are not related to the erosional response of the Himalayas to a cyclic rainfall; the time lags could originate from the delayed transport in the Ganges plains, for example. The temporary storage of sediments in the Indian plains is best described by a transport limited or diffusive model (Castelltort and Van Den Driessche, 2003). However, to fit the constraint provided by the two geochemical signals (i.e., that the time lag increases with the forcing period), the diffusivity parameter needs to be scaled in an ad hoc fashion with the period of fluctuations, which is difficult to justify. Alternatively and if we recall that the value of the $m$ and $n$ exponents is poorly constrained and remains the subject of much debate (see the recent review paper by Lague (2014) on this subject), the observed time lags could be regarded as new, independent constraints on the value of the stream power law parameters.

Our best fitting models have values for $m$ and $n$ that are either very large, if imposed to be in the accepted ratio of 2, or that are not in this accepted ratio. If $n$ is indeed large, the response of the erosional system to changes in slopes is strong. Interestingly, it has been recently demonstrated that the exponent $n$ may depend on the variability of river discharge, and thus climate (Lague and Hovius, 2005). In a variable climate,

$n$ and $m$ should have low values, with $n$ being close to unity, whereas, in locations where the climate is "steady", $n$ could be as large as 3 or 4 (Lague and Hovius, 2005). Alternatively, it could be that the ratio $m/n$ is not close to 0.5, but this is difficult to reconcile with the very numerous observations of the steady-state concavity of river profiles (see Whipple and Tucker (1999), for example), unless one calls into question the existence of steady-state conditions between uplift and erosion.

## 6   Conclusions

Based on both analytical and numerical solutions to the stream power law, we have shown that it is a natural behavior of this equation to produce a time lag between cycling climate forcing and the resulting erosional response. The main finding is that the time lag depends on the forcing period. If the forcing period is small compared to the characteristic timescale of the tectonic system (i.e., the time it takes for the system to approach steady state between uplift and erosion), the time lag is small; conversely, if the forcing period is large, the time lag tends towards a quarter of the period (the response is exactly out of phase with the forcing). The second important finding is that the erosional response is amplified in comparison with the amplitude of the climate forcing in a direct proportion to the parameter $m$, the discharge exponent in the stream power law when the forcing period is small. For large value of the forcing period (in comparison with the characteristic timescale of the system), the amplification tends towards 0, which means that very long-term variations in rainfall do not affect the erosional response of an active mountain belt and thus cannot be recorded, in the sedimentary record for example.

We have also demonstrated, based on simple 1-D and 2-D numerical landscape evolution experiments, that the response to climatic variations of an actively eroding mountain river, if it obeys the stream power equation, is independent of the size of its drainage basins, implying that, within a mountain belt, all rivers should respond in phase with each other to a periodic rainfall perturbation and, consequently, contribute constructively to the integrated sedimentary record.

We have shown that the response of a rapidly uplifting and eroding mountain belt to rainfall variations at Milankovitch periods can lag the climatic forcing by several thousands of years. This theoretical prediction should be used to interrogate the geological record and, potentially, test the validity of the stream power law as an adequate parameterization of fluvial erosion in active mountain belts. We have finally shown how geochemical signals could be used to extract such potential offsets under the assumption that they are adequate proxies for climate variability and the resulting erosional response. Potentially, such data sets could provide interesting and independent constraints on the slope and area exponents in the stream power law. We have also shown that the sedimentary flux fluctuations resulting from periodic rainfall variations can be amplified if $m > 1$, which may explain the strong imprint that Milankovitch cycles have on the sedimentary record despite the relatively small changes in both solar insulation and temperature that are associated with the corresponding variations in the Earth's orbital parameters.

## Appendix A: Response of the stream power law to periodic rainfall

### A1   The stream power law

Fluvial erosion in high-relief terrain is commonly parameterized by the stream power law (Howard, 1994):

$$\frac{\partial h}{\partial t} = U - K v^m A^m \left(\frac{\partial h}{\partial x}\right)^n, \tag{A1}$$

where $h$ is the height of the bedrock, $U$ is rock uplift, $A$ is drainage area, $v$ is precipitation rate and $K$, $n$ and $m$ are constants. Drainage area is known to increase as a power of the distance to the water divide:

$$A \propto (L - x)^p, \tag{A2}$$

where $x = L$ corresponds to the position of the water divide with respect to the point, $x = 0$, where the stream is held at base level:

$$h(0) = 0. \tag{A3}$$

Combining Eq. (A1) to Eq. (A2) leads to

$$\frac{\partial h}{\partial t} = U - K v^m (L - x)^{mp} \left(\frac{\partial h}{\partial x}\right)^n. \tag{A4}$$

Let us note that, as drainage area tends towards zero at the divide, this equation is singular at $x = L$. This is commonly handled by defining a critical slope, $S_c$, beyond which the stream power law is no more valid and colluvial and hillslope processes become dominant to maintain slope at $S = S_c$ (Whipple and Tucker, 1999).

### A2   Steady-state solution to the uplift/stream power equation

At steady state, we can write

$$U = K v^m (L - x)^{mp} \left(\frac{\partial \bar{h}}{\partial x}\right)^n, \tag{A5}$$

and the steady-state profile is

$$\bar{h} = \left(\frac{U}{K v^m}\right)^{1/n} \frac{1}{1 - mp/n} \left[L^{1-mp/n} - (L - x)^{1-mp/n}\right]. \tag{A6}$$

The maximum height is

$$H = \left(\frac{U}{K v^m}\right)^{1/n} \frac{L^{1-mp/n}}{1 - mp/n}. \tag{A7}$$

If we introduce dimensionless variables,

$$x' = x/L, \ h' = h/H, \ t' = t/\tau \tag{A8}$$

where $\tau = H/U$, Eq. (A4) becomes

$$\frac{\partial h'}{\partial t'} = 1 - (1 - x')^{mp} \left(\frac{\partial h'}{\partial x'}\right)^n. \tag{A9}$$

The boundary condition becomes

$$h'(0) = 0, \tag{A10}$$

and the steady-state solution takes the dimensionless form:

$$\bar{h}' = 1 - (1 - x')^{1-mp/n}. \tag{A11}$$

A similar expression can be found in Whipple and Tucker (1999).

### A3   Small perturbation in precipitation rate

Let us consider how a small temporal perturbation in precipitation rate $\delta v(t)$ affects the steady-state solution. We assume that the solution scales linearly with the perturbation:

$$h = \bar{h} + \delta h \tag{A12}$$

while still respecting the boundary condition:

$$\delta h(0) = 0. \tag{A13}$$

Equation (A4) becomes

$$\frac{\partial \delta h}{\partial t} = U - K(\bar{v} + \delta v)^m (L - x)^{mp} \left(\frac{\partial \bar{h}}{\partial x} + \frac{\partial \delta h}{\partial x}\right)^n, \tag{A14}$$

where $\bar{v}$ is the precipitation rate at steady state. Because both $\delta v$ and thus $\delta h$ are small, we can write

$$\frac{\partial \delta h}{\partial t} \approx U - K\left(v^m + mv^{m-1}\delta v\right)(L - x)^{mp}$$
$$\left[\left(\frac{\partial \bar{h}}{\partial x}\right)^n + n\left(\frac{\partial \bar{h}}{\partial x}\right)^{n-1} \frac{\partial \delta h}{\partial x}\right]. \tag{A15}$$

Using Eq. (A5) and keeping the terms $O(\delta v)$ and $O(\delta h)$ only, we obtain

$$\frac{\partial \delta h}{\partial t} \approx -m \frac{\delta v}{\bar{v}} U - n K \bar{v}^m (L - x)^{mp} \left(\frac{\partial \bar{h}}{\partial x}\right)^{n-1} \frac{\partial \delta h}{\partial x}. \tag{A16}$$

Using Eq. (A5) again, we can write

$$\frac{\partial \delta h}{\partial t} \approx -m \frac{\delta v}{\bar{v}} U - n K^{1/n} \bar{v}^{m/n} U^{(n-1)/n}$$
$$(L - x)^{mp/n} \frac{\partial \delta h}{\partial x} \tag{A17}$$

and by using Eq. (A7), we obtain

$$\frac{\partial \delta h}{\partial t} \approx -m \frac{\delta v}{\bar{v}} U - \frac{nU}{H} \frac{L^{1-mp/n}}{1-mp/n} (L-x)^{mp/n} \frac{\partial \delta h}{\partial x}. \quad \text{(A18)}$$

Using again the dimensionless variables,

$$\delta h' = \delta h / H, \ \delta v' = \delta v / \bar{v}, \ t' = t/\tau, \ x' = x/L, \quad \text{(A19)}$$

we obtain

$$\frac{\partial \delta h'}{\partial t'} \approx -m \delta v' - \frac{n}{1-mp/n} (1-x')^{mp/n} \frac{\partial \delta h'}{\partial x'}. \quad \text{(A20)}$$

If we integrate this equation over the length of the channel and introduce

$$\hat{h}' = \int_0^1 \delta h'(x') \mathrm{d}x', \quad \text{(A21)}$$

we obtain

$$\frac{\partial \hat{h}'}{\partial t'} \approx -m \delta v' - \frac{n}{1-mp/n} \int_0^1 (1-x')^{mp/n} \frac{\partial \delta h'}{\partial x'} \mathrm{d}x' \quad \text{(A22)}$$

and, by integrating by parts, we obtain

$$\frac{\partial \hat{h}'}{\partial t'} \approx -m \delta v' - \frac{mp}{1-mp/n} \int_0^1 (1-x')^{mp/n-1} \delta h' \mathrm{d}x' \quad \text{(A23)}$$

as $\delta h'(0) = 0$.

Here we need to make a further assumption, which is unlikely to be valid in all situations, but it will allow us to derive an approximate solution which we will test numerically. Under the assumption that the ratio $mp/n$ is close to unity, we can neglect the term $(1-x')^{mp/n-1}$ under the integral sign. This leads to

$$\frac{\partial \hat{h}'}{\partial t'} \approx -m \delta v' - \frac{mp}{1-mp/n} \int_0^1 \delta h' \mathrm{d}x' \quad \text{(A24)}$$

and

$$\frac{\partial \hat{h}'}{\partial t'} \approx -m \delta v' - \frac{mp}{1-mp/n} \hat{h}'. \quad \text{(A25)}$$

## A4   Periodic perturbation in precipitation rate

Assuming a periodic perturbation in precipitation rate which we express as

$$\delta v' = \delta v'_0 \sin(\alpha t'), \quad \text{(A26)}$$

we obtain the following solution:

$$\hat{h}' = \frac{m \delta v'_0 \left( \alpha \cos(\alpha t') - \frac{mp}{1-mp/n} \sin(\alpha t') \right)}{\alpha^2 + \left( \frac{mp}{1-mp/n} \right)^2}$$
$$+ C_0 e^{-\frac{mp}{1-mp/n} t'}. \quad \text{(A27)}$$

The negative exponential term corresponds to the transient response of the system, which is not of interest to us here and we will neglect it from now on.

## A5   Predicted sedimentary flux

The perturbation to the normalized sedimentary flux leaving the channel is given by

$$\delta Q' = \frac{\partial \hat{h}'}{\partial t'}$$
$$= -\frac{\alpha m \delta v'_0 \left( \alpha \sin(\alpha t') + \frac{mp}{1-mp/n} \cos(\alpha t') \right)}{\alpha^2 + \left( \frac{mp}{1-mp/n} \right)^2}, \quad \text{(A28)}$$

which can also be expressed in the following form:

$$\delta Q' \approx \frac{\alpha m \delta v'_0}{\sqrt{\alpha^2 + \left( \frac{mp}{1-mp/n} \right)^2}} \sin(\alpha t' + \theta'), \quad \text{(A29)}$$

where the lag, $\theta'$, is given by

$$\theta' = \tan^{-1} \left( \frac{1}{\alpha} \frac{mp}{1-mp/n} \right) \quad \text{(A30)}$$

and the amplification ratio, or gain, $G$, between normalized sedimentary flux variations and normalized precipitation variations, is given by

$$G = \delta q' / \delta v'_0 = \frac{\alpha m}{\sqrt{\alpha^2 + \left( \frac{mp}{1-mp/n} \right)^2}}. \quad \text{(A31)}$$

Going back to the dimensional variables, we see that this solution corresponds to a signal that lags behind the forcing by a time:

$$\theta = \frac{P}{2\pi} \tan^{-1} \left( \frac{P}{2\pi \tau} \frac{mp}{1-mp/n} \right), \quad \text{(A32)}$$

where $P$ is the period of forcing/climate change and $\tau$ is the characteristic time of the erosive system, i.e., the time it takes for erosion to come to equilibrium with tectonic uplift.

**Acknowledgements.** We thank Sébastien Castelltort, Tom Coulthard and two anonymous reviewers for their very helpful comments on an earlier version of this manuscript.

Edited by: J. Willenbring

# References

Anderson, R. S.: Modeling the tor-dotted crests, bedrock edges, and parabolic profiles of high alpine surfaces of the Wind River Range, Wyoming, Geomorphology, 46, 35–58, 2002.

Armitage, J. J., Duller, R. A., Whittaker, A. C., and Allen, P. A.: Transformation of tectonic and climatic signals from source to sedimentary archive, Nat. Geosci., 4, 1–5, 2011.

Boggart, P., van Balen, R., Kasse, C., and Vandenberghe, J.: Process-based modeling of fluvial system response to rapid climate change – I: model formulation and generic applications, Quaternary Sci. Rev., 22, 2077–2095, 2003.

Braconnot, P.: Modeling the last Glacial Maximum and mi-Holocene, Comptes Rendus Geoscience, 336, 711–719, 2004.

Braun, J. and Willett, S.: A very efficient $O(n)$, implicit and parallel method to solve the basic stream power law equation governing fluvial incision and landscape evolution, Geomorphology, 180–181, 170–179, 2013.

Castelltort, S. and Van Den Driessche, J.: How plausible are high-frequency sediment supply-driven cycles in the stratigraphic record?, Sediment. Geol., 157, 3–13, 2003.

Champagnac, J.-D., Molnar, P., Sue, C., and Herman, F.: Tectonics, climate, and mountain topography, J. Geophys. Res., 117, B02403, doi:10.1029/2011JB008348, 2012.

Clemens, S. C., Prell, W., Murray, D., Shimmield, G., and Weedon, G.: Forcing mechanism of the Indian-ocean monsoon, Nature, 353, 720–725, 1991.

Clemens, S., Murray, D., and Prell, W.: Nonstationary phase of the plio-pleistocene Asian monsoon, Science, 274, 943–948, 1996.

De Boer, P. L. and Smith, D. G.: Orbital forcing and cyclic sequences, 19, 1–14, 1994.

Demenocal, P. and Rind, S.: Sensitivity of Asian and African climate to variations in seasonal insolation, glacial cover, sea-surface temperature and Asian orography, J. Geophys. Res., 98, 7265–7287, 1993.

DiBiase, R. A. and Whipple, K. X.: The influence of erosion thresholds and runoff variability on the relationships among topography, climate, and erosion rate, J. Geophys. Res., 116, F04036, doi:10.1029/2011JF002095, 2011.

Dixon, J., Heimsath, A., Kaste, J., and Amundson, R.: Climate-driven processes of hillslope weathering, Geology, 37, 975–978, 2009.

Dosseto, A., Hesse, P., Maher, K., Fryirs, K., and Turner, S.: Climatic and vegetation control on sediment dynamics during the last glacial cycle, Geology, 38, 395–398, 2010.

Egholm, D., Nielsen, S., Pedersen, V., and Lesemann, J.-E.: Glacial effects limiting mountain height, Nature, 460, 884–887, 2009.

Godard, V., Tucker, G. E., Burch Fisher, G., Burbank, D. W., and Bookhagen, B.: Frequency-dependent landscape response to climatic forcing, Geophys. Res. Lett., 40, 859–863, doi:10.1002/grl.50253, 2013.

Gourlan, A., Meynadier, L., Allegre, C., Tapponier, P., Birck, J., and Joron, J.: Northern Hemisphere climate control of the Bengali rivers discharge during the past 4 Ma, Quaternary Sci. Rev., 29, 2482–2498, 2010.

Gourlan, A., Voisin, C., Chauvel, C., and Braun, J.: Delayed erosional response of the Hiamalayas to Milankovitch climate cycles, Geochem. Geophy. Geosy., submitted, 2014.

Hack, J. T.: Studies of longitudinal stream profiles in Virginia and Maryland, USGS Professional Paper 294, USGS, Washington, D.C., 97 pp., 1957.

Herman, F., Seward, D., Valla, P. G., Carter, A., Kohn, B., Willett, S. D., and Ehlers, T. A.: Worldwide acceleration of mountain erosion under a cooling climate, Nature, 504, 423–426, 2014.

Howard, A.: Equilibrium and time scales in geomorphology: application to sand-bed alluvial streams, Earth Surf. Proc. Land., 7, 303–325, 1982.

Howard, A.: A detachment–limited model of drainage basin evolution, Water Resour. Res., 30, 2261–2285, 1994.

Humphrey, N. F. and Heller, P. L.: Natural oscillations in coupled geomorphic systems: an alternative origin for cyclic sedimentation, Geology, 23, 499–502, 1995.

Hupp, C. R. and Osterkamp, W. R.: Riparian vegetation and fluvial geomorphic processes, Geomorphology, 14, 277–295, 1996.

Lague, D.: The stream power river incision model: evidence, theory and beyond, Earth Surf. Proc. Land., 39, 38–61, 2014.

Lague, D. and Hovius, N.: Discharge, discharge variability, and the bedrock channel profile, J. Geophys. Res., 110, F04006, doi:10.1029/2004JF000259, 2005.

Maher, K. and Chamberlain, C.: Hydrologic regulation of chemical weathering and the geologic Carbon cycle, Science, 343, 1502–1504, 2014.

Milankovitch, M.: Kanon der Erdbestrahlung und seine Andwendung auf das Eiszeiten-problem, Royal Serbian Academy, Belgrade, 1941.

Molnar, P. and England, P.: Late Cenozoic uplift of mountain ranges and global climate change: chicken and egg?, Nature, 346, 29–34, 1990.

Montgomery, D.: Valley incision and the uplift of mountain peaks, J. Geophys. Res., 99, 13913–13921, 1994.

Montgomery, D., Balco, G., and Willett, S.: Climate, tectonics, and the morphology of the Andes, Geology, 29, 579–582, 2001.

Paola, C., Heller, P. L., and Angevine, C. L.: The large-scale dynamics of grain-size variation in alluvial basins, 1: Theory, Basin Res., 4, 73–90, 1992.

Prell, W. and Kutzbach, J.: Monsoon variability over the past 150,000 years, J. Geophys. Res., 92, 8411–8425, 1987.

Rinaldo, A., Dietrich, W. E., Rigon, R., Vogel, G. K., and Rodríguez-Lturbe, I.: Geomorphological signatures of varying climate, Nature, 374, 632–635, 1995.

Roe, G., Stolar, D., and Willett, S.: The sensitivity of a critical wedge orogen to climatic and tectonic forcing, in: Tectonics, Climate, and Landscape Evolution, edited by: Willett, S., Hovius, N., Brandon, M., and Fisher, D., GSA Special Publication 398, 227–239, 2005.

Rossignolstrick, M.: African Monsoons, an immediate climate response to orbital insolation, Nature, 304, 46–49, 1983.

Simpson, G. and Castelltort, S.: Model shows that rivers transmit high-frequency climate cycles to the sedimentary record, Geology, 40, 1131–1134, 2012.

Strecker, M. R., Alonso, R. N., Bookhagen, B., Carrapa, B., Hilley, G. E., Sobel, E. R., and Trauth, M. H.: Tectonics and Climate of the Southern Central Andes, Annu. Rev. Earth Pl. Sc., 35, 747–787, 2007.

Thomson, S., Brandon, M., Tomkin, J., Reiners, P. W., Vasquez, C., and Wilson, N.: Glaciation as a destructive and constructive control on mountain building, Nature, 467, 313–317, 2010.

Tucker, G. and Slingerland, R.: Drainage basin responses to climate change, Water Resour. Res., 33, 2031–2047, 1997.

Valla, P., Shuster, D., and van der Beek, P.: Significant increase in relief of the European Alps during mid-Pleistocene glaciations, Nat. Geosci., 4, 688–692, 2011.

Whipple, K. X.: The influence of climate on the tectonic evolution of mountain belts, Nat. Geosci., 2, 97–104, 2009.

Whipple, K. X. and Meade, B.: Orogen response to changes in climatic and tectonic forcing, Earth Planet. Sc. Lett., 243, 218–228, 2006.

Whipple, K. X. and Tucker, G.: Dynamics of the stream-power incision model: implications for height limits of mountain ranges, landscape response timescales and research needs, J. Geophys. Res., 104, 17661–17674, 1999.

Whipple, K. X., Kirby, E., and Brocklehurst, S.: Geomorphic limits to climate-induced increases in topographic relief, Nature, 401, 39–43, 1999.

Willett, S., Beaumont, C., and Fullsack, P.: Mechanical model for the tectonics of doubly-vergent compressional orogens, Geology, 21, 371–374, 1993.

Willett, S., Schlunegger, F., and Picotti, V.: Messinian climate change and erosional destruction of the central European Alps, Geology, 34, 613–616, 2006.

Willgoose, G.: A statistic for testing the elevation characteristics of landscape simulation models, J. Geophys. Res., 99, 13987–13996, 1994.

Willgoose, G.: Mathematical Modeling of Whole Landscape Evolution, Annu. Rev. Earth Pl. Sc., 33, 443–459, 2005.

Wobus, C., Tucker, G., and Anderson, R.: Does climate change create distinctive patterns of landscape incision?, J. Geophys. Res., 115, F04008, doi:10.1029/2009JF001562, 2010.

Zhang, P., Molnar, P., and Downs, W.: Increased sedimentation rates and grain sizes 2–4 Myr ago due to the influence of climate change on erosion rates, Nature, 410, 891–897, 2001.

# 3

# Climate, tectonics or morphology: what signals can we see in drainage basin sediment yields?

T. J. Coulthard[1] and M. J. Van de Wiel[2]

[1] Department of Geography, Environment and Earth Sciences, University of Hull, UK
[2] Department of Geography, University of Western Ontario, London, Ontario, Canada

*Correspondence to:* T. J. Coulthard (t.coulthard@hull.ac.uk)

**Abstract.** Sediment yields from river basins are typically considered to be controlled by tectonic and climatic drivers. However, climate and tectonics can operate simultaneously and the impact of autogenic processes scrambling or shredding these inputs can make it hard to unpick the role of these drivers from the sedimentary record. Thus an understanding of the relative dominance of climate, tectonics or other processes in the output of sediment from a basin is vital. Here, we use a numerical landscape evolution model (CAESAR) to specifically examine the *relative* impact of climate change, tectonic uplift (instantaneous and gradual) and basin morphology on sediment yield. Unexpectedly, this shows how the sediment signal from significant rates of uplift (10 m instant or 25 mm a$^{-1}$) may be lost due to internal storage effects within even a small basin. However, the signal from modest increases in rainfall magnitude (10–20 %) can be seen in increases in sediment yield. In addition, in larger basins, tectonic inputs can be significantly *diluted* by regular delivery from non-uplifted parts of the basin.

## 1 Introduction

Sediment yields from upland basins are driven by tectonics and climate. The magnitude of sediment delivered by this "erosional engine" (Whittaker et al., 2009) is a major control on the size and location of sedimentary units found in depositional basins, and thus creates an opportunity to invert sedimentary records to establish climate and tectonic histories of the source basin.

However, the presence of two major external forcings (climate and tectonics) as well as the internal autogenic processing of these signals (e.g. Jerolmack and Paola, 2010; Van De Wiel and Coulthard, 2010) leads to a plurality of possible interpretations for each sedimentary record. The difficulty of inverting this plurality to establish whether individual or identifiable combinations of forcings can be determined is a fundamental limitation to our present capability to determine past climates and landscape histories from sedimentary records.

At its simplest, the sedimentary system can be split into three components, an upland production "erosional engine" (Whittaker et al., 2009), transfer of sediment, and deposition in a basin or store. Historically, research has focused on controls in the depositional setting; only more recently have researchers examined how tectonic and climatic changes can alter sediment production. For example, Willgoose et al. (1991), Whipple and Tucker (2002), Tucker and Whipple (2002), and others have demonstrated with numerical models how following uplift there is an increase in basin sediment discharge associated with a "wave" of incision that migrates upwards through a basin. Densmore et al. (2007) showed that for small fans the timing and amplitude of sediment flux to basins is controlled by changes in fault slip rates. Most recently, Armitage et al. (2011) modelled a coupled fan basin system and showed how fault slip rates altered grain size trends – with grain size changing spatially and temporally away from the fan apex. Other researchers have shown how climate changes can also change or increase sediment delivery (Tucker and Slingerland, 1997; Coulthard et al., 2002) through mechanisms such as the extension of the stream network and increased stream powers. Allen and

Densmore (2000) in their simulations from uplifting catchments suggest that sedimentary records more faithfully represent climate forcings than those from fault variability. Furthermore, physical experiments (Bonnet and Crave, 2003) also showed how erosional landscapes responded sharply to climate change. However, Armitage et al. (2013) found that sediment response was relatively insensitive to short-term climate variability.

One common theme from the above studies and others (Allen, 2008; Humphrey and Heller, 1995; Métivier and Gaudemer, 1999) is that basin response to external forcings is complex and strongly contingent upon the internal basin processing of these forcings. Therefore, there is clearly a need to disentangle the relative impacts of climate, tectonics and autogenics on sediment delivery and how these may manifest themselves in the sedimentary record.

In addition, comparatively little research has examined the transport of sediment from uplands to basin, with many theoretical and modelling studies assuming a direct link between the two. Yet in numerous natural settings there is a considerable length and width of alluvial "conduit" between areas of sediment erosion and deposition and therefore this conduit must play a key role in how signals are transmitted from uplands to basin (Castelltort and Van Den Driessche, 2003). Geomorphological research indicates that this conduit operates in a non-linear way as part of a complex response (e.g. Schumm, 1979) acting to buffer sediment supply signals (Castelltort and Van Den Driessche, 2003; Métivier, 1999), and more recent research has suggested that (geologically) short-term storage in floodplains may "shred" any upstream signals of forcings (Jerolmack and Paola, 2010). Simpson and Castelltort (2012) used a reach-based model to simulate how sediment pulses and water discharge perturbations may be transmitted to depositional settings. Their simulations showed that changes in discharge input and sediment storage (reflected in increased valley floor gradients) led to spikes in sediment associated with increased water inputs. Coulthard et al. (2005) calculated individual sediment budgets for reaches of a medium sized drainage basin from basin simulations carried out over a 9000 yr period. Their research showed that during wetter periods (increased rainfall magnitudes) there was a dramatic increase in sediment yields and this was largely sourced from the lower, valley floor sections of the catchment. In short, wetter climates mined out the valley floor of sediment. Clearly, the behaviour of the conduit between upland erosional areas and the ultimate depositional setting can have a major impact on sediment delivery.

In this paper we apply a numerical landscape evolution model to demonstrate that there are unexpected and major differences between the relative impacts of climate and tectonics on sediment delivery, and illustrate how the shape of the basin and length of conduit imparts an important control on this relationship.

## 2   Methods

### 2.1   The CAESAR model

The simulations were carried out with the CAESAR landscape evolution model (Coulthard et al., 2002; Van de Wiel et al., 2007). The main features of CAESAR are a combined hydrological and hydraulic flow model that operates on a sub-event time step, with multi-grain size erosion and deposition as well as slope processes (diffusive creep and landslides). CAESAR was initially developed to examine the relative roles of climate and land cover change on geomorphology and sediment yield and has been applied to a range of real drainage basins with outputs successfully compared to independent field data. These examples include patterns of sedimentation in Alpine environment (Welsh et al., 2009) sediment yields and longer term lowering rates from Northern Australia (Hancock et al., 2010), comparisons to field plot experiments (Coulthard et al., 2012a), predicting patterns of contaminated sediment dispersal (Coulthard and Macklin, 2003) and simulating 9000 yr of drainage basin evolution in the UK (Coulthard and Macklin, 2001).

CAESAR models landscape evolution by routing water over a grid of regular sized cells and changing elevations according to erosion and deposition from fluvial and slope processes. Caesar can be run in two ways: firstly in catchment mode (as used here), where there are no external in-fluxes other than rainfall; and secondly in reach mode, with one or more points at which sediment and water are input into the system. For both modes of operation CAESAR requires several parameters or initial conditions including surface elevation, grain sizes and rainfall (catchment mode) or a flow input (reach mode). In concept, the operation of Caesar is simple, where precipitation falling on the modelled surface drives fluvial and hillslope processes that determine the erosion and deposition for the modelled time step. This changes the topography, which then becomes the starting point for the following time step. Outputs of the model are the elevation changes across the whole modelled topography as well as water discharges and sediment fluxes at the outlet(s) over time. There are four main components in Caesar, a hydrological model, a surface flow model, fluvial erosion and deposition and slope processes.

When running in catchment mode, Caesar uses rainfall precipitation input to generate runoff over the catchment using an adaptation of TOPMODEL (Beven and Kirkby, 1979). TOPMODEL contains a lumped soil moisture store that creates surface runoff when it exceeds a threshold value. The movement of the surface runoff is then simulated using a surface flow model.

CAESAR's flow model uses a "flow-sweeping" algorithm, which calculates a steady-state, uniform flow approximation of the 2-D flow field. In this water discharge is distributed to all cells within a 2–5 cell range in front of a cell according to the differences between the water surface elevation of the donor cell and bed elevations of the receiving cells. Flow depths and velocity are calculated from discharges between cells using Manning's equation. These flow depths and velocities are then used to simulate the transport and deposition of sediment. Caesar calculates sediment transport over nine grain size fractions, which can be transported either as bed load or as suspended load, depending on the user specification. Caesar provides two options to calculate sediment transport, using Einstein (1950) or the Wilcock et al. (2003) equations. These equations are used as they calculate sediment transport for each size fraction that allows easy integration with the multiple grain size model components described later. For this study we have chosen to use the Einstein (1950) method where the calculation of sediment transport for each size fraction $i$ involves first determining the balance between the forces moving and restraining a particle:

$$\psi = \frac{(\rho_s - \rho) D_i}{\rho dS},  \tag{1}$$

where $\rho_s$ is the sediment density, $\rho$ water density, $D_i$ the grain size of the $i$ fraction, $d$ flow depth and $S$ slope. In the above equation the term $\rho dS$ is replaced by $\tau/g$, where $\tau$ denotes shear stress and $g$ gravitational acceleration. A dimensionless bedload transport rate, $\phi$, can then be estimated from $\psi$ using the relationship established by Einstein (1950):

$$\phi = 40 (1/\psi)^3.  \tag{2}$$

The value of $\phi$ is then used to rearrange the following equation to estimate $q_i$, the rate of sediment transport ($m^3\,s^{-1}$):

$$\phi = q_i \sqrt{\frac{\rho}{(\rho_s - \rho) g D_i^3}}.  \tag{3}$$

Deposition of sediments differs between bed load and suspended load. At each iteration all transported bed load is deposited in the receiving cells, whereas the deposition of suspended sediments is determined from fall velocities and concentrations for each suspended sediment fraction.

This procedure allows the selective erosion, transport and deposition of the different size fractions resulting in spatially variable sediment size distributions. Since this variability is expressed not only horizontally, but also vertically, CAESAR requires a method of storing sub-surface sediment data. This is enabled with a system of layers comprising an active layer representing the stream bed; multiple buried layers (strata); a base layer; and, if required, an immovable bedrock layer. These strata layers have a fixed thickness and their position is fixed relative to the bedrock layer and up to 20 strata can be stored at any cell on the grid. The active layer represents the river bed and has a variable thickness between 25 % and 150 % of the strata thickness. Erosion removes sediment and causes the active layer thickness to decrease. If the thickness becomes less than a threshold value, then the upper strata

layer is incorporated in the active layer to form a new, thicker active layer. Conversely, deposition adds material to the active layer, causing it to grow. If the active layer becomes greater than a set value, a new stratum is created leaving a thinner active layer.

Slope processes are also modelled, with mass movement represented as an instantaneous removal process. When the slope between adjacent cells exceeds a threshold (e.g. 0.5 or 45 degrees) material is moved from the uphill cell to the one below until the angle is lower than the threshold. As a small slide in a cell at the base of a slope may trigger more movement uphill, the model uses an iterative procedure to check the adjacent cells until there is no more movement.

Soil creep is also modelled by CAESAR using the equation below, where $C_{rate}$ is the user-specified rate of soil creep ($m\,yr^{-1}$), $T$ = time (years) and $Dx$ grid-cell size. This represents diffusion-like processes whereby sediment flux is linearly proportional to surface slope (Carson and Kirkby, 1972).

$$Creep = \frac{SC_{rate} T}{Dx}  \tag{4}$$

These mass movement and soil creep formulations allow material from slopes to be fed into the fluvial system as well as the input from landslides (both large scale and small – e.g. bank collapse). After the fluvial erosion/deposition and slope process amounts are calculated, the elevations and grain size properties of the cells are updated simultaneously.

CAESAR required modification to simulate uplift. A simple uplift algorithm was added which raises the elevations of designated cells by a specified amount, $\Delta Z_u$, at a specified time.

## 2.2 Model set-up and experimental procedure

Here, CAESAR was applied to the River Swale located in the North of England (Fig. 1). The section of the Swale modelled in this study lies upstream of Catterick Bridge, consisting of a total basin area of 490 km$^2$. The average elevation within the basin is 357 m, the altitudinal range is 68–712 m and the average river gradient is 0.0064. At the headwaters the landscape is characterized by steep valleys and the geology is Carboniferous limestone and millstone grit (Bowes et al., 2003). Further downstream, valleys become less pronounced and the underlying geology becomes Triassic mudstone and sandstone (Bowes et al., 2003).

For the numerical experiments, three different size sub-basins of the Swale were used (Fig. 1), herein termed small, medium and large, and located upstream of points S1, S2 and S3, respectively. The initial DEM for the simulations was obtained from the Ordnance Survey "OpenData" DEM library. The climatic inputs into the hydrological component of the model are a 10 yr hourly rainfall data set for the area generated by the UKCP09 weather generator (see Coulthard et al., 2012b), repeated for the length of the simulations. Basin

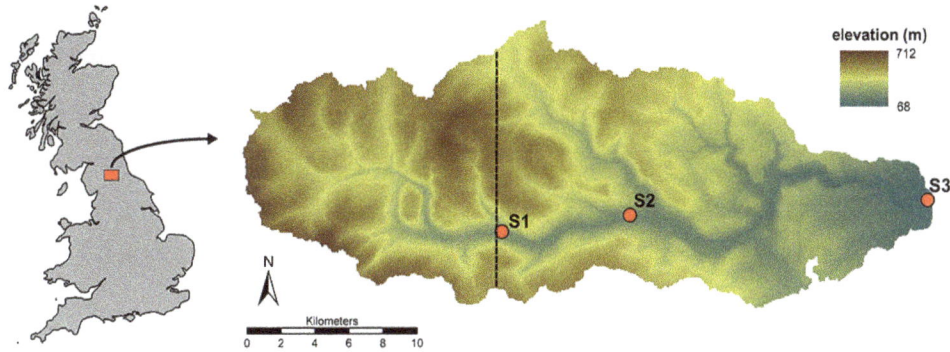

**Figure 1.** Location map (left) and elevation map (right) of the Swale basin. The area to the west of the fault line (dashed black line) was uplifted in the uplift simulations. Points S1, S2 and S3 indicate three locations where simulated sediment yields were recorded, respectively corresponding to a small, medium and large upstream sub-basin. Point S1 is located immediately downstream of the fault line. Points S2 and S3 are 10 and 30 km downstream from the fault line.

**Table 1.** Sediment yields in uplift scenarios for 100 yr simulations.

| scenario | S1 | | S2 | | S3 | |
|---|---|---|---|---|---|---|
| | cum. yield ($10^6$ m$^3$) | rel. change (%) | cum. yield ($10^6$ m$^3$) | rel. change (%) | cum. yield ($10^6$ m$^3$) | rel. change (%) |
| base | 2.18 | 0.0 | 1.63 | 0.0 | 16.16 | 0.0 |
| 1 m uplift | 2.23 | 2.5 | 1.65 | 1.2 | 16.24 | 0.5 |
| 2.5 m uplift | 2.46 | 13.0 | 1.68 | 3.1 | 16.23 | 0.4 |
| 5 m uplift | 2.98 | 36.9 | 1.71 | 4.8 | 16.41 | 1.5 |
| 10 m uplift | 3.96 | 81.8 | 2.02 | 24.1 | 16.58 | 2.6 |
| 25 m uplift | 6.74 | 209 | 2.67 | 63.7 | 17.78 | 10.0 |
| 50 m uplift | 11.82 | 442 | 3.75 | 130 | 19.90 | 23.1 |
| 100 m uplift | 21.03 | 865 | 6.12 | 275 | 23.76 | 47.0 |
| 250 m uplift | 40.67 | 1767 | 12.78 | 684 | 30.64 | 89.6 |

hydrology parameters were held constant during all simulated periods. There is no bedrock representation in these simulations, so all incision is into multi-grain sized material.

Nine groups of numerical experiments were conducted (E1 to E9), with each experiment consisting of several simulations in which either climate or uplift were increased over a range of values (Fig. 2). An additional simulation with no change provides a base reference to assess the impact of the changes in external driving conditions. Different experiments were conducted to investigate comparatively short-term impacts, over a period of 100 yr (E1, E2, E6, E7), and longer term impacts, over a period of 900 yr (E3, E4, E5, E7, E8). In these simulations, a year is defined as 365 days, that is, the extra day in leap years is ignored. Although CAESAR runs on sub-daily or sub-hourly time steps, output for the simulations is aggregated to facilitate analysis and interpretation. For the 100 yr simulations, data are aggregated in daily intervals. For the 900 yr simulations a 10 day interval was used. Additional analysis on annual and decadal intervals was done on further aggregation of these daily data.

In all experiments, the changes in rainfall or uplift were initiated at 50 yr into the simulation. For all uplift experiments (E1, E3, E5, E6, E8), only the area to the west of point S1 was uplifted (Fig. 2). Uplift is instantaneous in our simulations, with the exception of experiment E5 where a continued gradual uplift is applied every year for the last 850 yr of the simulation. For the climate experiments (E2, E4, E7, E9), the changes in climate were defined as a percentage increase in the base rainfall. This increase is incurred after 50 yr into the simulation, is sustained throughout the remainder of simulations and affects the entire basin (Fig. 2).

## 3 Results

### 3.1 Impacts of climate and tectonics on sediment totals

To investigate the role of tectonics we conducted a series of simulations (experiment E1) on the medium basin by instantaneously uplifting the upper section (i.e. the area west of S1; Figs. 2 and 3) after 50 yr of simulation, and then continuing the simulation for a further 50 yr. This was to establish whether there was any rapid/instant response as well as short-term (decadal) changes as the basin adjusts to its new topography. The simulations were repeated several times

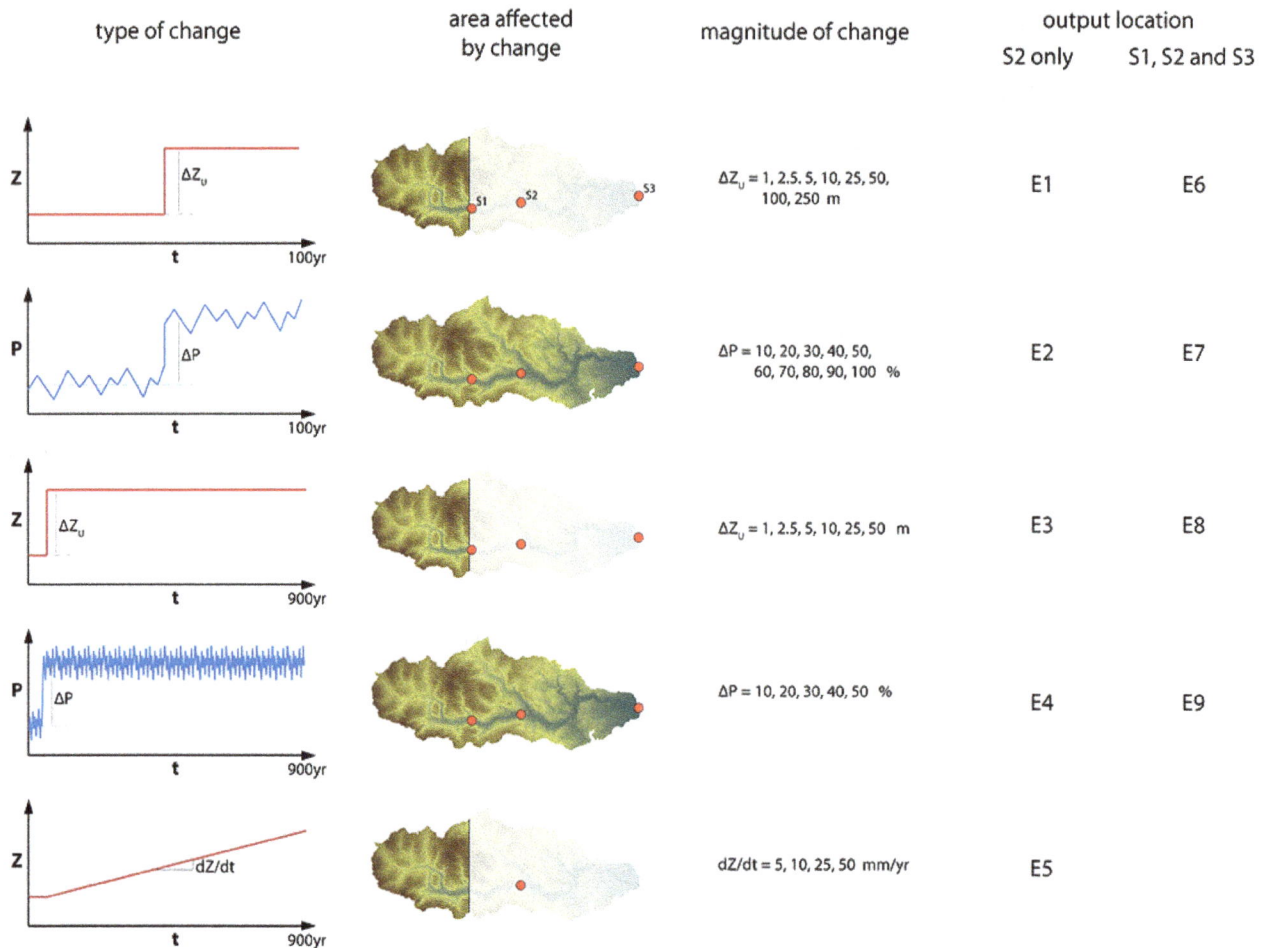

**Figure 2.** Schematic representation of experiment design. Nine numerical experiments were carried out (E1 to E9). Each experiment consists of a group of 4 to 10 simulations, where either rainfall (blue) or uplift (red) was altered after 50 yr into the simulation. In uplift experiments, only the western part of the basin is uplifted. Changes in rainfall affect the entire basin. Red dots on the maps indicate position of output points S1, S2 and S3.

with differing amounts of uplift ($\Delta Z_u = 1\,m, 2.5\,m, 5\,m, 10\,m,$ $25\,m, 50\,m, 100\,m, 250\,m$). Figure 3a shows cumulative sediment yields from the basin measured at point S2 and Table 1 contains cumulative yields and percentage increases in sediment yield. The red lines are from simulations with different instantaneous uplift amounts and generally show an increase in sediment yield following the uplift events. However, uplift of 10 m or more is required to create a noticeable shift in the cumulative line.

Using the same medium basin but disabling the uplift, we then investigated the role of climate by increasing the magnitude of the rainfall input to the hydrological model after 50 yr (E2). The simulations were run with increased rainfall rates ($\Delta P = 10$ to $100\,\%$, in $10\,\%$ intervals). The cumulative sediment yields from these simulations are plotted as blue lines on Fig. 3a. Increasing rainfall magnitude creates a very similar effect as tectonic uplift, though with a small but identifiable instantaneous response (a vertical increase

in the cumulative). Overall, comparatively small increases in rainfall magnitude result in relatively large increases in sediment yield (e.g. $20\,\%$ increase in rainfall increases sediment yield by $53\,\%$; Table 1, Fig. 3a). To achieve a similar increase of sediment yield, a large amount of instantaneous uplift is needed (e.g. 25 m of uplift results in $64\,\%$ increase in yield). In Fig. 4a, we compare the sediment yield totals after 100 yr from all the tectonic and climate simulations that comprised E1 and E2. Relatively modest rises in rainfall magnitudes ($+20\,\%$, $+30\,\%$) result in 50 to $100\,\%$ sediment yield increase, which is matched only by large levels of instantaneous uplift (i.e. 25 m or more). At larger rainfall increases, the difference becomes even starker. For example, approximately 90 m of uplift is required to achieve the cumulative sediment output from a $50\,\%$ increase in rainfall ($5.6 \times 10^6\,m^3$).

These simulations, however, are only over short 100 yr timescales. To investigate longer term trends we extended a

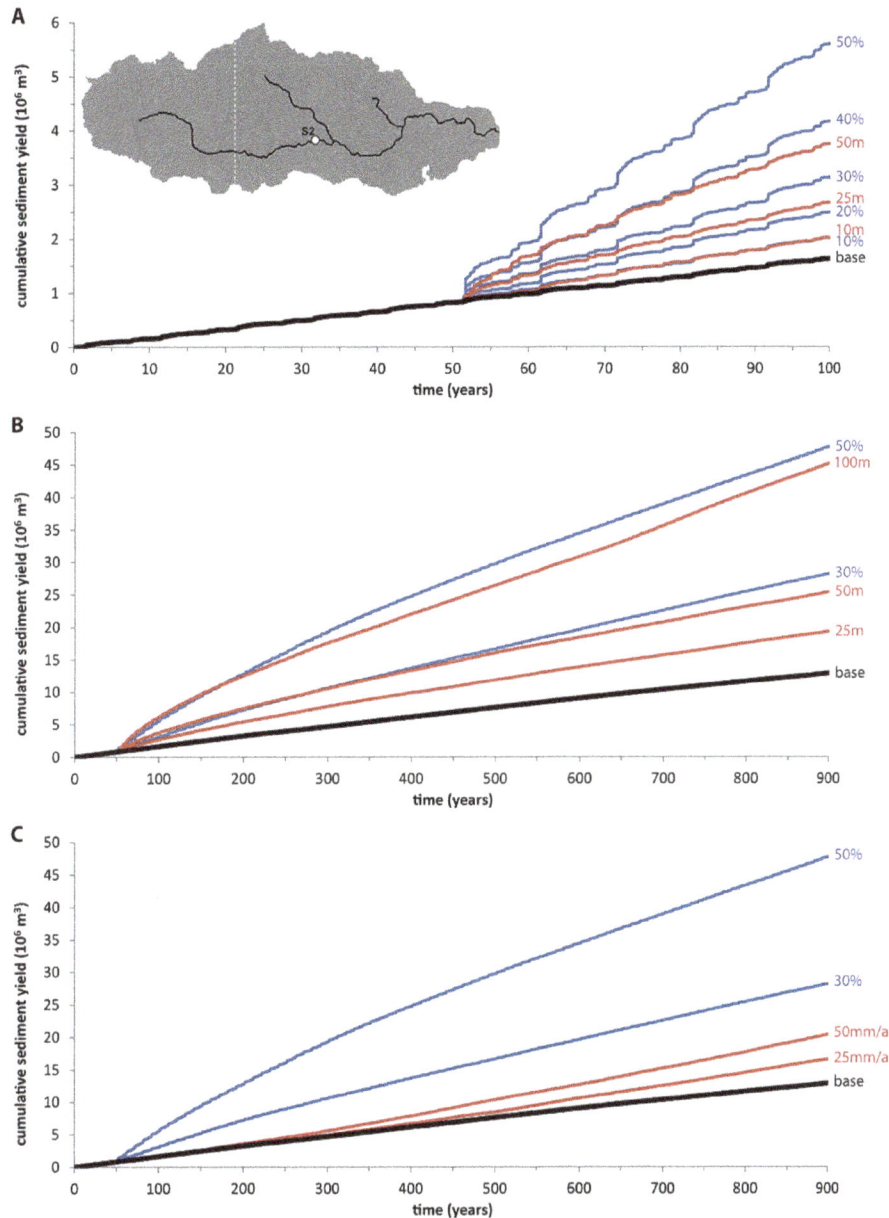

**Figure 3.** Simulated cumulative sediment yields at point S2 for a range of different scenarios from E1–E5. (**A**) 100 yr scenarios for rainfall increase and instantaneous uplift (E1, E2). (**B**) 900 yr scenarios for rainfall increase and instantaneous uplift (E3, E4). (**C**) 900 yr scenarios for rainfall increase and gradual uplift (E3, E5). The uplift or increase in rainfall occurred at 50 yr. The base scenario (no uplift, no rainfall increase) is shown as a thick black line, rainfall scenarios are shown as blue lines, and tectonic scenarios are shown as red lines. Inset shows location of fault line and measurement point S2.

select number of these runs to continue for 900 yr (E3, E4). Cumulative sediment yields show a very similar response to the shorter term simulations (Fig. 3b). There is a slight decay in the rate of increase in sediment yields after ca. 200 yr from both uplift and climate change – which is in response to locally increased gradients (tectonics) and expanding drainage areas (climate) leading to an initial surge from readily available sediment. Clearly, the trends observed in Fig. 3a are not transient conditions and persist over longer time periods.

We also investigated how gradual uplift ($dZ/dt = 5$, 10, 25 and 50 mm per year) compared to changes in climate (E5). As per previous results we can see that even large rates of gradual uplift are superseded by modest increases in climate (Fig. 3c).

To establish how basin shape and the length of channel between zones of uplift and deposition (the conduit) alter sediment delivery, an additional experiment (E6) was carried out uplifting the same upper section of the basin, but increasing

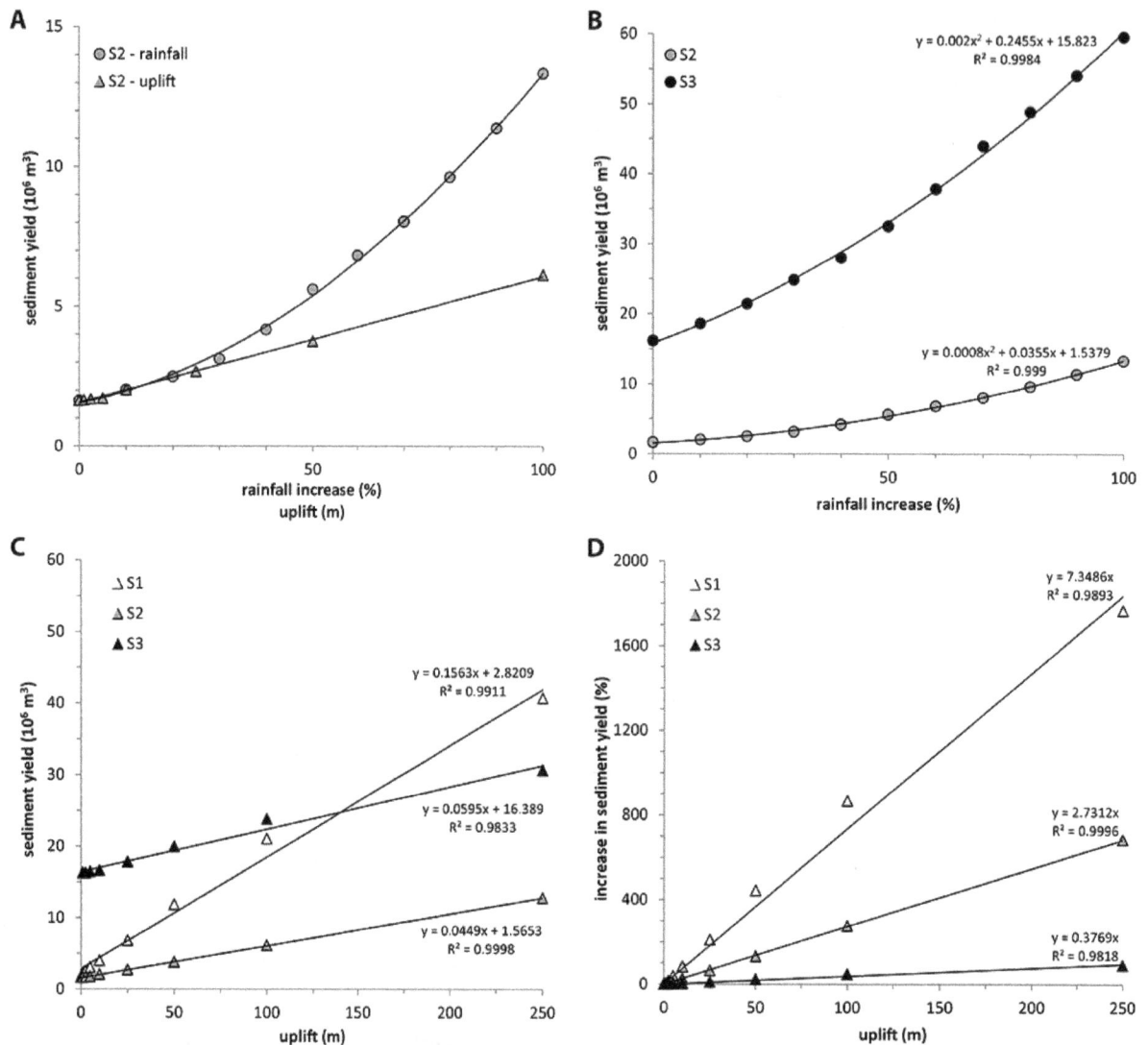

**Figure 4.** Impact of rainfall increase or tectonic uplift on simulated sediment yields after 100 yr, at point S1, S2 and S3 (see Fig. 1 for locations). (**A**) Impact of rainfall increase and tectonic uplift on sediment yield after 100 yr, at S2. (**B**) Total sediment yield for rainfall scenarios, at S2 and S3. (**C**) Total sediment yields for instantaneous uplift scenarios, at S1, S2 and S3. (**D**) Percentage change in sediment yield, relative to the base scenario, for instantaneous uplift scenarios, at S1, S2, S3.

the distance between uplift and basin outlet. Three size basins were used, with outlets at S1, S2 and S3, respectively 500 m, 10 km and 30 km downstream of the uplifted area (Figs. 1, 2). As per the previous experiments, uplift was instantaneous and added after 50 yr of simulation. As the temporal response was similar as previous simulations, only the impacts on total sediment yields are shown (Fig. 4c–d). These data demonstrate that the percentage increase in sediment yield due to uplift is highest in the small basin, and progressively smaller in the medium and large basins (Fig. 4d).

In absolute values, sediment yield increases linearly with uplift (Fig. 4c). In the larger basins a noticeable increase in sediment yield only occurs after 25 m of uplift, and is driven by the relatively substantial changes uplift causes in

basin relief (without uplift basin relief is 500 m). That aside, however, there are two other processes in operation. Firstly, adding a downstream section of valley floor to accommodate storage of sediment, as is the case in the medium and large basins, removes or "shreds" the part of the signal from the uplift events (Fig. 4c). The physical mechanism for this is evident studying the surface morphology of the simulations, where an alluvial fan forms immediately downstream of the uplift locus (Fig. 5). However, the larger basin does not seem to be more "effective" at shredding this signal than the medium basin, as the curves rise at very similar rates, albeit offset (Fig. 4c). Yet for the larger basin, the percentage change in sediment yield is far less (Fig. 4d). This demonstrates the second process, where the uplift signal is simply

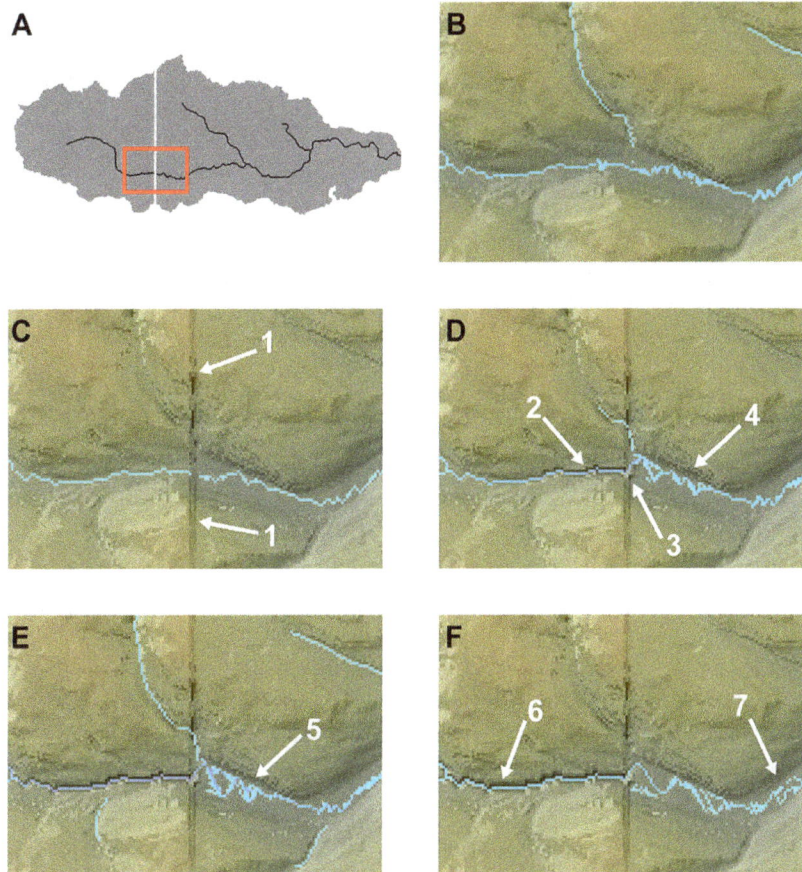

**Figure 5.** Geomorphic changes occurring after the 50 m instantaneous uplift scenario. (**A**) location of fault (solid white line) and area selected for elevation maps (red rectangle). (**B–F**) Elevation maps of selected area, respectively after 47.9, 50.1, 55.5, 63.5 and 99.2 yr of simulation. Arrows indicate specific features: 1. Escarpment after uplift; 2. Incision of channel upstream of fault line; 3. Formation of small alluvial fan; 4. Braided river pattern downstream of fault line; 5. Expansion of braiding; 6. Continued incision upstream of fault line; 7. Downstream propagation of braided river pattern.

*diluted* by sediment being added from the non-uplifted parts of the larger basin as the ratio of uplifted area to total basin area decreases.

Next we compared increases in sediment yield from uplift to those from rainfall in different size basins (E7). Unlike the uplift scenarios, where only part of the basin is uplifted, increases in rainfall affect the whole basin (Fig. 2), giving 2 to 3 times larger sediment yields for the larger basin, reflecting its larger area (Fig. 4b).

Figure 5 shows the impacts of 50 m of uplift on the geomorphology of the simulations around the area uplifted. This shows the before and ,at four time points, after uplift. Here there are clear changes in the nature of the channel both upstream and downstream. Upstream there is an entrenching of the main stream with a nick point migrating upstream ultimately leaving a river terrace behind. Downstream there is an immediate change in the channel pattern from single thread to braided (indicative of increased sediment yields) and the development of a small alluvial fan.

## 3.2 Impacts of climate and tectonics on sediment grain size

As well as providing information on sediment yields, the multiple grain sizes incorporated within CAESAR allow changes in grain size to be simulated in response to climate and tectonic forcings. Here we focus our discussion on the median grain size, D50, for experiment E6, that is, for the 100 yr simulations with uplift, for the S1, S2 and S3 basins (Fig. 6). These results show the annual mean D50 for all events within a year, and they thus increase and decrease with years that have fewer and greater large flood events. Daily data show similar patterns, but are omitted here as the yearly totals present a clearer picture. Results for D84 show similar trends and are not discussed.

For S1, the 5 m uplift initially leads to a 20 yr increase in the D50 (Fig. 6a). The initial coarsening agrees with field observations (e.g. Whittaker et al., 2009) associated with higher stream powers due to increased channel gradients (Whittaker et al., 2009). However, this effect begins to attenuate after

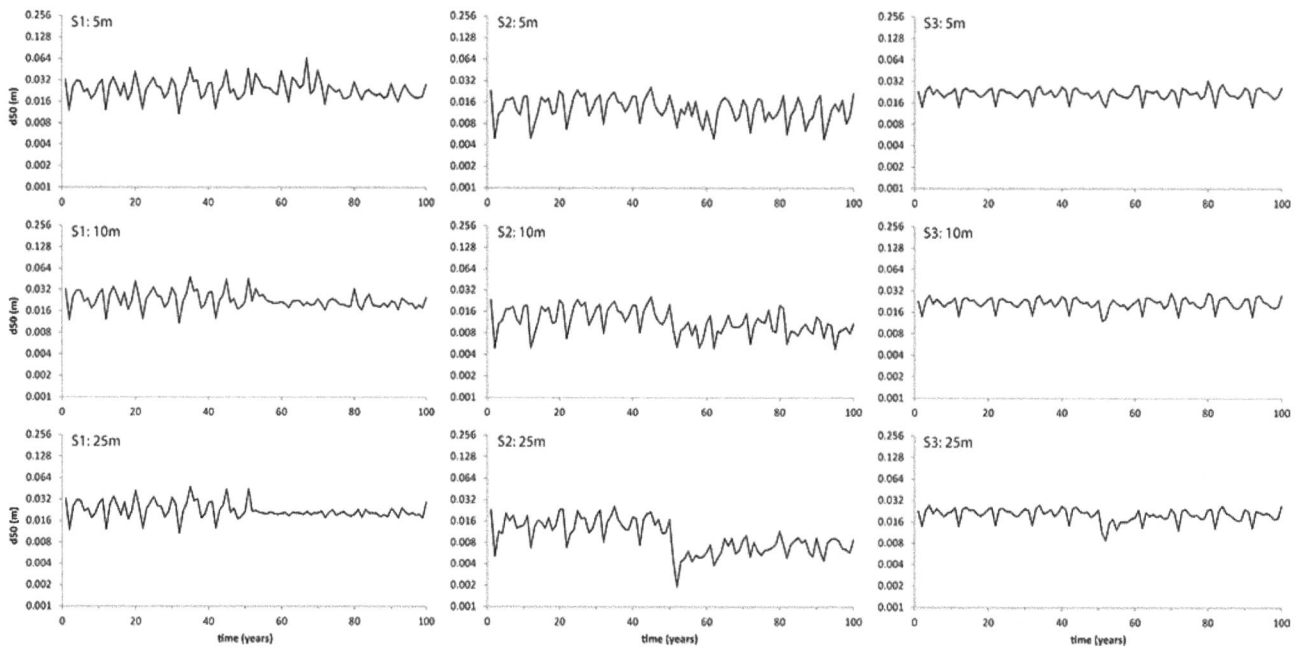

**Figure 6.** Median annual grain sizes for S1, S2 and S3 for 5, 10 and 25 m uplift simulations. Values are calculated by summing total daily sediment yields for each of 9 grain sizes output by CAESAR, over a simulated year. This was carried out in preference to calculating daily median grain sizes and then averaging this over a year, which would skew the median grain size by over-weighting the contribution of days with small daily sediment totals.

20 yr with the D50 diminishing to slightly less than that prior to uplift. For 10 m and 25 m of uplift there is a very different reaction with both a reduction in D50 and a smoothing of the signal from different annual events. For 10 m uplift there is a partial recovery in this signal after 30 yr and possibly beginning at 50 yr for 25 m uplift. The simulated fining associated with uplift is clearly different from expected results of coarsening (Whittaker et al., 2009) and we offer two explanations for this. Firstly, the 10 and 25 m uplift events we have simulated are very unusual in the field – and therefore it is quite possible we are simulating an effect not observed in the field. Secondly, our model has a high level of connectivity between channel, floodplain and hillslopes. Within our high uplift scenarios significant fluvial incision occurs (see Fig. 5), leading to the immediate introduction of fresh sediment from the collapse of river banks that will be sustained as a wave of incision passes up through the valley floor (as indicated in Fig. 5). The bank and slope material has a relatively higher concentration of fines than sediments in the river bed. In addition, this effect will also translate into tributary streams leading to a sustained input of finer sediment. The reduction in the annual variation in D50 in the 10 m and 25 m uplift simulations is a different example of signal shredding. Here the system has been overloaded with finer sediment creating a situation where climatic variations are no longer observed in the D50 (although they are in the total sediment yield).

At S2 all simulations show a fining of the sediment post uplift, with the effect becoming more amplified with greater amounts of uplift (Fig. 6). This shows that some of the finer sediment released from S1 after uplift is being transported through the reach – though, since the grain size reduction is reduced from S1 to S2, a substantial volume and proportionally more of the coarser material is being stored between S1 and S2. The reduction in interannual variability of the D50 that was observed at S1 is less noticeable at S2, although there still is some reduction in the variability as the amount of uplift increases.

For S3 there is little or no variation in D50 pattern in the 5 m and 10 m uplift scenarios. Only in the 25 m uplift scenario is there an observable impact on the D50, namely that the variability decreases. Location S3 is thus sufficiently far downstream of the fault-line for grain size not be affected by the uplift of 5 and 10 m with only a small impact from the 25 m uplift. In other words, the valley between S1 and S3 absorbs the effects of the uplift on transported sediment sizes.

Figure 7 describes the impacts of climate on grain size for basins S1, S2 and S3 (experiment E7). Unlike uplift overall there is a far weaker response in the grain size signal to increased wetness. For S1, S2 and S3 there is negligible change in grain size after the 10 % increase in wetness, with any variations indiscernible from variations prior to the increase. However, for larger increases in wetness drops or dips in grain size are amplified. These drops in grain size correspond to wetter years indicating that these correspond to the

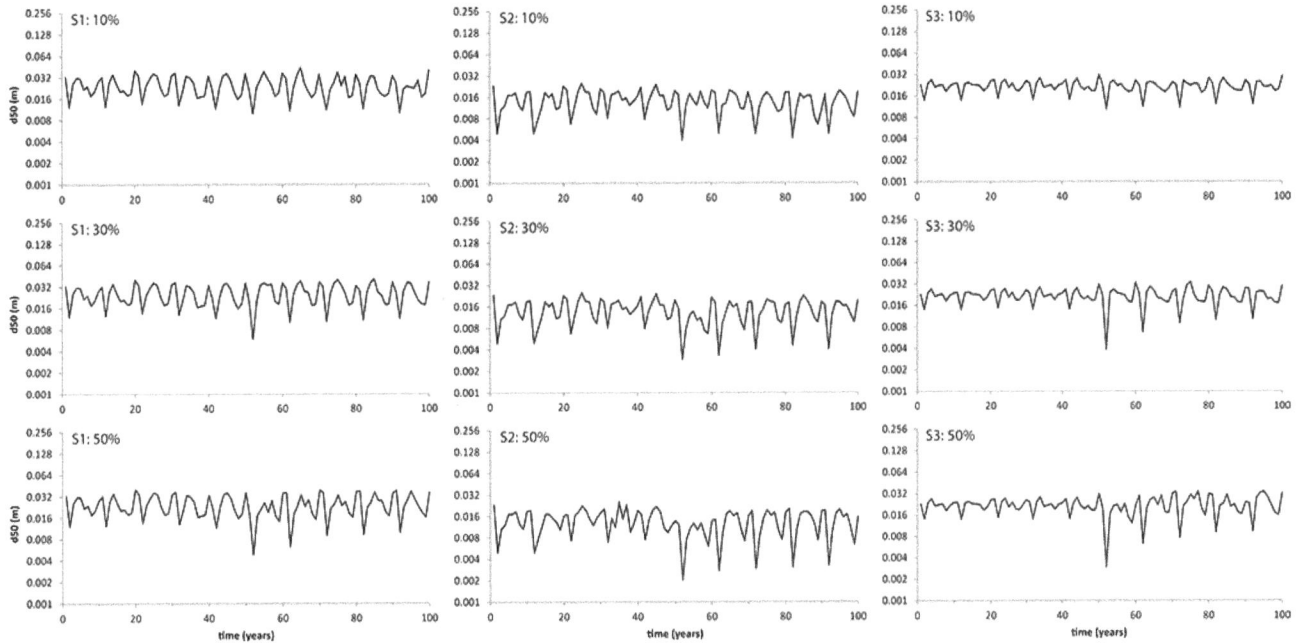

**Figure 7.** Grain size changes after rainfall changes for S1, 2 and 3 with 10, 30 and 50 % increases in precipitation magnitude after 50 simulated years. Median grain sizes were calculated as per the method outlined in Fig. 6.

increased delivery of fine sediments (from slopes, banks and also incision and expansion of tributaries) during wetter periods.

### 3.3 The role of autogenic processes

Within our results, we believe we can see evidence of autogenic, or internal processes that are creating signals within the basin outputs. In Fig. 8, we have plotted the annual and decadal sediment outputs from the 900 yr uplift simulations (E8), with actual and relative values (relative values are normalised to the mean pre-uplift sediment output).

Signals generated in post uplift sediment curve of S2 and to a lesser extent S3 are of similar (10 m) and greater (5 m) magnitude than peaks immediately following the uplift itself. Therefore, at both annual and decadal timescales the S2 and S3 basins are capable of generating sediment yield peaks equivalent to or greater than those seen immediately after uplift. There is also evidence that uplift changes the nature of sediment outputs from the system. For example, at S2, sediment yield variability increases following uplift (Table 2). This effect is more pronounced and longer lasting as the amount of uplift increases. At S3, only the 25 m uplift has an impact on the sediment yield variability (Table 2). This is also visible in Fig. 8, where the relative changes (right side graphs) also show there is a greater indication of the uplift signal in S2 than S3 (supporting earlier findings).

**Table 2.** Variability of relative sediment yield in uplift scenarios for 900 yr simulations.

| time period | S2 | | | S3 | | |
|---|---|---|---|---|---|---|
| | 5 m | 10 m | 25 m | 5 m | 10 m | 25 m |
| pre-uplift: 1–50 yr | 0.709 | 0.709 | 0.709 | 0.558 | 0.558 | 0.558 |
| post-uplift: 51–100 yr | 0.775 | 0.952 | 1.626 | 0.497 | 0.520 | 0.600 |
| post-uplift: 451–500 yr | 0.570 | 0.669 | 0.810 | 0.333 | 0.346 | 0.341 |
| post-uplift: 851–900 yr | 0.558 | 0.592 | 0.685 | 0.307 | 0.312 | 0.317 |

Variability is shown as standard deviations of relative annual sediment yield over the specified time period. Relative annual yields are normalized to the mean yield of the 50 yr pre-uplift period.

### 4 Discussion

These results provide us with considerable insight into how a drainage basin processes different external forcings. In these experiments climate changes clearly generate greater increases in sediment yield than uplift. In this study we have deliberately explored extreme values of climate and uplift, but if we restrict the results to more reasonable values (e.g. instantaneous uplift up to 10 m and rainfall/climate increases of 30 %) then in systems where the sedimentary deposits are *not* proximal to the sediment source, climate changes clearly have a far greater impact on sediment delivery. In our simulations elevated sediment yields from 10 m of uplift are equivalent to a 10 % increase in rainfall magnitude. Even looking at longer timescales and continual uplift rates of 25 mm yr$^{-1}$ equate to a 10 % increase in rainfall.

However, the converse is apparent when examining grain size. Here (especially at S1 and S2) there is a mixed reaction to uplift that generates an increase or decrease in D50

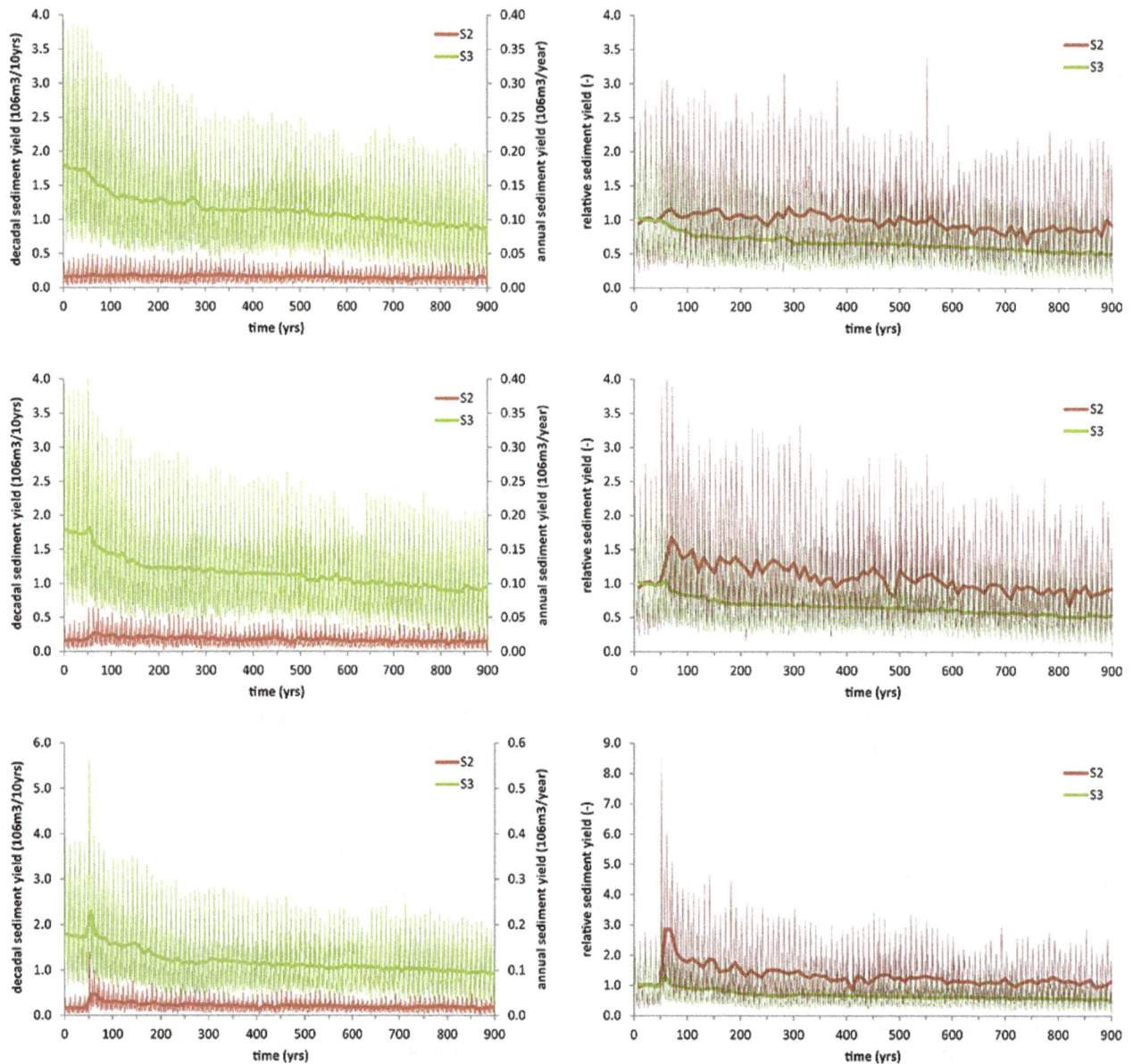

**Figure 8.** Annual and decadal sediment yields for S2 and S3 over 900 simulated years for 5, 10 and 25 m of uplift respectively (top to bottom). The left side shows absolute values and the right side relative values (normalised to pre-uplift average).

according to the magnitude of the change (increases associated with smaller amounts of uplift up to 5 m). For smaller (< 10 m) amounts of uplift our findings largely agree with those of Armitage et al. (2011) who note the deposition of larger sediment close to the uplift point (equivalent to our S1) with reductions in grain size further away (our S2 and S3). However, for larger uplift values, that is, 10 m and greater, we observed fining at S1 instead of coarsening. This would suggest a switch in sediment response above certain levels of uplift. Alterations in climate have a less apparent impact on D50 with the exception of the largest increases in rainfall magnitude (30 % and 50 %) leading to an increased drop in grain size during wetter years.

The role of the sedimentary system "shredding" input signals (Jerolmack and Paola, 2010) is also apparent – with the addition of less than 10 km of non-uplifted floodplain (the difference between S1 and S2) removing much of the uplift signal from both sediment volume and grain size. In Jerolmack and Paola's paper, they used a physical model of a rice pile – where the storage and (non-linear) release of rice in the pile led to the removal or shredding of any input signal in the output response. We would argue that a floodplain operates in a similar manner to the rice pile by storing and releasing sediment in a non-linear manner. Indeed, other workers have noted non-linear, self organising mechanisms in operation in floodplains with meandering and cut-offs (Hooke,

2003; Stølum, 1998). Thus, we would argue that for a given flood event the amount of sediment released from the end of a floodplain (bedload not suspended) may or may not bear any relation to the size of the flood event. Interestingly, our simulations show that as basin size increases the impact of "shredding" does not increase, thus indicating that only a short area of accommodation for storage and re-working of sediment is required. Furthermore, as the total area of basin relative to the area uplifted increases, the relative importance of shredding decreases as any signal from the uplift is diluted by sediment from tributaries in non-uplifted parts of the basin.

Our findings support previous work indicating long lag times from tectonic changes that are "buffered" by the fluvial system (Allen, 2008; Métivier and Gaudemer, 1999), in particular within areas of valley floor and floodplain (Castelltort and Van Den Driessche, 2003; Métivier and Gaudemer, 1999; Simpson and Castelltort, 2012) as found in our expanding catchment settings. It is worth noting that increases in sediment delivery are generated by *all* uplift and climate scenarios – but as catchment area and valley floor length grow, peaks in sediment were smoothed, or lost within the noise of the autogenic signals.

There may be important implications from this research for the interpretation of stratigraphy. Firstly, our results only generate sediment yields at the edge of our simulated drainage basins. We, therefore, do not account for any depositional settings or changes in accommodation space that may occur subsequently downstream. However, we suggest that the thickness of a sedimentary unit found in basins is more likely to represent changes in climate rather than any uplift history, except for where there is a direct and high level of connectivity between source and deposit. Closer to the point of uplift, coarsening in grain size is likely to be indicative of small/moderate uplift, but a fining in grain size indicates either large amounts of tectonic uplift or major increases in rainfall magnitude. In short, changes in the volume of sediment exported from a basin is more contingent on climate than tectonics, yet changes in grain size in this sediment are more likely to represent tectonic changes rather than climate. These findings re-enforce those from Simpson and Castelltort (2012) and of Allen and Densmore (2000) who suggest that discharge/climate variability may be better represented in the sediment signal than tectonic events and comment on the difficulties of inverting the sedimentary record (Simpson and Castelltort, 2012).

These results also indicate that short-term peaks in sediment volume are more likely to be generated by increases in rainfall than uplift events (as per the physical experiments of Bonnet and Crave, 2003). Therefore, marked increases in the thickness of sedimentary units – if representing increases in sediment supply – are more likely to indicate climatic changes rather than tectonic. It is important to note that this "short-term" sediment peak equates to a year or more of increased sediment delivery, rather than that from individual flood events, which may be very difficult to identify (Van de

Wiel and Coulthard, 2010). Furthermore, for moderate levels of uplift, the autogenic factors generate annual and even decadal peaks in sediment of equivalent or greater magnitude than uplift. In other words, the noise generated within the system is greater than the signal from the input.

We are aware that the timescales we have simulated here are considerably shorter than those used in other studies (e.g. Armitage et al., 2011, 2013). However, for sedimentary records over the Holocene and Quaternary scales, we would argue that the changes we are simulating are highly relevant. There are several studies that link increases in Holocene and Quaternary sedimentation rates to shorter term environmental forcings (e.g. Macklin et al., 2006) and even to individual events (e.g. Hinderer and Einsele, 1997; Hinderer, 2001; Larsen et al., 2013). We believe our findings are highly relevant in this context. Conversely, for records spanning tens of millions of years our findings may be less relevant and could just be part of the "noise" of the system. It is also important to note that our simulations are not run to any form of landscape steady state where there is equilibrium between (for example) uplift and erosion, and at these shorter timescales the catchments are always in a state of transience.

An important control on long-term basin development is bedrock – yet in these simulations we have not included its role. Within CAESAR, bedrock can be represented as a layer that is difficult/impossible to erode beneath layers of regolith/alluvium. However, when designing these experiments incorporating this would then introduce another parameter (the depth to bedrock) and we wished to keep the experiments as parsimonious as possible.

It is important to remember that whilst there is an unlimited depth of regolith in our simulations, this is of different grain sizes interacting through an active layer system. Therefore, the model will behave in a supply limited manner. This means that during the model operation the model will preferentially erode the finer fractions from channels and "wear down" any obstructions or high spots in the channel network (termed spinning up or conditioning, Coulthard et al., 2012b). As a result of this conditioning process the basin is left with an armoured channel bed that is relatively stable under the range of events that it formed under (i.e. the topography and rainfall events that formed it). If you perturb the basins driving forces (e.g. via increased precipitation or uplift) there follows an increase in sediment yield as the channel works through the armour layer that it had already developed. Therefore, in many respects, this scenario is identical to a the behaviour of a bedrock channel that experiences uplift or a change in climate, except the reaction speed of the bedrock basin will be far slower. For example, Godart et al. (2013) examine how basin lithology alters the response time of basins to climate changes with different "resonance periods" for harder and softer lithologies. For example a basin with a highly resistant bedrock would not have enough time to respond to short changes in climate – yet softer lithologies could respond (and thus generate sediment flux) to such changes (Godart et al., 2013). In addition, bedrock controls will have an increasing importance as the duration of simulations increases – so for the comparatively short times simulated here (100–900 yr) the effects may be less important.

Further complications of including bedrock include the position of channel bedrock sections relative to the point of uplift. For example if the uplift were located where bedrock was on the stream bed, then uplift would have a minimal impact on sediment delivery – as the controlling base level (the bedrock) would rise with the section of the catchment uplifted. However, if at the point of uplift bedrock were 2 m below the alluvial bed then we would expect a similar response to these simulations – equivalent to 2 m of uplift. Therefore, the introduction of bedrock below a surface would probably have the impact of reducing any spikes or pulses in sediment yield associated with uplift in effect limiting them to the depth to bedrock. Bedrock at a certain depth could also reduce the sediment output from increases in climate though the reduction would be smaller than for tectonics as climate change would impact across the whole basin.

Vegetation is another factor that could alter the relative result of climate and tectonic change on sediment yield. We did not model vegetation changes in these experiments, but in CAESAR, a simple vegetation model can act to bind the surface together until an erosional threshold is exceeded. This acts to reduce basin sediment output but would have a greater role on reducing the impact of climate change than tectonics – as increased sediment from climate is driven by channel network extension (and incision) into what would be vegetated slopes. Whereas tectonic increases in sediment are driven by local gradient changes that would largely be within non-vegetated (and thus unaffected) channels. The responses of a basin to bedrock and vegetation change are certainly worthy of further research, especially contrasting over the shorter (as per here) and longer timescales.

There are also limitations with the location and model parameterization. Firstly, the Swale is not a tectonically active basin, but was chosen as CAESAR has been extensively evaluated and validated on the Swale over decadal to centennial timescales. For determining how the distance from uplift affected any tectonic or climatic change signal the Swale has a relatively straightforward valley floor and is typical of many upland basins. Importantly, we chose a natural basin over an artificial landscape (e.g. Van De Wiel and Coulthard, 2010) as we wanted to include topographic heterogeneity (tributaries, floodplains, alluvial fans) as well as actual rainfall records and grain sizes. In addition, finding a field site with enough available data (in particular initial conditions e.g. topography) and with a tectonic and climatic history was difficult (as discussed later). We deliberately manipulated uplift rates and climate changes in a rather unrealistic manner, but this is to establish the outer limits of the relationships between climate, uplift and morphology on sediment yield.

Further modes of uplift were tried (fore tilt, back tilt, sideways tilt, gradual and instant) and all gave similar results.

Considering the model and parameterization, compared to alternative landscape evolution models, CAESAR has a different level of process representation. Therefore, some of the phenomena we have simulated here (e.g. different grain size responses to varying uplift) may be missed by different model configurations and parameterizations. For example the recent work by Armitage et al. (2011, 2013) is based on a one-dimensional model that whilst accounting for slope processes does not explicitly include them within a two-dimensional framework. It is also worth considering that in our study and that of Simpson and Castelltort (2012) hydraulics and flow processes are simulated and both models show a sensitivity of sediment output/throughput to climate. Whereas Armitage et al. (2013) do not simulate the hydraulics and show an insensitivity to climate (essentially impacting water discharge).

These observations raise the question as to what level of process detail, spatial and temporal resolution is required to simulate landscape dynamics. If high levels of detail are required, this may expose a particular problem for long-term landscape modelling, as field data to drive simulations (e.g. initial landscapes, grain sizes, climates) and validate model runs (e.g. stratigraphy, topographic data) are especially hard to come by. In addition, a circularity can develop whereby stratigraphy may be used to validate models, as well as generate driving data – yet could be highly variable and not truly reflect drivers or products of the basins dynamics (e.g. Coulthard et al., 2007; Van De Wiel and Coulthard, 2010).

Finally, assessing the reliability of numerical landscape evolution models is difficult given many of the issues outlined above (initial landscape, grain sizes, past climates etc.). As stated in the model description, CAESAR has successfully simulated landscape evolution in a range of environments (Coulthard and Macklin, 2001, 2003; Coulthard et al., 2012a; Hancock et al., 2011; Welsh et al., 2009), but it is important to question whether the non-linear response of the model is simply a model by-product or a representation of actual basin dynamics? CAESAR has a long history of modelling the non-linear reaction of catchments from 1998 through to 2010 (Coulthard and Van De Wiel, 2007; Coulthard et al., 1998, 2005; Van De Wiel and Coulthard, 2010) and given that similar non-linear dynamics have been well documented in fluvial systems (Cudden and Hoey, 2003; Gomez and Phillips, 1999; Hooke, 2003; Stølum, 1998), we consider the simulation of non-linear sediment dynamics in CAESAR genuine. In addition, CAESAR is certainly not the only model to show non-linear sediment responses. Examples of this include the Lapsus model (Temme and Van De Wiel, 2012), the results of Simpson and Castelltort (2012), and the Zscape model (Allen and Densmore, 2000). The notion that different numerical and physical models, with very different underlying algorithms produce non-linear responses indicates that what we simulate is a fundamental process of sediment transport dynamics rather than a model side effect.

## 5 Conclusions

A series of numerical experiments were carried out, in which the impacts of climate change and tectonic uplift on catchment sediment yield was evaluated. Our results indicate that both have an impact on the sediment yield, but the nature of that impact is different. Climate changes are more likely to impact the total volume of the sediment yield. Tectonic uplift, on the other hand is more likely to affect the grain size distribution of the sediment yield. In our simulations, the impacts of tectonic uplift on the volume sediment yield are pronounced immediately downstream of the uplift zone, but considerably less notable the further downstream you check. In effect the tectonic signal is diluted by storage of sediment in the floodplain.

In addition, our results also indicate that autogenic variability of sediment yield, due to temporary storage and release of sediment within the basin, can be of the same magnitude as the spikes in sediment yield associated with the external disturbance, that is, climate change or tectonic uplift.

All of these findings have implications for the reliability and meaning of inverting sedimentary records for determining past environmental and tectonic conditions.

**Acknowledgements.** M. J. Van de Wiel was supported by a NSERC Discovery Grant. We are very grateful for review comments from Arnaud Temme, Kevin Norton and Greg Hancock. In addition Sebastien Castelltort provided invaluable commentary on certain parts of the manuscript. The CAESAR model and source code is freely available from https://code.google.com/p/caesar-lisflood/.

Edited by: D. Lundbek Egholm

## References

Allen, P. A.: Time scales of tectonic landscapes and their sediment routing systems, Geological Society, London, Special Publications, 296, 7–28, doi:10.1144/SP296.2, 2008.

Allen, P. A. and Densmore, A. L.: Sediment flux from an uplifting fault block, Basin Res., 12, 367–380, doi:10.1046/j.1365-2117.2000.00135.x, 2000.

Armitage, J. J., Duller, R. A., Whittaker, A. C., and Allen, P. A.: Transformation of tectonic and climatic signals from source to sedimentary archive, Nat. Geosci., 4, 231–235, doi:10.1038/ngeo1087, 2011.

Armitage, J. J., Dunkley Jones, T., Duller, R. A., Whittaker, A. C., and Allen, P. A.: Temporal buffering of climate-driven sediment flux cycles by transient catchment response, Earth Planet. Sc. Lett., 369–370, 200–210, doi:10.1016/j.epsl.2013.03.020, 2013.

Beven, K. J. and Kirkby, M. J.: A physically based, variable contributing area model of basin hydrology / Un modèle à base physique de zone d'appel variable de l'hydrologie du bassin versant, Hydrological Sciences Bulletin, 24, 43–69, doi:10.1080/02626667909491834, 1979.

Bonnet, S. and Crave, A.: Landscape response to climate change: Insights from experimental modeling and implications for tectonic versus climatic uplift of topography, Geology, 31, 123, doi:10.1130/0091-7613(2003)031<0123:LRTCCI>2.0.CO;2, 2003.

Bowes, M. J., House, W. A., and Hodgkinson, R. A.: Phosphorus dynamics along a river continuum, Sci. Total Environ., 313, 199–212, doi:10.1016/S0048-9697(03)00260-2, 2003.

Carson, M. A. and Kirkby, M. J.: Hillslope Form and Process, Cambridge University Press, New York, 1972.

Castelltort, S. and Van Den Driessche, J.: How plausible are high-frequency sediment supply-driven cycles in the stratigraphic record?, Sediment. Geol., 157, 3–13, doi:10.1016/S0037-0738(03)00066-6, 2003.

Coulthard, T. J. and Macklin, M. G.: How sensitive are river systems to climate and land-use changes? A model-based evaluation, J. Quaternary Sci., 16, 347–351, doi:10.1002/jqs.604, 2001.

Coulthard, T. J. and Macklin, M. G.: Modeling long-term contamination in river systems from historical metal mining, Geology, 31, 451, doi:10.1130/0091-7613(2003)031<0451:MLCIRS>2.0.CO;2, 2003.

Coulthard, T. J. and Van De Wiel, M. J.: Quantifying fluvial non linearity and finding self organized criticality? Insights from simulations of river basin evolution, Geomorphology, 91, 216–235, doi:10.1016/j.geomorph.2007.04.011, 2007.

Coulthard, T. J., Kirkby, M. J., and Macklin, M. G.: Non-linearity and spatial resolution in a cellular automaton model of a small upland basin, Hydrol. Earth Syst. Sci., 2, 257–264, doi:10.5194/hess-2-257-1998, 1998.

Coulthard, T. J., Macklin, M. G., and Kirkby, M. J.: A cellular model of Holocene upland river basin and alluvial fan evolution, Earth Surf. Proc. Land., 27, 269–288, doi:10.1002/esp.318, 2002.

Coulthard, T. J., Lewin, J., and Macklin, M. G.: Modelling differential catchment response to environmental change, Geomorphology, 69, 222–241, doi:10.1016/j.geomorph.2005.01.008, 2005.

Coulthard, T. J., Hicks, D. M., and Van De Wiel, M. J.: Cellular modelling of river catchments and reaches: Advantages, limitations and prospects, Geomorphology, 90, 192–207, doi:10.1016/j.geomorph.2006.10.030, 2007.

Coulthard, T. J., Hancock, G. R., and Lowry, J. B. C.: Modelling soil erosion with a downscaled landscape evolution model, Earth Surf. Proc. Land., 37, 1046–1055, doi:10.1002/esp.3226, 2012a.

Coulthard, T. J., Ramirez, J., Fowler, H. J., and Glenis, V.: Using the UKCP09 probabilistic scenarios to model the amplified impact of climate change on drainage basin sediment yield, Hydrol. Earth Syst. Sci., 16, 4401–4416, doi:10.5194/hess-16-4401-2012, 2012b.

Cudden, J. R. and Hoey, T. B.: The causes of bedload pulses in a gravel channel: the implications of bedload grain-size distributions, Earth Surf. Proc. Land., 28, 1411–1428, doi:10.1002/esp.521, 2003.

Densmore, A. L., Allen, P. A., and Simpson, G.: Development and response of a coupled catchment fan system under changing tectonic and climatic forcing, J. Geophys. Res., 112, F01002, doi:10.1029/2006JF000474, 2007.

Einstein, H. A.: The Bed-load Function for Sediment Transportation in Open Channel Flows, in: Technical Bulletin No. 1026, USDA Soil Conservation Service, p. 71, US Department of Agriculture, 1950.

Godard, V., Tucker, G. E., Burch Fisher, G., Burbank, D. W., and Bookhagen, B.: Frequency-dependent landscape response to climatic forcing, Geophys. Res. Lett., 40, 859–863, doi:10.1002/grl.50253, 2013.

Gomez, B. and Phillips, J. D.: Deterministic Uncertainty in Bed Load Transport, J. Hydraul. Eng., 125, 305–308, 1999.

Hancock, G., Lowry, J., Coulthard, T., Evans, K. and Moliere, D.: A catchment scale evaluation of the SIBERIA and CAESAR landscape evolution models, Earth Surf. Proc. Land., 35, 863–875, doi:10.1002/esp.1863, 2010.

Hancock, G. R., Coulthard, T. J., Martinez, C., and Kalma, J. D.: An evaluation of landscape evolution models to simulate decadal and centennial scale soil erosion in grassland catchments, J. Hydrol., 398, 171–183, doi:10.1016/j.jhydrol.2010.12.002, 2011.

Hinderer, G. and Einsele, M.: Terrestrial sediment yield and the lifetimes of reservoirs, lakes, and larger basins, Geol. Rundsch., 86, 288–310, 1997.

Hinderer, M.: Dénudation quaternaire récente dans les Alpes, remplissage des vallées et des lacs, charge solide des rivières actuelles, Geodin. Acta, 14, 231–263, doi:10.1016/S0985-3111(01)01070-1, 2001.

Hooke, J.: River meander behaviour and instability: a framework for analysis, T. I. Brit. Geogr., 28, 238–253, doi:10.1111/1475-5661.00089, 2003.

Humphrey, N. F. and Heller, P. L.: Natural oscillations in coupled geomorphic systems: An alternative origin for cyclic sedimentation, Geology, 23, 499–502, doi:10.1130/0091-7613(1995)023, 1995.

Jerolmack, D. J. and Paola, C.: Shredding of environmental signals by sediment transport, Geophys. Res. Lett., 37, 1–5, doi:10.1029/2010GL044638, 2010.

Larsen, A., Bork, H.-R., Fuelling, A., Fuchs, M., and Larsen, J. R.: The processes and timing of sediment delivery from headwaters to the trunk stream of a Central European mountain gully catchment, Geomorphology, doi:10.1016/j.geomorph.2013.06.022, in press, 2013.

Macklin, M. G., Benito, G., Gregory, K. J., Johnstone, E., Lewin, J., Michczyńska, D. J., Soja, R., Starkel, L., and Thorndycraft, V. R.: Past hydrological events reflected in the Holocene fluvial record of Europe, Catena, 66, 145–154, doi:10.1016/j.catena.2005.07.015, 2006.

Métivier, F.: Diffusive like buffering and saturation of large rivers, Phys. Rev. E, 60, 5827–5832, doi:10.1103/PhysRevE.60.5827, 1999.

Métivier, F. and Gaudemer, Y.: Stability of output fluxes of large rivers in South and East Asia during the last 2 million years: implications on floodplain processes, Basin Res., 11, 293–303, doi:10.1046/j.1365-2117.1999.00101.x, 1999.

Schumm, S. A.: Geomorphic Thresholds: The Concept and Its Applications, T. I. Brit. Geogr., 4, 485–515, 1979.

Simpson, G. and Castelltort, S.: Model shows that rivers transmit high-frequency climate cycles to the sedimentary record, Geology, 40, 1131–1134, doi:10.1130/G33451.1, 2012.

Stølum, H.: Planform geometry and dynamics of meandering rivers, Geol. Soc. Am. Bull., 11, 1485–1498, 1998.

Temme, A. J. A. M. and Van De Wiel, M. J.: Filtering signals?: complexity experiments with LAPSUS, Geophysical Research Abstracts, EGU General Assembly, 14, 5299, 2012.

Tucker, G. E. and Slingerland, R.: Drainage basin responses to climate change, Water Resour. Res., 33, 2031, doi:10.1029/97WR00409, 1997.

Tucker, G. and Whipple, K.: Topographic outcomes predicted by stream erosion models: Sensitivity analysis and intermodel comparison, J. Geophys. Res., 107, 2179, doi:10.1029/2001JB000162, 2002.

Van De Wiel, M. J. and Coulthard, T. J.: Self-organized criticality in river basins: Challenging sedimentary records of environmental change, Geology, 38, 87–90, doi:10.1130/G30490.1, 2010.

Van de Wiel, M. J., Coulthard, T., Macklin, M., and Lewin, J.: Embedding reach-scale fluvial dynamics within the CAESAR cellular automaton landscape evolution model, Geomorphology, 90, 283–301, doi:10.1016/j.geomorph.2006.10.024, 2007.

Welsh, K. E., Dearing, J. A., Chiverrell, R. C., and Coulthard, T. J.: Testing a cellular modelling approach to simulating late-Holocene sediment and water transfer from catchment to lake in the French Alps since 1826, The Holocene, 19, 785–798, doi:10.1177/0959683609105303, 2009.

Whipple, K. and Tucker, G.: Implications of sediment-flux-dependent river incision models for landscape evolution, J. Geophys. Res., 107, 1–20, 2002.

Whittaker, A. C., Attal, M., and Allen, P. A.: Characterising the origin, nature and fate of sediment exported from catchments perturbed by active tectonics, Basin Res., 22, 809–828, doi:10.1111/j.1365-2117.2009.00447.x, 2009.

Wilcock, P. R., Crowe, J. C., and Wilcock, P. Crowe, J.: Surface-based transport model for mixed-size sediment, J. Hydraul. Eng., 129, 120–128, doi:10.1061/(ASCE)0733-9429(2003)129:2(120), 2003.

Willgoose, G., Bras, R. L., and Rodriguez-Iturbe, I.: A coupled channel network growth and hillslope evolution model 1. Theory, Water Resour. Res., 27, 1671–1684, doi:10.1029/91WR00935, 1991.

# 4

# Short Communication: Humans and the missing C-sink: erosion and burial of soil carbon through time

T. Hoffmann[1], S. M. Mudd[2], K. van Oost[3], G. Verstraeten[4], G. Erkens[5], A. Lang[6], H. Middelkoop[5], J. Boyle[6], J. O. Kaplan[7], J. Willenbring[8], and R. Aalto[9]

[1]Department of Geography, University of Bonn, Meckenheimer Allee 166, 53115 Bonn, Germany

[2]School of Geosciences, University of Edinburgh, Drummond Street, Edinburgh EH8 9XP, UK

[3]KU Leuven – University of Leuven, Department of Earth and Environmental Sciences, Celestijnenlaan 200e, 3001 Leuven, Belgium

[4]Department of Earth and Environmental Sciences, University of Leuven, Celestijnenlaan 200e, 3001 Heverlee, Belgium

[5]Department of Physical Geography, University of Utrecht, Heidelberglaan 2, 3584 CS Utrecht, the Netherlands

[6]School of Environmental Sciences, University of Liverpool, Liverpool L69 3GP, UK

[7]Institute of Environmental Engineering, Ecole Polytechnique Fédérale de Lausanne, Station 2, 1015 Lausanne, Switzerland

[8]University of Pennsylvania, Department of Earth and Environmental Science, 240 S. 33rd Street, Philadelphia, PA 19104-6313, USA

[9]College of Life and Environmental Sciences, University of Exeter, Rennes Drive, Exeter EX4 4RJ, UK

*Correspondence to:* T. Hoffmann (thomas.hoffmann@uni-bonn.de)

**Abstract.** Is anthropogenic soil erosion a sink or source of atmospheric carbon? The answer depends on factors beyond hillslope erosion alone because the probable fate of mobilized soil carbon evolves as it traverses the fluvial system. The transit path, residence times, and the resulting mechanisms of C-loss or gain change significantly down-basin and are currently difficult to predict as soils erode and floodplains evolve – this should be a key focus of future research.

## 1 Introduction

A considerable fraction of human-induced carbon dioxide ($CO_2$) emissions due to fossil fuel combustion and land cover change is absorbed by the oceans and the terrestrial biosphere. Soils store about 80 % or 2500 Pg C of the total carbon (C) in the terrestrial biosphere, and thus contain three times more C than the atmosphere (Lal, 2004). Consequently, soil C represents a substantial and highly sensitive component within the global carbon cycle and small changes in the soil C may result in large changes of atmospheric $CO_2$ at timescales of $10^1$ to $10^3$ yr.

During the last decade, research has highlighted the importance of vertical exchanges of carbon between the atmosphere, biosphere and pedosphere (e.g. Ruddiman, 2003).

Little attention was devoted to the full "life cycle" of the eroded minerals from soil to sea including the lateral fluxes of organic carbon (OC) associated with human-induced soil erosion. In many regions of the world, soils have increasingly been eroded (Wilkinson and McElroy, 2007), owing to increasing agricultural activity during the last few thousand years, with negative impacts on soil fertility and productivity as well as strongly increased lateral sediment-burden carbon fluxes. Therefore, sediment-burden carbon fluxes potentially provide an important, yet unknown, component in the global carbon cycle. As yet, lateral C-fluxes on hillslopes (Quinton et al., 2010; Starr et al., 2000; Van Oost et al., 2007) and in river channels (Aufdenkampe et al., 2011; Battin et al., 2008, 2009; Cole et al., 2007) (Fig. 1) have received little

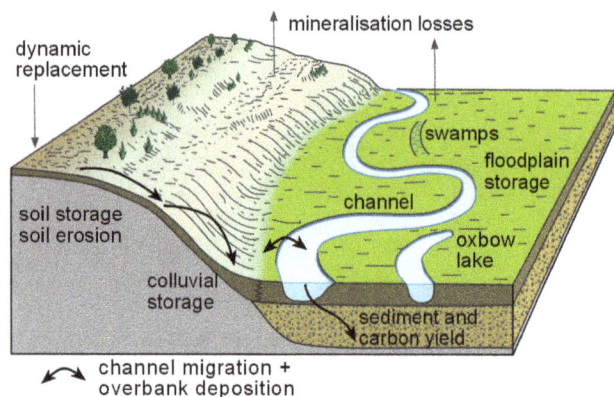

**Figure 1.** Storage compartments and lateral and vertical C-fluxes in agricultural landscapes.

attention, and their representation in contemporary carbon cycle models is rudimentary: soils are typically represented as spatially homogeneous and static entities, which does not reflect their dynamic nature in response to anthropogenic disturbance. Hillslopes are often regarded as simple engines that sequester or release carbon through soil formation and subsequent erosion. River channels are generally viewed as passive pipes that flush carbon from the hillslopes to the oceans (Battin et al., 2009), with floodplains that are mostly considered as wetlands that trap sediment and nutrients from the river, sustain productive riparian vegetation and hence induce higher $CH_4$ fluxes (Petrescu et al., 2010).

Sediment mobilized in river catchments by the introduction of agriculture has important implications for lateral carbon fluxes in fluvial systems. Soil erosion and carbon emissions are tightly linked through OC-stabilization by organo–mineral associations, requiring a profound understanding of the coupling between sediment and OC-erosion, as well as transport and storage. Three key mechanisms determine the coupling between sediment and carbon fluxes: (i) OC fixation to surfaces of mineral particles eroded on hillslopes and transported in river systems, (ii) increased mineralization of OC due to aggregate breakdown during transport, and (iii) stabilization of OC through burial (Fig. 1) (Van Oost et al., 2007). While soil scientists, geomorphologists and geochemists generally agree these are all key mechanisms, markedly different assumptions have been made about their relative importance. This has resulted in an intense debate as the inter-relation of these key mechanisms decides the significance of erosion-induced terrestrial carbon as sink or source (Berhe et al., 2007; Lal and Pimentel, 2008; Van Oost et al., 2007): strong mineralization acts as an atmospheric $CO_2$ source, while OC-fixation and burial provide a major atmospheric $CO_2$ sink. Because erosion, transport and depositional processes change along the sediment and carbon transport path from hillslopes to large river systems, the relative contributions of erosion induced OC-fixation, mineralization and OC-protection through burial change over space and

time. Recent studies have focused on decadal timescale perturbations to the C-cycle via soil erosion (Jacinthe and Lal, 2001; Ritchie and Rasmussen, 2000; Van Oost et al., 2007), yet it has emerged that sediments are stored on hillslopes and floodplains for several thousand years, meaning that human-induced changes to sediment-driven carbon fluxes can be delayed and altered over millennial timescales (Hoffmann et al., 2009a; Van Oost et al., 2012). Accounting for the non-steady-state C-dynamics along the flow path from hillslopes to river channels and then into the oceans is thus pertinent to understanding both the past and future global C-cycle. Here, we review the current understanding of sediment and carbon dynamics on hillslopes and in river channels and encourage an integrated view of long-term sediment-associated carbon dynamics.

## 2  Hillslope erosion and deposition

Hillslopes are major source areas of sediment and carbon. Estimates of global soil erosion on agricultural land vary between 28.1 and $150\,\text{Pg}\,\text{a}^{-1}$. Most recently Quinton et al. (2010) estimated the total contemporary erosion rate (including water, tillage and wind erosion) at $35 \pm 10\,\text{Pg}\,\text{a}^{-1}$, which corresponds to a carbon erosion rate of $0.5 \pm 0.15\,\text{Pg}\,\text{a}^{-1}$. These rates change through time as a result of population density, cultivation techniques and climate (Hoffmann et al., 2009a; Notebaert et al., 2011; Trimble, 1999). Eroded sites are usually characterised by lower soil fertility, crop productivity and C-contents than non-eroded sites supporting the notion that soil erosion is a major source of atmospheric $CO_2$ (Jacinthe and Lal, 2001; Lal, 2005; Victoria et al., 2012). Yet, OC-loss is partially balanced by OC-input through plants and fertilization and it may thus be argued that OC-fixation changes in concert with erosion rates at timescales covering the period of agriculture.

Estimates of soil and carbon erosion rates on regional to global scales in agriculturally developed river systems either rely on erosion plot studies or on measured sediment discharges. Plot studies, however, measure the amount of soil that is moved on the fields, and thus overestimate the "loss" of soils to river channels, which is only a small proportion of the eroded soil. In contrast, sediment discharges at river gauging stations estimate the sediment or OC-yield (efflux) that is transported beyond the gauging station, but do not quantify any sediment deposition upstream. Major flux differences found between these two approaches indicate that a large fraction of the detached soil is transported only a limited distance (Parsons et al., 2006) and that a substantial amount of sediment is stored at the foot of hillslopes: i.e. the transition between the hillslopes and the channels (Houben, 2008; Verstraeten et al., 2009). Thus, considering hillslopes simply as sources of sediment and carbon neglects (i) the differences between soil erosion and sediment yield (Dotterweich, 2008; Trimble, 1999; Verstraeten et al., 2009) and (ii) the complexity of the internal sediment dynamics at the hillslope scale (Lang and Honscheidt, 1999; Cerdà et al., 2012). This simplification can account for the major discrepancies in the current estimates of the effects of soil erosion: Lal (2005) indicate that 20–30 % of eroded OC is released through erosion-induced mineralization, whereas Van Oost et al. (2007) derive an order of magnitude lower mineralization rates of only $\sim 2$ % based on $^{137}$Cs inventories within fields.

During the last two decades, long-term sediment budgets (Hoffmann et al., 2010; Brown et al., 2009), which can account for some of the sources, pathways and sedimentary sinks, have increasingly shed light on the response of hillslope systems to human-induced soil erosion (Lang et al., 2003). These budgets provide an organizing framework to answer the following open research questions: How much eroded carbon is stored in colluvial sediments, and what are typical residence times of carbon and sediment on hillslopes? Sediment redistribution across hillslopes (as estimated by soil erosion plots) does not affect the downstream channel system, but does it have a significant impact on the global carbon cycle? How far must sediment be transported until it is depleted in carbon? Possible approaches to answer these questions are presented by Wang et al. (2010), who suggest OC-enrichment of exported sediment due to grain size sorting and the association of OC to fine grain fraction (e.g. clays), and by Van Oost et al. (2012) stating that 50 % of colluvial OC is decomposed within 500 yr. Yet, major uncertainties on the role of the hillslope sediment dynamics in the global carbon cycle remain (Aufdenkampe et al., 2011; Brantley et al., 2011; Yoo et al., 2011).

## 3 Fluvial transport and deposition

Recent studies have highlighted the role of rivers (including small streams, lakes, artificial reservoirs and wetlands) not only in transporting the C exported from terrestrial ecosystems but also in metabolising and burying significant amounts of C (Aufdenkampe et al., 2011; Battin et al., 2008, 2009; Cole et al., 2007; Tranvik et al., 2009). Globally, rivers receive about 2.9 Pg C each year, a quantity that represents the differences between global annual terrestrial production and respiration (Tranvik et al., 2009; Aufdenkampe et al., 2011). A major part of this carbon is associated with sediments mobilized through surface runoff. Only a fraction of this carbon is transported directly into the oceans (0.9 Pg C a$^{-1}$) while the majority is mineralized or outgassed to the atmosphere (1.4 Pg C a$^{-1}$), or buried in lakes and reservoirs (0.6 Pg a$^{-1}$) (Tranvik et al., 2009). Still, long-term C-burial in floodplains and subsequent outgassing are generally not considered (Battin et al., 2009; Cole et al., 2007; Tranvik et al., 2009) due to the limited availability of data on global floodplain extent, sedimentation rates, duration of inundation

and gas exchange velocities between floodplains and the atmosphere.

First estimates of global C-burial on hillslopes and floodplains (Aufdenkampe et al., 2011) range between 0.5–1.5 Pg C a$^{-1}$. These numbers are derived from the difference between global soil erosion (50–150 Pg C a$^{-1}$) (Wilkinson and McElroy, 2007) and sediment delivery to the ocean (12.6 Pg a$^{-1}$) (Syvitski et al., 2005), which is multiplied by an average C-concentration of eroded and exported sediment ($\sim 1$ %) (Aufdenkampe et al., 2011). Although informative, these numbers are not direct measurements, and account for neither temporal storage nor remobilization within the fluvial system. More explicit representation of floodplains and their impact on the global carbon cycle is essential because floodplains represent a major depositional environment (Aalto et al., 2003; Hoffmann et al., 2007; Noe and Hupp, 2005; Verstraeten et al., 2009). Based on a Holocene sediment budget of the Rhine basin, Hoffmann et al. (2013) estimated that $1.1 \pm 0.5$ Pg C is stored in the floodplains of the non-alpine part of the Rhine basin (i.e. 125 000 km$^2$), equivalent to a long-term OC-sequestration rate of 5.3 to 17.7 g C m$^{-2}$ a$^{-1}$. While these rates are time-integrated Holocene averages, several lines of evidence support a significant increase of organic-rich overbank deposition during the late Holocene (Hoffmann et al., 2009a; Verstraeten et al., 2009). Pre-human background rates of overbank deposition in the Rhine are $\sim 0.5$ mm a$^{-1}$, corresponding to a C-sequestration of 8.3 g C m$^{-2}$ a$^{-1}$. In contrast, maximum sedimentation rates during the last 300 yr indicate at least an order of magnitude increase to 15 mm a$^{-1}$ or 166 g C m$^{-2}$ a$^{-1}$. These high rates generally coincide with increased hillslope erosion resulting from agricultural intensification (Notebaert and Verstraeten, 2010; Hoffmann et al., 2009a).

High sequestration rates of 100 g C m$^{-2}$ a$^{-1}$ associated with soil formation in freshly deposited overbank deposits (Zehetner et al., 2009) imply that a large fraction of floodplain OC is not the result of sediment-burden carbon fluxes, but represents in-situ OC-formation. These high rates are maintained during the initial 100 yr after sediment deposition, and thus strongly conditioned by the input of fresh sediments that provide abundant mineral surface area for complexation of OC on timescales > 100 yr. As shown by studies on OC-burial in marine fans and shelves (Hilton et al., 2008; Galy et al., 2007), high sediment input not only favours OC-sequestration, but also increases the burial efficiency of OC in marine deposits.

The residence time and the stability of OC in floodplains are strongly related to the geomorphological and hydrological floodplain dynamics, depending on the dominant grain size of floodplain sediments and the structure and age of riparian vegetation. High energy, non-cohesive floodplains of headwater streams are characterised by coarse sediments, mobile channels, open vegetation cover, and large groundwater fluctuations. The high sediment transport capacity causes a low storage potential, and thus short residence times of floodplain OC. In contrast, low energy, cohesive floodplains are characterised by stable channel banks with massive and cohesive overbank deposits, and dense vegetation coverage. These represent major sedimentary sinks that are able to store sediment-burden carbon for several thousand years (Battin et al., 2008). Further important links between geomorphic dynamics and floodplain OC include changes of the groundwater tables, hydrological connectivity and river incision caused by river engineering and land cover change (Hupp et al., 2009; Noe and Hupp, 2005; Osterkamp et al., 2012).

The comparison of the burial efficiency (given by the rate of OC-accumulation minus the oxidation of OC within the sink) in different depositional settings suggests that sequestration of OC within floodplains exceeds that of lakes, artificial impounds and afforestation of catchments (Fig. 2). The ubiquitous prevalence of floodplains and their comparable burial efficiency to peats highlights both their importance and their insufficient representation within global C-budgets. Despite growing awareness of the multiple sources of floodplain OC and the feedbacks between sediment and vegetation dynamics in floodplains, quantitative understanding of the functioning of floodplains as C-sources or sinks remains elusive, with specific implications for the global C-cycle are hardly considered.

## 4   Towards integrated biogeochemical and geomorphological approaches

Recently, major progress has been made in understanding land use and climate impacts on sediment dynamics through the study of Holocene sediment budgets (Hoffmann et al., 2010). These studies provided essential information on (i) the sensitivity of hillslopes and channel systems to environmental change (Notebaert and Verstraeten, 2010; Verstraeten et al., 2009); (ii) the storage, residence time and remobilization of sediment along the flow path (Houben et al., 2009; Hoffmann et al., 2007; Aalto and Nittrouer, 2012); (iii) the connectivity between hillslopes and channels (Verstraeten et al., 2009; Lang et al., 2003); and (iv) the resulting non-linear dynamics between soil erosion and sediment yield (Erkens et al., 2011; Van De Wiel and Coulthard, 2010). While long-term budgets portrayed changing sediment dynamics through time and are of great value to reconstruct variable sediment burden OC-fluxes, their potential is not yet fully exploited. This becomes apparent when comparing estimates of sediment-associated OC-erosion during the last 50 yr with long-term OC-burial studies. In contrast to the limited impact of soil erosion on atmospheric-C during the last 50 yr (Van Oost et al., 2007), terrestrial sediment storage presents an important long-term atmospheric C-sink (Fig. 3) (Hoffmann et al., 2009b; Van Oost et al., 2012). The discrepancies between short- and long-term OC-budgets and their implications highlight that mechanisms associated with changing

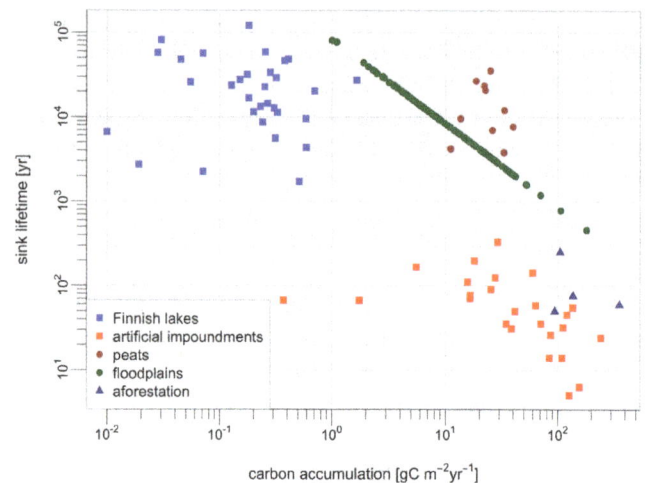

**Figure 2.** Lifetime versus sequestration rate of selected carbon pools. For details on data and calculations see Supplement.

conditions of OC-fixation and stabilization need to be considered more explicitly. These include the effects of changing hydrology, ecology (Osterkamp et al., 2012), and anthropogenically induced land cover change ALCC (Lechterbeck et al., 2009) during the Holocene. For instance, severe soil erosion in Mediterranean landscapes during the last Millennia changed soil covered regions into landscape of bare rock over large areas and thus resulted in almost irreversible changes of the ecological conditions (Fuchs et al., 2004; Lowdermilk, 1948). Dusar et al. (2011) and Marselli and Trincardi (2013), for instance, suggest that sediment yields in many Mediterranean landscapes declined during the last 1–2 ka as a consequence of the widespread soil depletion. Of major importance with respect to future global C-cycle is the declining capacity of the remaining soil to replace eroded OC as the extent and severity of soil degradation and desertification increases (Lal, 2009). In addition, the decay of buried OC in depositional settings is decreasing under drier climatic conditions as projected for the future decades (Solomon et al., 2007). Furthermore, deforestation increases not only soil erosion and sediment flux into sedimentary sinks, but also transforms the morphology of river channels (Walter and Merritts, 2008). Changing channel morphology involves, for example, (i) transformation of floodplains from stable channels in cohesive deposits to coarse sediments and mobile channels, and (ii) transitions from higher to low groundwater levels with corresponding changes from stabilized floodplain OC to destabilization and emission of large amounts of organic C to the atmosphere.

These conceptual considerations highlight the influence of indirect links between geomorphological processes and OC-fluxes that act on century-to-Holocene timescales. We believe that the on-going discussion about whether anthropogenic soil erosion is a sink or source of atmospheric carbon will not be solved until we synthesize biogeochemical

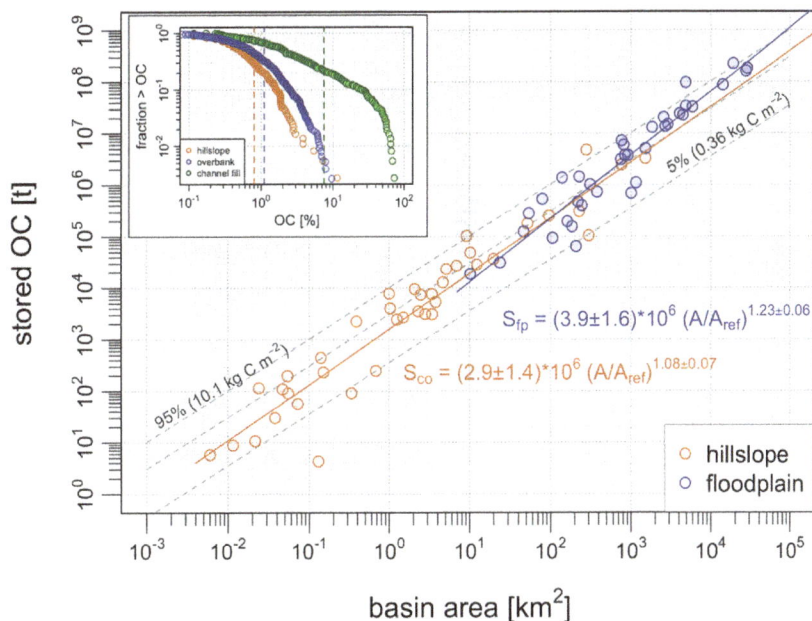

**Figure 3.** Holocene carbon storage on hillslopes and in floodplains as a function of basin size (Hoffmann et al., 2013). Lines represent predicted hillslope ($S_{co}$) and floodplain ($S_{fp}$) storage using power law regression with normalized basin area: $S = \alpha \times (A \times A_{ref}^{-1})^{\beta}$. The scaling coefficient $\alpha$ represent OC-storage [t] at the reference basin area ($A_{ref} = 1000 \, km^2$) and mainly reflect the greater OC-concentration of floodplain deposits (i.e. overbank and channel fills) compared to hillslope deposits (see inset: cumulative frequency distributions of OC-inventory from hillslope, and floodplain sediments). The scaling exponent of floodplain storage $\beta_{fp} = 1.23 \pm 0.06$ indicates a greater increase of floodplain OC with basin than hillslope OC, which is given by $\beta_{co} = 1.08 \pm 0.07$ slightly larger than one. These results are in accordance with an increasing accommodation space for floodplain storage with basin size. For details see Supplement.

and geomorphological approaches, thereby properly considering the fate of mobilized OC as it travels from the hillslopes through the river network at corresponding timescales of $10^0$ to $10^4$ yr. Such detailed studies are increasingly feasible using improved techniques for high-resolution geochronology of colluvial and fluvial deposits over timescales for decades to millennia (Aalto and Nittrouer, 2012; Chiverrell et al., 2008; Hobo et al., 2010).

To put our estimates of hillslope- and floodplain-OC storage into perspective (Fig. 3), we extracted C-emissions as a result of ALCC for the Rhine basin from the global modelling scenarios described in Kaplan et al. (2011, 2012) (Fig. 4). Our sediment storage data indicates that hillslopes and floodplains could have sequestered an amount of OC similar to the cumulative C-emissions from anthropogenic land cover change during the last 8000 yr. Thus, in this region, soil erosion associated with human-induced land-cover changes potentially offsets the effects of C-emissions caused by long-term ALCC. Furthermore, high accumulation rates and long residence times indicate that floodplains and peats represent the dominant terrestrial OC-sink, while OC-uptake in lakes, reservoirs and forests is limited due to either lower accumulation rates or shorter residence times in these pools (Fig. 2). Yet, the timescales of protection of eroded organic soil material are insufficiently understood (including the complex mechanisms of ensuing reburial and then fur-

ther destabilization due to changing environmental conditions on hillslopes and in floodplains). Thus, whether anthropogenic soil erosion is a sink or source of atmospheric carbon is not dependent on hillslopes alone. Instead, the fate of mobilized soil carbon changes as it travels through the river network. The transit path and the resulting C-loss or gain dynamically changes and is complicated to predict as soils erode and floodplains evolve. In fact, hillslopes and floodplains are important components of the "boundless carbon cycle" (Battin et al., 2009), that introduce legacy effects into contemporary C-dynamics (Van Oost et al., 2012). Their functioning within the Holocene's boundless carbon cycle is heavily dependent on ALCC and catchment management, both of which require an appropriate understanding of changing environmental conditions on mineral sediments and associated carbon fluxes across spatial and temporal scales. A better accounting of hillslopes and floodplains (Fig. 3) as factors of global change demands integrated geomorphological and biogeochemical studies on long-term sediment-burden carbon fluxes under different environmental conditions (e.g. old vs. new world and temperate vs. semi-arid climate). Erosion-induced soil degradation and decline in biomass production should also be considered when evaluating the OC-budget, particularly in regions of the world where long-term land degradation is observed. Such research is now technically possible. Potential benefits from the study

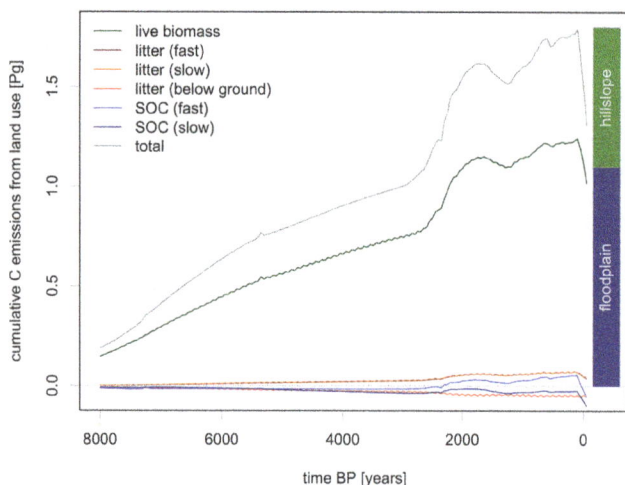

**Figure 4.** Cumulative carbon emissions as a result of anthropogenic land cover change in the non-alpine Rhine Basin. Emissions were calculated using the LPJ Dynamic Global Vegetation Model in the standard simulation using the KK10 anthropogenic land cover change described in Kaplan et al. (2011, 2012). Emissions accumulated steadily as a result of anthropogenic deforestation throughout most of the Holocene, with accelerated deforestation occurring during the Iron Age and Roman period around 500 BC–AD 500, followed by a period of land abandonment during the Migration Period, and accelerating deforestation during Medieval times. The period AD 1500–1900 is marked by relative stability in land use with little new deforestation emissions; during the 20th century land abandonment and afforestation lead to a strong uptake of carbon. Hillslope and floodplain storage are taken from Hoffmann et al. (2013) (compare Fig. 3).

of geomorphically coupled biogeochemical cycles would be substantial and help to (i) evaluate the wide-ranging implications of changing erosion and sediment dynamics for the global carbon budget, and (ii) assist policymakers to incorporate hillslopes and floodplains into catchment management strategies to mitigate climate change.

**Acknowledgements.** This paper is a result of the open IGBP-PAGES-PHAROS workshop "Sediment and carbon fluxes under human impact and climate change", which was held in July 2011 in Bern, Switzerland. We would like to thank IGBP-PAGES for the financial support of the workshop. Furthermore, funding was provided for T. Hoffmann through the DFG-SFB 806 "Our way to Europe".

Edited by: A. Cerdà

## References

Aalto, R., Maurice-Bourgoin, L., Dunne, T., Montgomery, D. R., Nittrouer, C. A., and Guyot, J. L.: Episodic sediment accumulation on Amazonian flood plains influenced by El Nino/Southern Oscillation, Nature, 425, 493–497, 2003.

Aalto, R. and Nittrouer, C. A.: 210Pb geochronology of flood events in large tropical river systems, Philos. T. R. Soc. Lond. A, 370, 2040–2074, 2012.

Aufdenkampe, A. K., Mayorga, E., Raymond, P. A., Melack, J. M., Doney, S. C., Alin, S. R., Aalto, R. E., and Yoo, K.: Riverine coupling of biogeochemical cycles between land, oceans, and atmosphere, Front. Ecol. Environ., 9, 53–60, 2011.

Battin, T. J., Kaplan, L. A., Findlay, S., Hopkinson, C. S., Marti, E., Packman, A. I., Newbold, J. D., and Sabater, F.: Biophysical controls on organic carbon fluxes in fluvial networks, Nat. Geosci., 1, 95–100, 2008.

Battin, T. J., Luyssaert, S., Kaplan, L. A., Aufdenkampe, A. K., Richter, A., and Tranvik, L. J.: The boundless carbon cycle, Nat. Geosci., 2, 598–600, 2009.

Berhe, A. A., Harte, J., Harden, J. W., and Torn, M. S.: The significance of the erosion-induced terrestrial carbon sink, Bioscience, 57, 337–346, 2007.

Brantley, S. L., Megonigal, J. P., Scatena, F. N., Balogh-Brunstad, Z., Barnes, R. T., Bruns, M. A., Van Cappellen, P., Dontsova, K., Hartnett, H. E., Hartshorn, A. S., Heimsath, A., Herndon, E., Jin, L., Keller, C. K., Leake, J. R., McDowell, W. H., Meinzer, F. C., Mozdzer, T. J., Petsch, S., Pett-Ridge, J., Pregitzer, K. S., Raymond, P. A., Riebe, C. S., Shumaker, K., Sutton-Grier, A., Walter, R., and Yoo, K.: Twelve testable hypotheses on the geobiology of weathering, Geobiology, 9, 140–165, 2011.

Brown, A. G., Carey, C., Erkens, G., Fuchs, M., Hoffmann, T., Macaire, J. J., Moldenhauer, K. M., and Walling, D. E.: From sedimentary records to sediment budgets: Multiple approaches to catchment sediment flux, Geomorphology, 108, 35–47, 2009.

Cerdà, A., Brazier, R., Nearing, M. A., and de Vente, J.: Preface: scales and erosion, Catena, 102, 1–2, 2012.

Chiverrell, R. C., Foster, G. R., Thomas, G. S. P., Marshall, D., and Hamilton, D.: Robust chronologies for landform development, Earth Surf. Proc. Land., 34, 319–328, 2008.

Cole, J. J., Prairie, Y. T., Caraco, N. F., McDowell, W. H., Tranvik, L. J., Striegl, R. G., Duarte, C. M., Kortelainen, P., Downing, J. A., Middelburg, J. J., and Melack, J.: Plumbing the global carbon cycle: Integrating inland waters into the terrestrial carbon budget, Ecosystems, 10, 171–184, doi:10.1007/s10021-006-9013-8, 2007.

Dotterweich, M.: The history of soil erosion and fluvial deposits in small catchments of central Europe: Deciphering the long-term interaction between humans and the environment – A review, Geomorphology, 101, 192–208, 2008.

Dusar, B., Verstraeten, G., Notebaert, B., and Bakker, J.: Holocene environmental change and its impact on sediment dynamics in the Eastern Mediterranean, Earth-Sci. Rev., 108, 137–157, 2011.

Erkens, G., Hoffmann, T., Gerlach, R., and Klostermann, J.: Complex fluvial response to Late Glacial and Holocene allogenic forcings in the Lower Rhine embayment (Germany), Quaternary Sci. Rev., 30, 611–627, 2011.

Fuchs, A., Lang, A., and Wagner, G. A.: The history of soil erosion and landscape degradation in the Phlious basin, NE Peleponnes, Greece, The Holocene, 14, 334–345, 2004.

Galy, A., France-Lanord, C., Beyssac, O., Faure, P., Kudrass, H., and Palhol, F.: Efficient organic carbon burial in the Bengal fan sustained by the Himalayan erosional system, Nature, 450, 407–410, 2007.

Hilton, R. G., Galy, A., Hovius, N., Chen, M. C., Horng, M. J., and Chen, H.: Tropical-cyclone-driven erosion of the terrestrial biosphere from mountains, Nat. Geosci., 1, 759–762, 2008.

Hobo, N., Makaske, B., Wallinga, J., and Middelkoop, H.: Reconstruction of sedimentation rates in embanked floodplains, a comparison of different methods, Earth Surf. Proc. Land., 35, 1499–1515, 2010.

Hoffmann, T., Erkens, G., Dikau, R., Houben, P., Seidel, J., and Cohen, K. M.: Holocene floodplain sediment storage and hillslope erosion within the Rhine catchment, The Holocene, 17, 105–118, 2007.

Hoffmann, T., Erkens, G., Gerlach, R., Klostermann, J., and Lang, A.: Trends and controls of Holocene floodplain sedimentation in the Rhine catchment, Catena, 77, 96–106, 2009a.

Hoffmann, T., Glatzel, S., and Dikau, R.: A carbon storage perspective on alluvial sediment storage in the Rhine catchment, Geomorphology, 108, 127–137, 2009b.

Hoffmann, T., Thorndycraft, V. R., Brown, A. G., Coulthard, T., Damnati, B., Kale, V. S., Middelkoop, H., Notebaert, B., and Walling, D. E.: Human impact on fluvial regimes and sediment flux during the Holocene: Review and future research agenda, Global Planet. Change, 72, 87–98, 2010.

Hoffmann, T., Schlummer, M., Verstraeten, G., and Notebaert, B.: Significance of sediment and carbon storage on hillslopes and floodplains, Global Biogeochem. Cy., 27, 22013, doi:10.1002/gbc.20071, 2013.

Houben, P.: Scale linkage and contingency effects of field-scale and hillslope-scale controls of long-term soil erosion: Anthropogeomorphic sediment flux in agricultural loess watersheds of Southern Germany, Geomorphology, 101, 172–191, 2008.

Houben, P., Wunderlich, J., and Schrott, L.: Climate and long-term human impact on sediment fluxes in watershed systems, Geomorphology, 108, 1–7, 2009.

Hupp, C. R., Pierce, A. R., and Noe, G. B.: Floodplain geomorphic processes and environmental impacts of human alteration along coastal plain rivers, USA, Wetlands, 29, 413–429, 2009.

Jacinthe, P. A. and Lal, R.: A mass balance approach to assess carbon dixoide evolution during erosion events, Land Degrad. Dev., 12, 329–339, 2001.

Kaplan, J. O., Krumhardt, K., Ellis, E. C., Ruddiman, W. F., Lemmen, C., and Klein Goldewijk, K.: Holocene carbon emissions as a result of anthropogenic land cover change, The Holocene, 21, 775–791, 2011.

Kaplan, J. O., Krumhardt, K. M., and Zimmermann, N. E.: The effects of land use and climate change on the carbon cycle of Europe over the past 500 years, Glob. Change Biol., 18, 902–914, 2012.

Lal, R.: Soil Carbon Sequestration Impacts on Global Climate Change and Food Security, Science, 304, 1623–1627, 2004.

Lal, R.: Soil erosion and carbon dynamics, Soil Till. Res., 81, 137–142, 2005.

Lal, R.: Sequestering carbon in soils of arid ecosystems, Land Degrad. Dev., 20, 441–454, 2009.

Lal, R. and Pimentel, D.: Soil erosion: A carbon sink or source? Science, 319, 1040–1041, 2008.

Lang, A. and Honscheidt, S.: Age and source of colluvial sediments at Vaihingen-Enz, Germany, Catena, 38, 89–107, 1999.

Lang, A., Bork, H.-R., Mäckel, R., Preston, N., Wunderlich, J., and Dikau, R.: Changes in sediment flux and storage within a fluvial system: some examples from the Rhine catchment, Hydrol. Process., 17, 3321–3334, 2003.

Lechterbeck, J., Kalis, A. J., and Meurers-Balke, J.: Evaluation of prehistoric land use intensity in the Rhenish Loessboerde by canonical correspondence analysis – A contribution to LUCIFS, Geomorphology, 108, 138–144, 2009.

Lowdermilk, W. C.: Conquest of the Land through 7,000 Years, Soil Conservation Service, Misc. Pub. No. 32, 1948.

Maselli, V. and Trincardi, F.: Man made deltas, Scientific Reports, 3, 1926, doi:10.1038/srep01926, 2013.

Noe, G. B. and Hupp, C. R.: Carbon, nitrogen, and phosphorus accumulation in floodplains of Atlantic Coastal Plain rivers, USA, Ecol. Appl., 15, 1178–1190, 2005.

Notebaert, B. and Verstraeten, G.: Sensitivity of West and Central European river systems to environmental changes during the Holocene: A review, Earth-Sci. Rev., 103, 163–182, 2010.

Notebaert, B., Verstraeten, G., Ward, P. J., Renssen, H., and Van Rompaey, A.: Modeling the sensitivity of sediment and water runoff dynamics to Holocene climate and land use changes at the catchment scale, Geomorphology, 126, 18–31, 2011.

Osterkamp, W. R., Hupp, C. R., and Stoffel, M.: The interactions between vegetation and erosion: new directions for research at the interface of ecology and geomorphology, Earth Surf. Proc. Land., 37, 23–36, 2012.

Parsons, A. J., Brazier, R., Wainwright, J., and Powell, D. M.: Scale relationships in hillslope runoff and erosion, Earth Surf. Proc. Land., 31, 1384–1393, 2006.

Petrescu, A. M. R., Van Beek, L. P. H., Van Huissteden, J., Prigent, C., Sachs, T., Corradi, C. A. R., Parmentier, F. J. W., and Dolman, A. J.: Modeling regional to global CH4 emissions of boreal and arctic wetlands. Global Biogeochem. Cy., 24, GB4009, doi:10.1029/2009GB003610, 2010.

Quinton, J. N., Govers, G., Van Oost, K., and Bardgett, R. D.: The impact of agricultural soil erosion on biogeochemical cycling, Nat. Geosci., 3, 311–314, doi:10.1038/ngeo838, 2010.

Ritchie, J. C. and Rasmussen, P. E.: Application of 137Cesium to estimate erosion rates for understanding soil carbon loss on long-term experiments at Pendleton, Oregon, Land Degrad. Dev., 11, 75–81, 2000.

Ruddiman, W. F.: The anthropogenic greenhouse era began thousands of years ago, Climatic Change, 61, 261–293, 2003.

Solomon, S., Qin, D., Manning, M., Chen, Z., Marquis, M., Averyt, K. B., Tignor, M., and Miller, H. L.: Climate Change 2007 – The Physical Science Basis. Contribution of Working Group I to the Fourth Assessment Report of the Intergovernmental Panel on Climate Change, Cambridge University Press, Cambridge, 2007.

Starr, G. C., Lal, R., Malone, R., Hothem, D., Owens, L., and Kimble, J. M.: Modeling soil carbon transported by water erosion processes, Land Degrad. Dev., 11, 83–91, 2000.

Syvitski, J. P. M., Vorosmarty, C. J., Kettner, A. J., and Green, P.: Impact of humans on the flux of terrestrial sediment to the global coastal ocean, Science, 308, 376–380, 2005.

Tranvik, L. J., Downing, J. A., Cotner, J. B., Loiselle, S. A., Striegl, R. G., Ballatore, T. J., Dillon, P., Finlay, K., Fortino, K., Knoll, L. B., Kortelainen, P. L., Kutser, T., Larsen, S., Laurion, I., Leech, D. M., McCallister, S. L., McKnight, D. M., Melack, J. M., Overholt, E., Porter, J. A., Prairie, Y., Renwick, W. H., Roland, F., Sherman, B. S., Schindler, D. W., Sobek, S., Tremblay, A., Vanni, M. J., Verschoor, A. M., von Wachenfeldt, E., and Weyhenmeyer, G. A.: Lakes and reservoirs as regulators of carbon cycling and climate, Limnol. Oceanogr., 54, 2298–2314, 2009.

Trimble, S. W.: Decreased rates of alluvial sediment storage in the Coon Creek Basin, Wisconsin, 1975–93, Science, 285, 1244–1246, 1999.

Van De Wiel, M. J. and Coulthard, T.: Self-organized criticality in river basins: Challenging sedimentary records of environmental change, Geology, 38, 87–90, 2010.

Van Oost, K., Quine, T. A., Govers, G., De Gryze, S., Six, J., Harden, J. W., Ritchie, J. C., McCarty, G. W., Heckrath, G., Kosmas, C., Giraldez, J. V., da Silva, J. R. M., and Merckx, R.: The impact of agricultural soil erosion on the global carbon cycle, Science, 318, 626–629, 2007.

Van Oost, K., Verstraeten, G., Doetterl, S., Notebaert, B., Wiaux, F., Broothaerts, N., and Six, J.: Legacy of human-induced C erosion and burial on soil-atmosphere C exchange, P. Natl. Acad. Sci., 109, 19492–19497, 2012.

Verstraeten, G., Rommens, T., Peeters, I., Poesen, J., Govers, G., and Lang, A.: A temporarily changing Holocene sediment budget for a loess-covered catchment (central Belgium), Geomorphology, 108, 24–34, 2009.

Victoria, R., Banwart, S., Black, H., Ingram, J., Joosten, H., Milne, E., and Noellemeyer, E.: The benefits of soil carbon. Managing soils for multiple economic, societal and environmental benefits, in: The UNEP years books 2012, edited by: UNEP, 19–33, 2012.

Walter, R. C. and Merritts, D. J.: Natural streams and the legacy of water-powered mills, Science, 319, 299–304, 2008.

Wang, Z., Govers, G., Steegen, A., Clymans, W., Van den Putte, A., Langhans, C., Merckx, R., and Van Oost, K.: Catchment-scale carbon redistribution and delivery by water erosion in an intensively cultivated area, Geomorphology, 124, 65–74, 2010.

Wilkinson, B. H. and McElroy, B. J.: The impacts of humans on continental erosion and sedimentation, Geol. Soc. Am. Bull., 119, 140–156, 2007.

Yoo, K., Ji, J., Aufdenkampe, A. K., and Klaminder, J.: Rates of soil mixing and associated carbon fluxes in a forest versus tilled agricultural field: Implications for modeling the soil carbon cycle, J. Geophys. Res.-Biogeo., 116, GB3003, doi:10.1029/2004GB002271, 2011.

Zehetner, F., Lair, G. J., and Gerzabek, M. H.: Rapid carbon accretion and organic matter pool stabilization in riverine floodplain soils, Global Biogeochem. Cy., 23, GB4004, doi:10.1029/2009GB003481, 2009.

# 5

# Threshold effects of hazard mitigation in coastal human–environmental systems

**E. D. Lazarus**

Environmental Dynamics Laboratory, Earth Surface Processes Research Group, School of Earth & Ocean Sciences, Cardiff University, Main Building, Park Place, Cardiff CF10 3AT, UK

*Correspondence to:* E. D. Lazarus (lazarusED@cf.ac.uk)

**Abstract.** Despite improved scientific insight into physical and social dynamics related to natural disasters, the financial cost of extreme events continues to rise. This paradox is particularly evident along developed coastlines, where future hazards are projected to intensify with consequences of climate change, and where the presence of valuable infrastructure exacerbates risk. By design, coastal hazard mitigation buffers human activities against the variability of natural phenomena such as storms. But hazard mitigation also sets up feedbacks between human and natural dynamics. This paper explores developed coastlines as exemplary coupled human–environmental systems in which hazard mitigation is the key coupling mechanism. Results from a simplified numerical model of an agent-managed seawall illustrate the nonlinear effects that economic and physical thresholds can impart into coastal human–environmental system dynamics. The scale of mitigation action affects the time frame over which human activities and natural hazards interact. By accelerating environmental changes observable in some settings over human timescales of years to decades, climate change may temporarily strengthen the coupling between human and environmental dynamics. However, climate change could ultimately result in weaker coupling at those human timescales as mitigation actions increasingly engage global-scale systems.

## 1 Introduction

Beach nourishment, artificial dune construction, and shoreline armoring with rock revetments, bulkheads, and sea walls (Fig. 1) are methods of mitigating coastal hazards in developed coastal zones worldwide. Coastal engineering is an old science, and seawalls are an especially old technology: the Phoenicians used them to protect their ports (Marriner et al., 2006), as did early Chinese dynasties (Qingzhou, 1989). In the UK, coastal defences, including many built during the 19th-century Victorian heyday of English seaside resorts (Tunstall and Penning-Rowsell, 1998), now extend along approximately 44 % of the coastline in England and Wales (DEFRA, 2010; BGS, 2012). In the US, shoreline hardening was a common but localized practice that boomed with the post-war housing market of the 1950s, rapidly transforming much of the mid-Atlantic seaboard (Pilkey and Wright, 1988). However, even with this legacy, developed coastlines illustrate a confounding paradox in the modern science of natural hazards and extreme events: that despite "improved…understanding of the physical processes underlying natural hazards and the complexities of social decision-making before, during, and after disasters, …troubling questions remain about why more progress has not been made in reducing dollar losses" (Mileti, 1999). The UK, the Netherlands, and Belgium still recall "the Big Flood" event of January, 1953, a North Sea storm surge that devastated the English east coast and caused an estimated equivalent GBP 5 billion in damage there (Summers, 1978; Johnson et al., 2005; Lumbroso and Vinet, 2011). The same storm was so catastrophic to the Netherlands that it prompted the now iconic Delta Plan, a massive national investment in flood-control infrastructure (Gerritsen, 2005). The Big Flood again made recent headlines when the *Guardian* reported a GBP 1 billion funding gap in UK flood-control infrastructure, including coastal defenses, with the UK Environment Agency calling for a year-on-year funding increase of GBP 20 million just to maintain current protection (Guardian, 2012). On the other side of the Atlantic, Hurricane Katrina, in August 2005, and Hurricane Sandy, in October 2012, rank as the two most expensive weather-related disasters on record for the US (NOAA, 2013). How is it that coastal disasters, which human ingenuity has been trying to ward off for millennia, are presently both better understood and more costly than ever before?

As a thought experiment, "Mileti's paradox" sets up three hypotheses. First, that extreme weather events, whether through an increasing mean or increasing variability, are becoming more frequent. There is compelling evidence of rising trends in temperature extremes (Rahmstorf and Coumou, 2011) and perhaps in other weather phenomena (Emanuel, 2005; Lubchenco and Karl, 2012). Second, that vulnerable infrastructure and hazard mitigation around the globe is more expensive, beyond inflation, than ever before, and thus the higher financial cost of disasters is independent of any trend toward greater extremes in natural systems. Recent growth in high-value development is apparent on coastlines worldwide (Cooper and McKenna, 2009). In the US, the per-cubic-yard cost of sand for beach nourishment has risen sevenfold since the 1970s (Seabrook, 2013). And third, that a fundamental consequence of hazard mitigation is to filter out small-scale hazard events at the greater expense of infrequent, large ones (Werner and McNamara, 2007). Assessing the risk of a natural hazard involves accounting for the economic value of infrastructure or activities vulnerable to a hazard event, and the probability that an event of a given magnitude will occur. Infrastructural value changes with markets, demographics, land use, and even hazard protection, thus changing the risk associated with a given hazard (Mileti, 1999; Smith, 2013). More difficult to anticipate is the effect of hazard mitigation on the magnitude frequency distribution of the hazard itself, which can change even with a stationary climate.

This third hypothesis arises from considering human activities an "anthropic force" of landscape change (Hooke, 1994, 2000; Haff, 2003) that in some cases results in coupled human–environmental systems: contexts in which human activities and the natural physical environment are dynamically linked, such that the state and behavior of each becomes a function of the other. Note that the logic of the second hypothesis regarding Mileti's paradox, to be independent of the third, assumes that coastal development growth is unrelated to mitigation interventions. In fact, property values and development pressures tend to increase with investment in engineered protection against natural hazards, especially along coastlines; in the absence of hazard mitigation, property values in vulnerable places would be completely different (Mileti, 1999; Werner and McNamara, 2007; Smith, 2013). Parsing the complex dynamics of coupled human–environmental systems (also called coupled human–natural systems, coupled human–landscape systems, and

coupled social–ecological systems) is a grand challenge in the physical and social sciences (Kates et al., 2001; Haff, 2003; Liu et al., 2007a; Ostrom et al., 2007; Murray et al., 2009; NRC, 2002, 2010; Ostrom, 2010). This paper explores Mileti's paradox in the context of shoreline protection, a setting in which coupled-system dynamics can manifest over human timescales of years to decades. Drawing on recent advances in coastal morphodynamics involving interactions between human activities and shoreline processes, I present a numerical model of a seawalled shoreline as a human–environmental system governed by economic and physical thresholds, with implications for the function of threshold structures in managed landscapes more generally.

## 2 Recent advances in understanding coastal coupled systems

### 2.1 Beach nourishment

A body of recent numerical modeling research examines the coupled economic–beach dynamics of developed shorelines, with particular attention to beach nourishment (Slott et al., 2008, 2010; Smith et al., 2009; Slott et al., Lazarus et al., 2011; McNamara et al., 2011; Murray et al., 2013; Ells and Murray, 2012; Jin et al., 2013; McNamara and Keeler, 2013; Williams et al., 2013). The work frames beach nourishment as a cumulative cost–benefit optimization problem (Smith et al., 2009), adopting an environmental economics approach typically applied to a renewable resource like timber (e.g., Hartman, 1976). Trees take a certain amount of time to grow into a mature stand. Consequently, in commercial forestry, there is an optimal interval at which trees may be harvested. Harvest them too soon and the timber is worth less money; wait too long and the cumulative return on that patch of forest diminishes. In a coastal town, the width of the beach is analogous to standing timber: the beach constitutes a resource of natural capital (Smith et al., 2009; Gopalakrishnan et al., 2011). A town experiencing shoreline erosion, which depletes that natural capital, may maintain the value of a wide beach through cyclical beach nourishment. Theoretically, like timber harvesting, that nourishment cycle has an optimal frequency: nourishing too often is unnecessarily expensive; waiting too long to nourish results in a narrow beach that negatively affects the town's economic capital. The dynamics of this optimization problem change when a series of towns share the spatial context of a continuous shoreline, such that the management actions of one town begin to affect the beach widths and corresponding management actions of the others. Model scenarios suggest that uncoordinated beach replenishment among neighboring coastal towns may make shoreline erosion rates and mitigation actions more unpredictable, and the use of sand resources more inequitable, as nonlocal effects become more pronounced or unstable in response to forcing conditions associated with climate change, such as sea-level rise and increased storminess.

## 2.2 Engineered structures and management decisions as thresholds

Foundational to these beach-nourishment models are two fully coupled, agent-based, dynamic landscape models, one describing the evolution of New Orleans, Louisiana (USA), as a city on a major river delta prone to flooding (Werner and McNamara, 2007), and the other the evolution of Ocean City, Maryland (USA), as a resort town on an eroding barrier island prone to storm-driven overwashing (McNamara and Werner, 2008a, b). Both models demonstrate a boom-and-bust cycle of development and disaster that is an emergent consequence of the human–environmental coupling rather than an intrinsic characteristic of either the economic or physical components of the system. Moreover, thresholds in states of landscape stability, and in development and hazard-mitigation actions by human agents, play an integral role.

In the New Orleans case, artificial levee construction is economically driven; levee height increases in response to flood events that destroy city infrastructure and private property. When flooding destroys property worth more than the cost of levee reconstruction, the levee gets repaired and the local economic market drives property redevelopment. Meanwhile, incremental channelization of the river drives gradual deltaic inundation, increasing the severity of subsequent flood events (e.g., Criss and Shock, 2001). Similarly, in the Ocean City example, resort development and cyclical beach nourishment driven by tourism economics restricts natural barrier dynamics by inhibiting the ability of island width and height to change with sea-level rise. When erosion mitigation eventually becomes economically untenable and beach nourishment becomes too infrequent to hold the barrier island in place, the vulnerable resort is destroyed by a storm event and developer agents site new construction on a more stable part of the island where projected economic return is higher.

A key implication of the New Orleans scenario is that "the long-time-scale dynamics of the modeled system appears to be characterized by an attractor with emergent dynamics in which small-scale floods are filtered out at the expense of amplifying the impact of large floods to be significant disasters, because protection from small-scale floods facilitates development in areas prone to disaster and increased channelization causes an increase in flood size that results in enhanced damage from the low-frequency flood events" (Werner and McNamara, 2007). This same feedback extends to the barrier-island resort scenario, in which "hazard-protection measures filter out high-frequency responses to storms and sea-level rise, but create long-period boom and bust cycles..." (McNamara and Werner, 2008a). The alternative states that arise in these coupled-system examples may be characterized as "undamaged" and "damaged", with long periods of the former punctuated by sudden episodes of the latter. Dynamical systems discourse defines thresholds in terms of transitions between alternative states (Abraham and

Shaw, 1988; Scheffer, 2009), a definition that ecology-based perspectives of human–environmental systems have tended to adopt (Beisner et al., 2003; Groffman et al., 2006; Liu et al., 2007a, b; Chin et al., 2013). But from a geomorphology-based perspective, engineered hazard-mitigation structures like artificial levees and sand dunes, beach nourishment, and seawalls also function as physical thresholds: imposed barriers that a hazard must erode, crest, or breach before it can interact (through flooding, erosion, sediment transport, deposition) with the otherwise sheltered landscape. The threshold between dynamical states is thus the scale, physical integrity, and indirect effects of the mitigating barrier itself, filtering the impact of high-frequency events but literally breaking down with low-frequency recurrence.

Formal definitions of coupled human–environmental systems emphasize the importance of feedbacks that link human activities and natural processes (Turner et al., 2003; Liu et al., 2007a; Chin et al., 2013), but none addresses feedback reciprocity and strength more specifically than the description by Werner and McNamara (2007), who write that coupling "should be strongest where fluvial, oceanic, or atmospheric processes render significant stretches of human-occupied land vulnerable to large changes and damage, and where market processes assign value to the land and drive measures to protect it from damage. These processes typically operate over the (human) medium scale of perhaps many years to decades, over which landscapes become vulnerable to change and over which markets drive investment in structures, evaluate profits from those investments, and respond to changes in conditions". This definition categorically distinguishes coupled systems from, for example, extractive resource activities that obviate or ignore preventative measures against damage (e.g., McDaniel and Gowdy, 2000). Furthermore, Werner and McNamara (2007) associate strong human–landscape coupling with environmental hazard and risk, which distinguishes their definition from others derived more from social complexity in common-pool resources (e.g., Ostrom, 2010). Coastal environments offer such accessible examples of coupled dynamics because risk exposure to natural hazard is arguably an inherent characteristic of developed coastlines (e.g., Nordstrom, 2000; Kelley and Brothers, 2009). Seawalls, for example, are a ubiquitous response to coastal hazard and risk associated with shoreline erosion and storm surge. Controversy regarding seawalls as a coastal management practice tends to hinge on whether a seawall exacerbates shoreline erosion (Kraus, 1988; Pilkey and Wright, 1998; Kraus and McDougal, 1996; Dean and Dalrymple, 2002), but more broadly, the role of seawalls may be interpreted in terms of a managed physical threshold within a coastal human–environmental system.

**Figure 1.** Three examples of typical coastal hazard mitigation practices: (**A**) beach replenishment in Monmouth, New Jersey (USA), nine months after Hurricane Sandy (photo: A. Coburn, Program for the Study of Developed Shorelines, http://www.psds-wcu.org/); (**B**) artificial dune reconstruction following an overwash event on Highway 12, North Carolina Outer Banks (USA) (photo: A. Coburn, PSDS); and (**C**) concrete seawall on the Channel Island of Jersey – note the pitting and fresh scour (light gray band) along the wall base just above the shingle toe.

## 3   Example: threshold dynamics in a seawall model

A deliberately simplified, one-dimensional numerical model demonstrates how a seawalled shoreline may exhibit system dynamics similar to those described for leveed rivers (Criss and Shock, 2001; Werner and McNamara, 2007) and artificial dune fronts on barrier islands (Magliocca et al., 2011). The model combines mechanical interaction between the beach and seawall with decisions by a coastal-manager agent regarding seawall construction and repair. Rather than simulating a particular place, the model represents systemic dynamics of seawalls in abstracted terms.

### 3.1   Landscape and damage

As a motivating analog, the evolution of cliffed coastlines involves a nonlinear relationship between fronting beach width and the rate of cliff erosion (Limber and Murray, 2011). The argument supposes that, comparable to mechanisms for regolith production (Anderson, 2002; Strudley et al., 2006), the beach functions as an erosive tool. The cliff has a natural background retreat rate that a fronting beach then accelerates, at least up to a critical width; beyond that critical width, the beach prevents wave action from reaching the cliff toe, insulating the cliff from erosion.

Here, I assume the seawall acts like a sea cliff (Fig. 2), with a background deterioration rate (in units of % yr$^{-1}$) that is (1) constant in the absence of a fronting beach; (2) increases with beach width (BW) up to a critical width; and (3) decreases with beach width exceeding the critical width, as a wider beach insulates the seawall from wear. Because the "shoreline" in this one-dimensional model is a single cell, I use a detrended, normalized Brownian time series (generated from

the cumulative sum of a white-noise time series) to represent temporally autocorrelated, year-to-year beach width (Fig. 2).

The seawall has two principal variables: wall strength ($W_S$, represented as a percentage) and wall height ($W_H$). Wall deterioration rate ($\rho$) goes by the function

$$\rho = \left( a \times (\text{BW} + b) \times e^{(-c \times W_S)} + 1 \right) / \rho_{max}, \tag{1}$$

where $a$, $b$, and $c$ are curve-tuning constants (here, $a = 40$, $b = 0.05$, and $c = -5$; these deliver a corresponding $\rho_{max}$ of 4.78). Although beach width sets the effective timescale of seawall deterioration, the seawall also experiences one "storm event" per year. Storm magnitude ($S$) here is analogous to a flood stage or surge height, and is sampled from a normalized, temporally uncorrelated time series of random values. Storm damage the wall sustains in a given year ($D_{storm}$) is calculated as the difference between the scale of the storm ($S$) and the product of wall height multiplied by wall strength:

$$D_{storm} = S - (W_H \times W_S). \tag{2}$$

Wall strength is adjusted by both the storm damage and annual deterioration related to beach width:

$$W_S^{t+1} = W_S^t - (\rho + D_{storm}), \tag{3}$$

where $t$ is the model time step (year). If $S$ is less than the product of $W_H$ and $W_S$, then $D_{storm} = 0$.

### 3.2   Hazard mitigation

Hazard mitigation actions affect wall height and wall strength. When the model begins, the initial wall height is set equal to the scale ($S$) of the first storm. Subsequent increases in wall height and repairs to wall strength are reactionary, lagging storm impacts. As a record of storm-driven

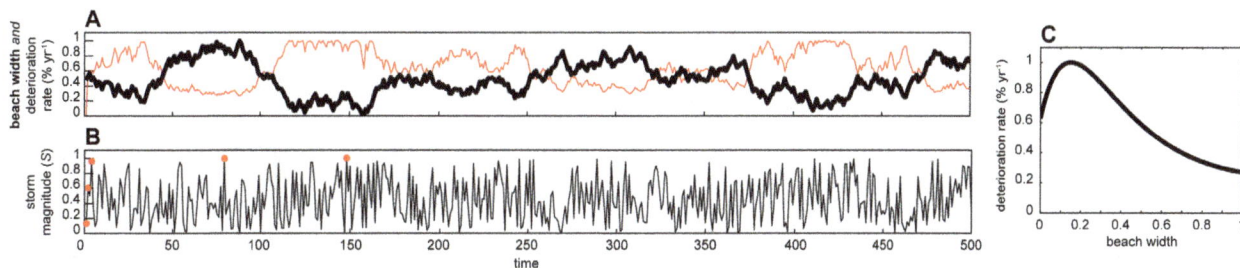

**Figure 2.** Forcing components for the seawall model. (**A**) Normalized Brownian signal representing beach width over time (bold) and the corresponding wall deterioration rate (red). (**B**) Randomized time series of storm events over the simulation period (500 yr); red dots mark record-setting storms. (**C**) Plot of wall deterioration rate as a function of beach width, motivated by Limber and Murray (2011), Strudley et al. (2006), and Anderson (2002).

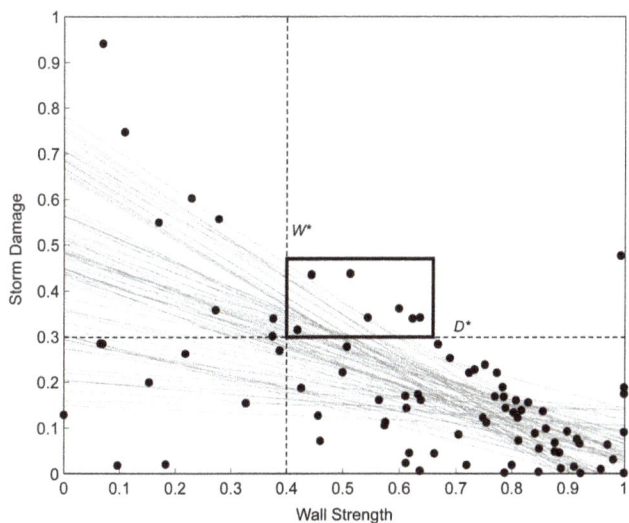

**Figure 3.** The manager agent uses a line of best fit relating storm damage to wall strength for the previous $N$ damaging storm events to determine whether wall repairs are cost effective, given imposed thresholds for damage tolerance ($D^*$) and economy of scale ($W^*$). In this plot ($N = 10$ events), damage must exceed $D^* = 0.3$ and at least 40 % of the wall ($W^* = 0.4$) must have deteriorated to warrant repair. Lines of best fit that satisfy these conditions pass through the box shown in bold.

wall damage develops, the coastal-manager agent interprets the record to make decisions about whether or not to repair the wall. When repaired, the wall is always restored to full strength ($W_S = 100\,\%$), and wall height is determined by the largest storm on record ($W_H = S^*$).

The manager's decision process goes as follows. Each year, the manager looks back over the previous $N$ storms that caused wall damage, plots the damage sustained ($D_{\mathrm{storm}}$) against wall strength at the time of the storm for those $N$ events, and calculates a best-fit linear trend through the data points (Fig. 3). The manager requires hindsight of $N > 1$ damage events in order to calculate a line, and the best-fit line needs a negative slope to be physically meaningful

(e.g., damage is high when wall strength is weak); hindsight of $N > 4$ prevents nonsensical (positive) trend calculations. Given a best-fit line, mitigation action then depends on two thresholds: a damage tolerance ($D^*$), meaning that damage exceeding $D*$ warrants mitigation, and a second threshold requiring that wall strength be degraded beyond a certain percentage ($W^*$) for repairs to be considered cost effective. In real coastal management settings, fixed costs associated with work crews, equipment, and permitting make capital-works projects like seawalls subject to economies of scale (e.g., Leafe et al., 1998; Smith et al., 2009). This component of the model reflects the relationship between scale of repair and cost distribution. The manager finds where the calculated best-fit line intersects the ordinate line $D^*$. If the abscissa of the intersection is greater than $W^*$, then wall repair is deemed cost effective. Indeed, if this condition is satisfied, then repairs to all wall conditions between $W^*$ and the abscissa intercept of the best-fit line are cost effective (Fig. 3). This window of cost effectiveness varies each time the manager recalculates the best-fit line. If the abscissa for the best-fit intersection with $D^*$ is less than $W^*$, the condition for cost effectiveness is not satisfied and the wall is not repaired that year. Both $D^*$ and $W^*$ are imposed constraints, and remain fixed for the duration of a model run.

## 3.3 Results

Figure 4 shows representative time series of key model parameters: the occurrence of record-setting storms, increases in seawall height, years punctuated by storm-driven damage, and the threshold of cost effectiveness for wall repair calculated by the manager agent. Because the model assumes a stationary climate (zero trend in storm magnitude or sea-level rise), more extreme events occur in the beginning of the time series: the likelihood of an unprecedented extreme (a maximum value in the time series) declines with $1/t$, where $t$ is the number of previous years in the time series (Rahmstorf and Coumou, 2011). This quick succession of impactful storms drives early investment in wall construction until wall height is nearly equal to the largest possible storm

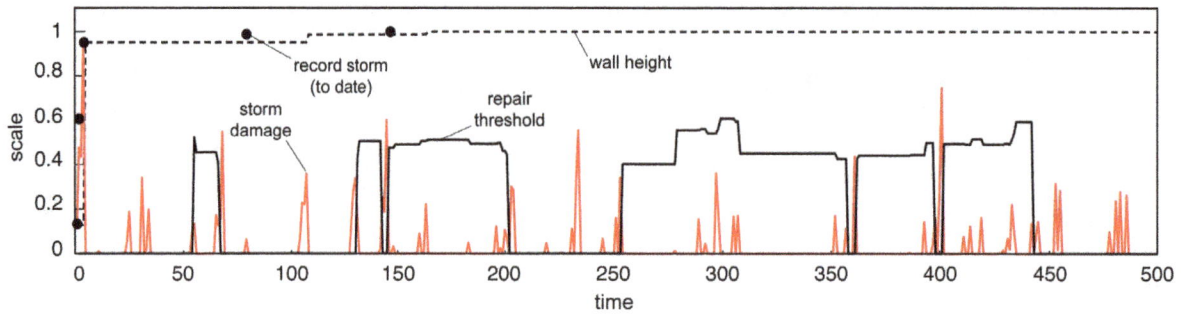

**Figure 4.** Representative time series of key model parameters (where $D^* = 0.3$, $W^* = 0.4$, and $N = 10$): record-setting storms (black dots), seawall height (dashed line), storm-driven damage (solid red line), and the threshold of cost effectiveness for wall repair calculated by the manager agent (solid black line).

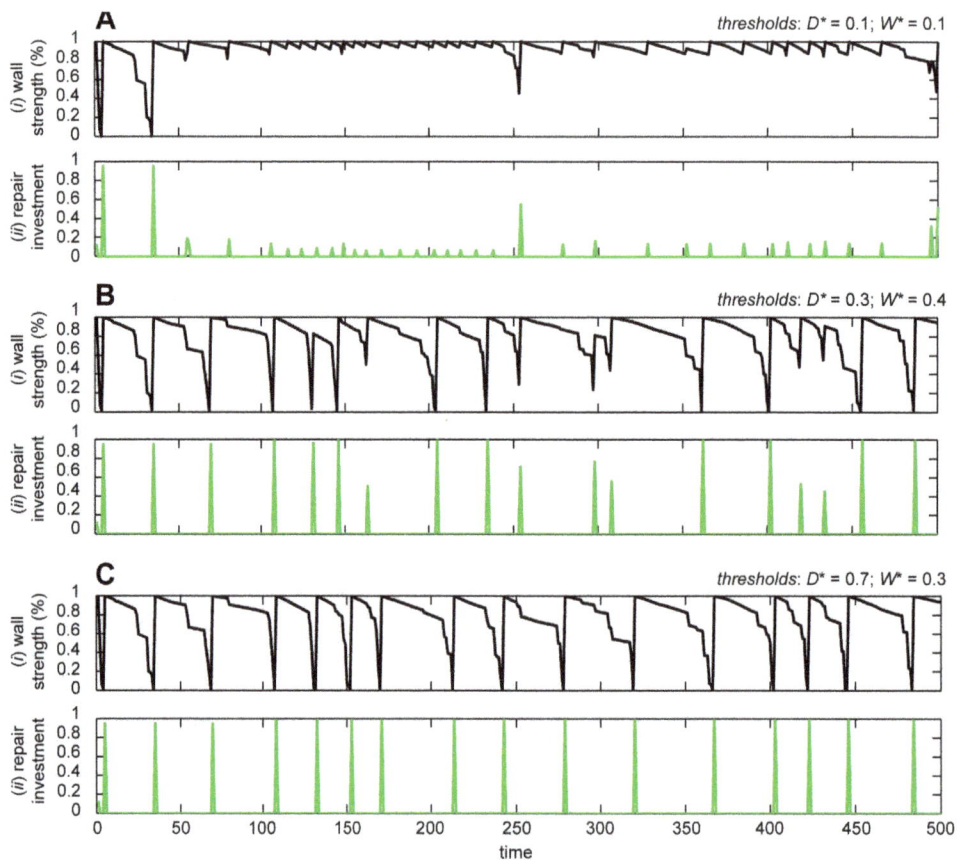

**Figure 5.** Time series of wall strength and repair investment for three representative pairs of threshold conditions in which: **(A)** thresholds for damage tolerance and economy of scale are both low ($D^* = 0.1$ and $W^* = 0.1$), **(B)** both thresholds are intermediate ($D^* = 0.3$ and $W^* = 0.4$), and **(C)** damage tolerance is high and the economy of scale is intermediate ($D^* = 0.7$ and $W^* = 0.3$). In all three cases shown, $N = 10$ events.

event. When the wall is low, storm damage is almost always extensive enough for the manager to opt for wall investment. Annual storm damage decreases as the wall gains height and serves as a more effective barrier against a greater range of storm events.

If the wall did not deteriorate, here the model would effectively stop. Oppositely, if left unrepaired (and in the absence

of major storms), the model seawall degrades within a century, a time frame consistent with lifespan estimates for real coastal defenses (e.g., Yokota and Komure, 2003). The model is insensitive to the specific Brownian time series of beach width (Fig. 2); the year-to-year details of the output differ but the model dynamics remain consistent under different forcing patterns. Likewise, adjusting the deterioration rate as a

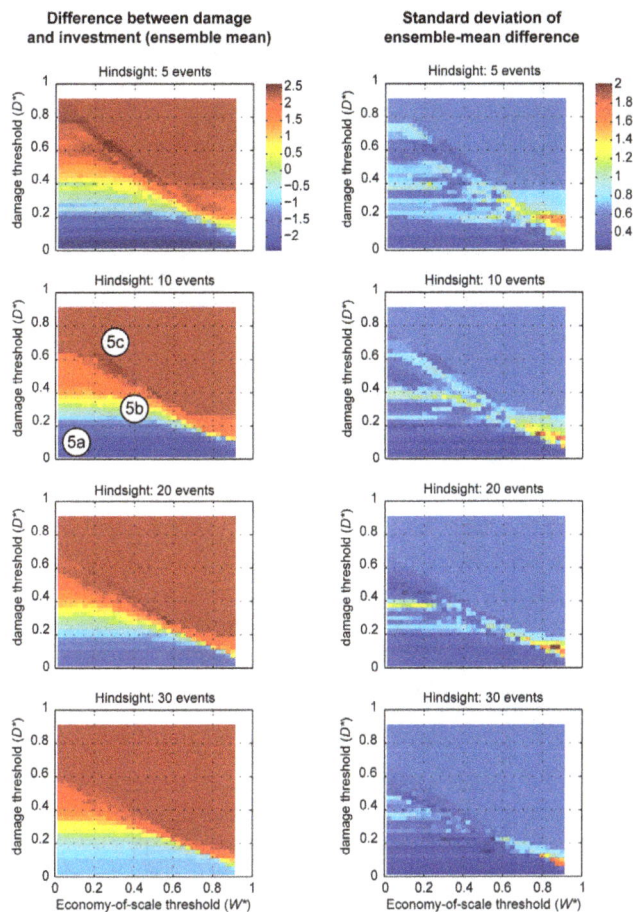

**Figure 6.** Parameter spaces of threshold pairs $D^*$ and $W^*$ for four hindsight conditions ($N = 5$, 10, 20, 30). In the left column, color represents the ensemble-mean difference between total storm damage and total investment in repair (derived from 10 different randomized forcing time series for beach width and storms, as in Fig. 2). In the right column, color represents standard deviation around each ensemble mean. Numbered circles indicate parameter pairs shown in Fig. 5.

function of beach width changes the inherent wall lifespan but does not affect the system dynamics. Therefore, maintenance action (including inaction) by the manager agent determines the cumulative record of storm-driven damage. This is emphasized in the temporal record of the manager's economy-of-scale threshold shown in Fig. 4. When the seawall is kept in good repair, data points begin to accumulate in the low-damage, high-wall-strength region of Fig. 3, gradually reducing the slope of the manager's calculated line of best fit. The slope of that line may get so shallow – again, precisely because the wall has been a strong protective barrier – that the manager finds repairs are not cost effective. Once the wall is allowed to degrade, even a storm of average size can result in a costly damage event.

Figure 5 shows indicative time series of the relationship between storm damage and wall investment under three rep-

resentative pairs of $D^*$ and $W^*$ thresholds. When both the damage and economy-of-scale thresholds are low (Fig. 5a), the manager determines that repairing even minor wall deterioration is cost effective: after the seawall is near maximum height, storm damage and associated repair costs remain low through time. When the manager tolerates greater storm damage and requires a greater economy of scale to initiate wall repair, investment in response to damage is erratic (Fig. 5b). Finally, when the manager tolerates extensive storm damage and maintains an economy of scale that requires major wall deterioration to warrant investment in repair, even relatively minor storm damage may necessitate total reconstruction of the seawall (Fig. 5c).

The other imposed parameter governing agent behavior is hindsight ($N$). Figure 6 shows the parameter space defined by damage tolerance and economy of scale for four different hindsight conditions (5, 10, 20, and 30 yr). Each square within each plot in Fig. 6 is the ensemble mean of 10 model trials per parameter pair under different randomized forcing conditions. In the left column, color represents the difference between total storm damage and total wall investment over the duration of a model run; in the right column, color represents standard deviation in the ensemble results. The totals in Fig. 6 illustrate in aggregate what the time series in Fig. 5 show in detail. Minimizing seawall deterioration in the model is an effective but expensive preventative measure against storm damage. Oppositely, infrequent wall repair costs less overall but comes at the expense of large amounts of damage. Moreover, when repairs do happen, they require maximum expenditure. The longer the data series ($N$) the manager agent uses in the decision calculations, the lower the variability in the system outcome: a best-fit line through 30 data points changes less from year to year than a line through 5 points, tempering the extremes of damage and investment costs over time.

Of course, this model does not simulate the fine-grained intricacy of real shoreline management. The premise assumes sustained managerial commitment to a seawall (e.g., Pilkey and Wright, 1988) and does not explore cost–benefit conditions for abandonment (e.g., McNamara and Keeler, 2013). Strangely enough, the manager's hindsight-based approach to decision making has a surprising, if dubious, recent precedent: in June 2012, the state senate in North Carolina, USA, passed a bill requiring that state agencies use linear extrapolations from historical data to project future sea-level rise, but the state's house of representatives subsequently rejected the measure (Phillips, 2012). Among the model's limitations, a single manager agent treats the seawall as a single managed unit, where a long seawall might be repaired in parts and can extend across adjacent municipalities. Also, the manager agent's decision-making process operates within the imposed bounds of fixed parameters for damage tolerance and economy of scale. But even if those thresholds varied as a function of property value, for example, the results shown still illustrate the kinds of patterns a resulting time series would comprise. A trend of increasing property value might drive increased investment in mitigation and thereby decrease total incurred damage (e.g., Fig. 5a). A decline in property value might make hazard mitigation less relevant, and the seawall might be left unrepaired for long periods (e.g., Fig. 5c). The value here of exploring the model's behavior given discrete pairs of bounding conditions is that the underlying mechanisms for system dynamics remain transparent.

Finally, in this construction of the model, wall strength ($W_S$) is the only dynamical variable. Changing the height or strength of the seawall does not affect any beach width or storm characteristics: the distribution of the environmental forcing does not change as a result of the hazard mitigation. Lacking a feedback that affects the environmental conditions, the model represents some but not all of the dynamics involved in the third hypothesis for Mileti's paradox. Here the seawall simply acts as a filter, one that can deteriorate (until it protects against only minor events in the environmental forcing signal) and be restored (filtering a wider range of event magnitudes). Moreover, because the deterioration rate ($\rho$) is a negative exponential function of wall strength, the seawall deteriorates exponentially faster the weaker it is. Punctuated, large damage events thus occur when the manager agent delays wall repairs, but those damage events get no worse with time because there is no feedback on an intermediate timescale (e.g., Werner and McNamara, 2007) to drive an emergent trend in either the coastal development or shoreline components of the system. With the coupling feedback left out, this model serves the expository purpose of illustrating the kind of dynamics that can derive strictly from variability in the forcing conditions imposed on a threshold-governed system.

## 4  Discussion

The high financial and economic costs of coastal erosion and flooding are expected to increase further with future sea-level rise and the cumulative effects of anthropogenic changes in coastal sediment fluxes (Mendelsohn and Neumann, 2004; Stern, 2007; Nicholls and Cazenave, 2010; Syvitsky and Kettner, 2011). Some argue that recent coastal development around the world, fueled by the housing market bubble behind the 2008 global financial crisis, has outstripped strategies for sustainable coastal management (Cooper and McKenna, 2009). For a brief period after Hurricane Sandy, US states even debated the prospect of Dutch-scale barrier engineering for parts of the US eastern seaboard (Higgins, 2012; Navarro, 2012). Insight into the dynamics of coastal vulnerability is therefore valuable to government agencies whose remits involve hazard assessment, impact forecasting, and environmental adaptation strategies (Thorne et al., 2007; Plant et al., 2010; MCCIP, 2010; DEFRA 2010). But if protected infrastructure is increasingly valuable or if storm

impacts are increasingly powerful, or if both conditions are true, then the cost of damage may go up regardless of how well people understand the hazard. In the presence of a rising trend in either the economic or natural component of the coastal coupled system, is Mileti's paradox inevitable?

Theoretically, situations for which long-term predictions are possible (or reasonable) should lend themselves to optimization, a standard analysis in resource economics that projects into the infinite future to maximize cumulative net benefits (Conrad, 2010; Smith et al., 2009). Ideally, a coastal manager could operate by a cyclical, economically optimal mitigation schedule and never need to deviate from it. For a coupled system in which small changes in environmental conditions might drive large changes in decision making, the more irregularity a manager introduces into a mitigation program, the farther the system's net benefits drift from the optimal outcome (e.g., Lazarus et al., 2011). The seawall model presented here illustrates increased systemic variability in response to frenetic management recalculations (Figs. 5 and 6), even under stationary forcing conditions. The time series shown in Fig. 5b, and the regions of the model parameter space where the difference between total damage and total investment hovers around zero in Fig. 6, represent a managed coastal system unwilling to tolerate damage but reluctant to invest in the scale of infrastructural maintenance necessary to prevent it. By contrast, of any managed coastline in the world, the best example of a long-term optimization strategy in practice must be the Netherlands, strategy in practice must be the Netherlands (Kabat et al., 2005, 2009), whose damage-versus-investment trajectory might look more like Fig. 5a. The Dutch Delta Plan has engineered Mileti's paradox into a moot point: the coastal hazard is well understood, the cost of maintenance is high but accepted, and the resulting cost of damage is small.

Although the Delta Plan is a physical threshold of such magnitude that impact from all but the rarest events are preventatively filtered, it is not a solution to sea-level rise. Therefore, the scale of intervention on the Dutch coast has not decoupled human dynamics from natural environmental processes so much as coupled them to the environment at a timescale significantly longer than those most governments treat as dynamically relevant. Such contrasting scales of intervention complicate the suggestion that "climate change, by accelerating the rates of landscape change, tends to strengthen the coupling with human dynamics" (Murray et al., 2013). With various technological innovations throughout history, humans have proven remarkably successful at weakening the strength of coupling between human and environmental dynamics. For example, land clearing in pre-Columbian Mesoamerica triggered intensive soil erosion, but agricultural terracing – another physical-threshold system – was so effective at preventing subsequent soil loss that the next spike in erosion only occurred after the landscape was abandoned and the terrace structures began failing from lack of maintenance (Fisher et al., 2003). Other technological innovations have made uninhabitable places habitable, resulting in ways of life that are functionally disconnected from surrounding natural systems: consider what air conditioning makes possible by masking outdoor heat and humidity, or that cities in arid basins can divert water from major river drainages over huge distances (e.g., Hundley, 2009). Technological interventions that disconnect human activities from local environmental conditions allow an anthropic "built layer... of artificial composition and structure" to be superimposed on the Earth's surface (Haff, 2003). Indeed, the frontier of innovative environmental interventions has reached the scale of geoengineering, the "intentional alteration" of natural planetary-scale processes (Caldiera et al., 2013). Mitigation technology, whether agricultural terracing, air conditioners, levees, or schemes for solar radiation management, is characteristically sensitive – strongly coupled – to environmental conditions. Such responsiveness, rather like buildings designed to tolerate earthquakes, is what makes them good interventions. Technological interventions that are sensitive to environmental conditions enable humans to build and live anywhere they choose, to be effectively insensitive to environmental change, with the potential to be insulated all but entirely from natural variability.

In light of this technological track record, and assuming that strongly coupled "human dynamics" refers to societal actions rather than to physical hazard-defense infrastructure, in the near term, climate change may strengthen environmental coupling with human dynamics in systems that are already strongly coupled: settings in which mitigation actions such as river levees (Chriss and Shock, 2001; Werner and McNamara, 2007), artificial dunes (Magliocca et al., 2011), beach nourishment (Lazarus et al., 2011), and seawalls match rather than overwhelm natural system processes. Given the conditions that contribute to strong coupling between human and environmental dynamics (Werner and McNamara, 2007), it follows that coupling strength is an inherently transient property of these systems. If institutions invest in mitigation infrastructures that function as physical thresholds on the scale of Earth's global systems, climate change could ultimately reduce, rather than increase, human–environmental coupling strength observable over human timescales.

**Acknowledgements.** My thanks to D. E. McNamara, P. K. Haff, B. T. Werner, and M. A. Ellis for comments and conversations that contributed to the development of the ideas presented here; to the School of Earth & Ocean Sciences and the Sustainable Places Research Institute at Cardiff University; and to G. Coco and the organizers of the Eighth Symposium on River, Coastal, and Estuarine Morphodynamics (RCEM) in Santander, Spain (June 2013).

Edited by: G. Coco

## References

Abraham, R. H. and Shaw, C. H.: Dynamics, the geometry of behavior: Part 4 – bifurcation behavior, Visual Mathematics Library, Addison-Wesley, 1988.

Anderson, R. S.: Modeling the tor-dotted crests, bedrock edges, and parabolic profiles of high alpine surfaces of the Wind River Range, Wyoming, Geomorphology, 46, 35–58, 2002.

Beisner, B. E., Haydon, D. T., and Cuddington, K.: Alternative stable states in ecology, Front. Ecol. Environ., 1, 376–382, 2003.

British Geological Survey (BGS) coastal erosion summary: http://www.bgs.ac.uk/research/climatechange/environment/coastal/caseStudies.html, last access: September 2013.

Caldeira, K. Bala, G., and Cao, L.: The science of geoengineering, Ann. Rev. Earth Pl. Sc., 41, 231–256, 2013.

Chin, A., Florsheim, J. L., Wohl, E., and Collins, B. D.: Feedbacks in human–landscape systems, Environ. Manage., 53, 28–41, 2013.

Conrad, J. M.: Resource economics (2nd Edn.), Cambridge University Press, 2010.

Cooper, J. A. G. and McKenna, J.: Boom and bust: The influence of macroscale economics on the world's coasts, J. Coast. Res., 25, 533–538, 2009.

Criss, R. E. and Shock, E. L.: Flood enhancement through flood control, Geology, 29, 875–878, 2001.

Dean, R. G. and Dalrymple, R. A.: Coastal processes with engineering applications, Cambridge University Press, 2002.

Department for Environment, Food, and Rural Affairs (DEFRA): Productive Seas – Coastal Defense: http://chartingprogress.defra.gov.uk/coastal-defence, last access: September 2013.

Ells, K. and Murray, A. B.: Long-term, non-local coastline responses to local shoreline stabilization, Geophys. Res. Lett., 39, L19401, doi:10.1029/2012GL052627, 2012.

Emanuel, K: Increasing destructiveness of tropical cyclones over the past 30 years, Nature, 436, 686–688, 2005.

Fisher, C. T., Pollard, H. P., Israde-Alcántara, I., Garduño-Monroy, V. H., and Banerjee, S. K.: A reexamination of human-induced environmental change within the Lake Patzcuaro Basin, Michoacan, Mexico, P. Natl. Acad. Sci. USA, 100, 4957–4962, 2003.

Guardian (Press Association): Warning of 'significant' flood risk, available at: http://www.guardian.co.uk/uk/feedarticle/10330403, 11 July 2012.

Gerritsen, H.: What happened in 1953? The Big Flood in the Netherlands in retrospect, Philo. T. Roy. Soc. A, 363, 1271–1291, 2005.

Gopalakrishnan, S., Smith, M. D., Slott, J. M., and Murray, A. B.: The value of disappearing beaches: a hedonic pricing model with endogenous beach width, J. Environ. Econ. Manag., 61, 297–310, 2011.

Groffman, P. M., Baron, J. S., Blett, T., Gold, A. J., Goodman, I., Gunderson, L. H., Levinson, B. M., Palmer, M. A., Paerl, H. W., Peterson, G. D., Poff, N. L., Rejeski, D. W., Reynolds, J. F., Turner, M. G., Weathers, K. C., and Wiens, J.: Ecological thresholds: the key to successful environmental management or an important concept with no practical application?, Ecosystems, 9, 1–13, 2006.

Haff, P. K.: Neogeomorphology, prediction, and the Anthropic Landscape, in: Prediction in Geomorphology, edited by:

Wilcock, P. R. and Iverson, R. M., AGU Monograph Series, 135, 15–26, 2003.

Hartman, R.: The harvesting decision when the standing forest has value, Econ. Inq., 14, 52–58, 1976.

Higgins, A.: Lessons for the U.S. from a flood-prone land, The New York Times, A6, 14 November 2012.

Hooke, R. LeB.: On the efficacy of humans as geomorphic agents, GSA Today, 4, 217 pp., 1994.

Hooke, R. LeB.: On the history of humans as geomorphic agents, Geology, 28, 843–846, 2000.

Hundley, N.: Water and the West: the Colorado River compact and the politics of water in the American West, University of California Press, Berkeley, 2009.

Jin, D., Ashton, A. D., and Hoagland, P.: Optimal responses to shoreline changes: an integrated economic and geological model with application to curved coasts, Nat. Resour. Model., 26, 572–604, 2013.

Johnson, C. L., Tunstall, S. M., and Penning-Rowsell, E. C.: Floods as catalysts for policy change: historical lessons from England and Wales, Water Res. Devel., 21, 561–575, 2005.

Kabat, P., van Vierssen, W., Veraart, J., Vellinga, P., and Aerts, L.: Climate proofing the Netherlands, Nature, 438, 283–284, 2005.

Kabat, P., Fresco, L. O., Stive, M. J. F., Veerman, C. P., van Alphen J. S. L. J., Parmet, B. W. A. H., Hazeleger, W., and Katsman, C. A.: Dutch coasts in transition, Nat. Geosci., 2, 450–452, 2009.

Kates, R. W., Clark, W. C., Corell, R., Hall, J. M., Jaeger, C. C., Lowe, I., McCarthy, J. J., Schellnhuber, H. J., Bolin, B., Dickson, N. M., Faucheux, S., Gallopin, G. C., Grübler, A., Huntley, B., Jäger, J., Jodha, N. S., Kasperson, R. E., Mabogunje, A., Matson, P., Mooney, H., Moore, B., O'Riordan, T., and Svedin, U.: Sustainability science, Science, 292, 641–642, 2001.

Kelley, J. T. and Brothers, L. L.: Camp Ellis, Maine: a small beach community with a big problem …its jetty, in: America's most vulnerable coastal communities, edited by: Kelley, J. T., Pilkey, O. H., and Cooper, J. A. G., Geol. S. Am. S., 460, 1–20, 2009.

Kraus, N. C.: The effects of seawalls on the beach: an extended literature review, J. Coastal Res., 4, 1–28, 1988.

Kraus, N. C. and McDougal, W. G.: The effects of seawalls on the beach: Part I, an updated literature review, J. Coastal Res., 4, 1–28, 1996.

Lazarus, E. D., McNamara, D. E., Smith, M. D., Gopalakrishnan, S., and Murray, A. B.: Emergent behavior in a coupled economic and coastline model for beach nourishment, Nonlin. Process. Geophys., 18, 989–999, 2011.

Leafe, R., Pethick, J., and Townend, I.: Realizing the benefits of shoreline management, Geog. J., 164, 282–290, 1998.

Limber, P. W. and Murray, A. B.: Beach and sea-cliff dynamics as a driver of long-term rocky coastline evolution and stability, Geology, 39, 1147–1150, 2011.

Liu, J., Dietz, T., Carpenter, S. R., Alberti, M., Folke, C., Moran, E., Pell, A. N., Deadman, P., Kratz, T., Lubchenco, J., Ostrom, E., Ouyang, Z., Provencher, W., Redman, C. L., Schneider, S. H., and Taylor, W. W.: Complexity of coupled human and natural systems, Science, 317, 1513–1516, 2007a.

Liu, J., Dietz, T., Carpenter, S. R., Folke, C., Alberti, M., Redman, C. L., Schneider, S. H., Ostrom, E., Pell, A. N., Lubchenco, J., Taylor, W. W., Ouyang, Z., Deadman, P., Kratz, T., and Provencher, W.: Coupled human and natural systems, AMBIO, 36, 639–649, 2007b.

Lubchenco, J. and Karl, T. R.: Predicting and managing extreme weather events, Phys. Today, 65, 31–37, 2012.

Lumbroso, D. M. and Vinet, F.: A comparison of the causes, effects and aftermaths of the coastal flooding of England in 1953 and France in 2010, Nat. Hazard. Earth Sys., 2321–2333, 2011.

Magliocca, N. R., McNamara, D. E., and Murray, A. B.: Long-term, large-scale morphodynamic effects of artificial-dune construction along a barrier-island coastline, J. Coast. Res., 27, 918–930, 2011.

Marine Climate Change Impacts Programme (MCCIP): Annual Report Card 2010–2011, edited by: Baxter, J. M., Buckley P. J., and Wallace, C. J., Lowestoft, available at: www.mccip.org.uk/arc, 2010.

Marriner, N., Morhange, C., Doumet-Serhal, C., and Carbonel, P.: Geoscience rediscovers Phoenicia's buried harbors, Geology, 34, 1–4, 2006.

McDaniel, C. N. and Gowdy, J. M.: Paradise for sale: a parable of nature, University of California Press, 2000.

McNamara, D. E., Murray, A. B., and Smith, M. D.: Coastal sustainability depends on how economic and coastline responses to climate change affect each other, Geophys. Res. Lett., 38, L07401, doi:10.1029/2011GL047207.

McNamara, D. E. and Keeler, A.: A coupled physical and economic model of the response of coastal real estate to climate risk, Nature Clim. Change, 3, 559–562, 2013.

McNamara, D. E. and Werner, B. T.: Coupled barrier island–resort model: 1. Emergent instabilities induced by strong human-landscape interactions, J. Geophys. Res., 113, F01016, doi:10.1029/2007JF000840, 2008a.

McNamara, D. E. and Werner, B. T.: Coupled barrier island–resort model: 2. Tests and predictions along Ocean City and Assateague Island National Seashore, Maryland, J. Geophys. Res., 113, F01017, doi:10.1029/2007JF000841, 2008b.

Mendelsohn, R. and Neumann, J. E. (Eds.): The impact of climate change on the United States economy, Cambridge University Press, 2004.

Mileti, D.: Disasters by design: A reassessment of natural hazards in the United States, Joseph Henry Press, 1999.

Murray, A. B., Lazarus, E., Ashton, A., Baas, A., Coco, G., Coulthard, T., Fonstad, M., Haff, P., McNamara, D., Paola, C., Pelletier, J., and Reinhardt, L.: Geomorphology, complexity, and the emerging science of the Earth's surface, Geomorphology, 103, 496–505, 2009.

Murray, A. B., Gopalakrishnan, S., McNamara, D. E., and Smith, M. D.: Progress in coupling models of human and coastal landscape change, Comput. Geosci., 53, 30–38, 2013.

National Oceanic and Atmospheric Administration (NOAA): Billion-dollar Weather/Climate Disasters, http://www.ncdc.noaa.gov/billions/events, last accessed: September 2013.

National Research Council (NRC): The drama of the commons, National Academy Press, 2002.

National Research Council (NRC): Landscapes on the edge: new horizons for research on Earth's surface, National Academy Press, 2010.

Navarro, M.: Weighing sea barriers as protection for New York, The New York Times, A21, 8 November 2012.

Nicholls, R. J. and Cazenave, A.: Sea-level rise and its impact on coastal zones, Science, 328, 1517–1520, 2010.

Nordstrom, K. F.: Beaches and dunes of developed coasts, Cambridge University Press, 2000.

Ostrom, E.: Beyond markets and states: polycentric governance of complex economic systems, Amer. Econ. Rev., 100, 1–33, 2010.

Ostrom, E., Janssen, M. A., and Anderies, J. M.: Going beyond panaceas, P. Natl. Acad. Sci. USA, 104, 15176–15178, 2007.

Phillips, L.: Sea versus senators: North Carolina sea-level rise accelerates while state legislators put the brakes on research, Nature, 486, p. 450, 2012.

Pilkey, O. H. and Wright, H. L.: Seawalls versus beaches, J. Coast. Res., 4, 41–64.

Plant, N. G., Stockdon, H. F., Sallenger Jr., A. H., Turco, M. J., East, J. W., Taylor, A. A., and Shaffer, W. A.: Forecasting hurricane impact on coastal topography, Eos Trans. AGU, 91, 65–66, 2010.

Qingzhou, W.: The Protection of China's Ancient Cities from Flood Damage, Disasters, 13, 193–227, 1989.

Rahmstorf, S. and Coumou, D. Increase of extreme events in a warming world, P. Natl. Acad. Sci. USA, 108, 17905–17909, 2011.

Scheffer, M.: Critical transitions in nature and society, Princeton University Press, 2009.

Seabrook, J.: The beach builders, The New Yorker, 42, 22 July 2013.

Slott, J. M., Murray, A. B., and Ashton, A. D.: Large-scale responses of complex-shaped coastlines to local shoreline stabilization and climate change, J. Geophys. Res. Earth Surf., 115, F03033, doi:10.1029/2009JF001486, 2010.

Slott, J. M., Smith, M. D., and Murray, A. B.: Synergies between adjacent beach-nourishing communities in a morpho-economic coupled coastline model, Coast. Manage., 36, 374–391, 2008.

Smith, K.: Environmental hazards: Assessing risk and reducing disaster (6th Edn.), Routledge, 2013.

Smith, M. D., Slott, J. M., McNamara, D. E., and Murray, A. B.: Beach nourishment as a dynamic capital accumulation problem. J. Environ. Econ. Manag., 58, 58–71, 2009.

Stern, N. N. H. (Ed.): The economics of climate change: the Stern review. Cambridge University Press, 2007.

Strudley, M. W., Murray, A. B., and Haff, P. K.: Emergence of pediments, tors, and piedmont junctions from a bedrock weathering–regolith thickness feedback, Geology, 34, 805–808, 2006.

Summers, D.: The east coast floods, David & Charles, 1978.

Syvitski, J. P. and Kettner, A.: Sediment flux and the Anthropocene, Philos. T. Roy. Soc. A, 369, 957–975, 2011.

Thorne, C. R., Evans, E. P., and Penning-Rowsell, E. C. (Eds.): Future flooding and coastal erosion risks, Thomas Telford Services Ltd, 2007.

Tunstall, S. M. and Penning-Rowsell, E. C.: The English beach: experiences and values, Geogr. J., 164, 319–332, 1998.

Turner, B. L., Matson, P. A., McCarthy, J. J., Corell, R. W., Christensen, L., Eckley, N., Hovelsrud-Broda, G. K., Kasperson, J. X., Kasperson, R. E., Luers, A., Martello, M. L., Mathiesen, S., Naylor, R., Polsky, C., Pulsipher, A., Schiller, A., Selin, H., and Tyler, N.: Illustrating the coupled human–environment system for vulnerability analysis: three case studies, P. Natl. Acad. Sci. USA, 100, 8080–8085, 2003.

Werner, B. T. and McNamara, D. E.: Dynamics of coupled human–landscape systems, Geomorphology, 91, 393–407, 2007.

Williams, Z. C., McNamara, D. E., Smith, M. D., Murray, A. B., and Gopalakrishnan, S.: Coupled economic–coastline modeling with suckers and free riders, J. Geophys. Res. Earth Surf., 118, 887–899, 2013.

Yokota, H. and Komure, K.: Estimation of structural deterioration process by Markov-chain and costs for rehabilitation, in: Life-cycle performance of deteriorating structures: assessment, design, and management, edited by: Frangopol, D. M., Brühwiler, E., Faber, M. H., and Adey, B., Life-cycle performance of deteriorating structures, Am. Soc. Civil Engin., 424–431, 2003.

# 6

# The linkages among hillslope-vegetation changes, elevation, and the timing of late-Quaternary fluvial-system aggradation in the Mojave Desert revisited

**J. D. Pelletier**

Department of Geosciences, University of Arizona, Gould-Simpson Building, 1040 East Fourth Street, Tucson, Arizona 85721-0077, USA

*Correspondence to:* J. D. Pelletier (jdpellet@email.arizona.edu)

**Abstract.** Valley-floor-channel and alluvial-fan deposits and terraces in the southwestern US record multiple episodes of late-Quaternary fluvial-system aggradation and incision. Perhaps the most well-constrained of these episodes took place from the latest Pleistocene to the present in the Mojave Desert. One hypothesis for this episode – i.e., the paleovegetation-change hypothesis (PVCH) – posits that a reduction in hillslope vegetation cover associated with the transition from Pleistocene woodlands to Holocene desert scrub generated a pulse of sediment that triggered a primary phase of aggradation downstream, followed by channel incision, terrace abandonment, and initiation of a secondary phase of aggradation further downstream. A second hypothesis – i.e., the extreme-storm hypothesis – attributes episodes of aggradation and incision to changes in the frequency and/or intensity of extreme storms. In the past decade a growing number of studies has advocated the extreme-storm hypothesis and challenged the PVCH on the basis of inconsistencies in both timing and process. Here I show that in eight out of nine sites where the timing of fluvial-system aggradation in the Mojave Desert is reasonably well constrained, measured ages of primary aggradation are consistent with the predictions of the PVCH if the time-transgressive nature of paleovegetation changes with elevation is fully taken into account. I also present an alternative process model for PVCH that is more consistent with available data and produces sediment pulses primarily via an increase in drainage density (i.e., a transformation of hillslopes into low-order channels) rather than solely via an increase in sediment yield from hillslopes. This paper further documents the likely important role of changes in upland vegetation cover and drainage density in driving fluvial-system response during semiarid-to-arid climatic changes.

## 1 Introduction and motivation

Quaternary deposits of the southwestern US are dominated by valley-floor-channel and alluvial-fan deposits and their associated terraces that record multiple regionally correlative episodes of aggradation, channel incision, and terrace abandonment (Christensen and Purcell, 1985; Bull, 1991; Harvey et al., 1999; Menges et al., 2001; McDonald et al., 2003; Anders et al., 2005). What drives these aggradation and incision episodes has been a fundamental question in the geomorphology and Quaternary geology of the southwestern US for decades. Given the approximate correlation between the

timing of fluvial-system aggradation events and semiarid-to-arid transitions recorded in paleoclimatic proxies, together with the correlative nature of Quaternary deposits and terraces across tectonically active and inactive regions, climate change has most often been invoked as the primary trigger for these episodes. How climate change drives episodes of aggradation and incision is debated, however.

In this paper I focus on the timing and mechanisms of fluvial-system aggradation and incision in the Mojave Desert portion of the southwestern US from the latest Pleistocene to the present. I focus on this study area and this time interval because the constraints on both fluvial-system behavior and its potential driving mechanisms are arguably better constrained than for any other area and any other time interval in the world that has experienced a semiarid-to-arid climatic change. For example, the timing of local paleovegetation changes is unusually well constrained; i.e., 87 dated packrat middens within the central Mojave Desert exist with woodland species (*Juniperus*) clearly present or absent from 17 to 0 ka. Also, dozens of state-of-the-art stratigraphic and surface-exposure ages have been obtained (e.g., Miller et al., 2010; Antinao and McDonald, 2013a, and references therein). Semiarid-to-arid climatic transitions are of particular interest given that semiarid landscapes may be particularly sensitive to climatic changes (e.g., Langbein and Schumm, 1958) and because large portions of earth's surface have the potential to transition from semiarid to arid climates in the future (e.g., Held and Soden, 2006; Lau et al., 2013). More broadly, drainage basin responses to climatic changes are mediated in large part through changes in vegetation cover, and understanding the feedbacks between vegetation cover and landscape evolution has emerged as a "grand challenge" problem in earth surface science (e.g., Murray et al., 2009; Reinhardt et al., 2010). As such, understanding the late-Quaternary record of fluvial-system response to climatic changes in the Mojave Desert has the potential to enhance our conceptual understanding of how vegetation cover and landforms coevolve in other process zones.

In his work on the paleovegetation-change hypothesis (PVCH) for fluvial-system response to climatic changes in the southwestern US, Bull (1991) argued that a reduction in vegetation cover during late-Quaternary semiarid-to-arid transitions led to sediment pulses characterized by an initial increase in sediment yield (resulting in aggradation in valley-floor channels and/or alluvial fans downstream) followed by a decrease in sediment yield as the reservoir of colluvium stored on hillslopes was depleted (resulting in channel incision and terrace abandonment). As Bull (1991) wrote, "when the climate changes from semiarid to arid, the concurrent decrease in vegetation cover results in a rapid increase in sediment yield. The sediment-yield maximum is attained quickly, after which the yield progressively decreases as the area of hillslope colluvium decreases and outcrop area increases." (p. 113). Bull (1991) also invoked changes in the frequency and/or intensity of extreme storms; i.e., "hillslope sediment yields were greatly increased – partly because of increased rainfall intensities associated with the return of monsoon thunderstorms" (p. 114). As such, Bull (1991) did not envision the PVCH working in isolation. Which of these two driving mechanisms (paleovegetation changes or an increase in the frequency and/or intensity of extreme storms) plays a greater role in driving fluvial-system aggradation and incision has been a subject of controversy ever since Bull's seminal work.

As the dating of fluvial-system deposits in the Mojave Desert has improved over the past 2 decades, numerous studies have used the apparently poor correlation between the

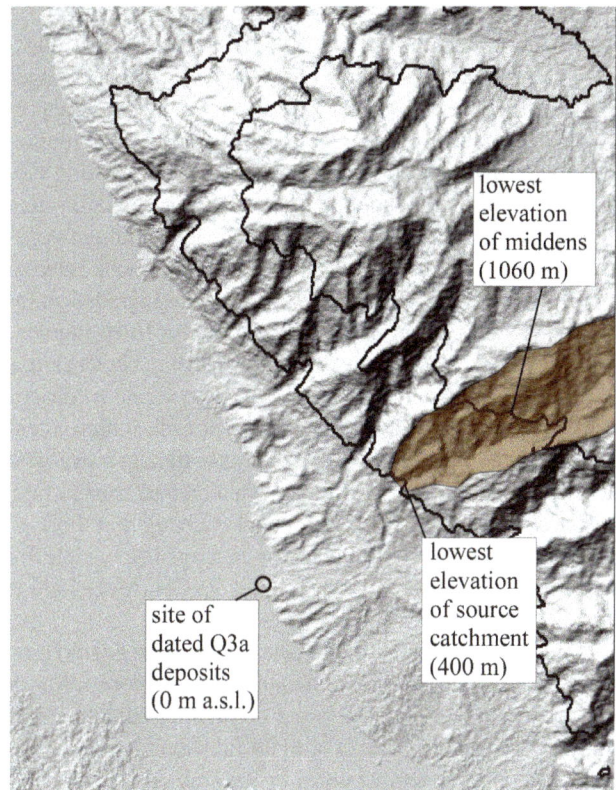

**Figure 1.** Illustration of the key elevations associated with the southern Death Valley fan-aggradation site, using a shaded-relief image as a base map. Location is shown in Fig. 4. Samples were collected by Sohn et al. (2007) at approximately 0 m a.s.l. The timing of the initiation of fan aggradation is controlled, according to the PVCH, by the timing of the retreat of the lower elevational limit of woodland vegetation through the lowest elevations of the source catchment; i.e., 400 m a.s.l. In their central Mojave subregion, Antinao and McDonald (2013a) used midden records with lowest elevations of 1060 m a.s.l. (Supplement of Antinao and McDonald, 2013a), resulting in an elevational gap of approximately 600 m. This gap produces a 6 ka cal (calibrated)time lag between assumed and actual paleovegetation changes in the lowest elevations of the source catchment.

timing of paleovegetation changes and alluvial-fan aggradation to argue against the PVCH. For example, in the central Mojave subregion of their study, Antinao and McDonald (2013a) tested the PVCH by comparing the timing of fan aggradation at a site (southern Death Valley – DV) sourced by an eroding catchment with lowest elevations of ≈ 400 m a.s.l. (above sea level) against the timing of paleovegetation changes constrained by a group of nearby packrat midden sites (which record the vegetation types within a 100 m range) with a lowest elevation of 1060 m a.s.l. (Granite Mountains) (Fig. 1; Supplement of Antinao and McDonald, 2013a). This approximately 600 m gap in elevation between the lowest elevations of the source catchment and the lowest elevation where the timing of paleovegetation changes was constrained is problematic because, as I demonstrate in Sect. 2, the transition from woodlands to desert scrub measured at 1060 m a.s.l. occurred 6 ka after vegetation changes in the lowest elevations of the source catchment (i.e., the portion of the catchment responsible for triggering the initiation of fan aggradation according to the PVCH). Antinao and McDonald (2013a) concluded that the onset of fluvial-system aggradation "began well before changes in catchment vegetation cover" and, as a result, that "the ambiguous relation between vegetation change and alluvial fan aggradation indicates that vegetation had a reduced role in LPH aggradation". To test the PVCH most comprehensively, however, it is necessary to constrain the timing of the woodland-to-desert-scrub transition at the lowest elevations of each source catchment because vegetation changes are time-transgressive with elevation (occurring first at low elevations and later at high elevations) and hence it is the vegetation changes at the lowest elevations of source catchments that are responsible for triggering aggradation according to the PVCH (McDonald et al., 2003; Sect. 2 in this paper).

Miller et al. (2010) concluded that alluvial-fan aggradation in the Mojave Desert was principally due to more frequent and/or more intense storms based on an approximate correlation between the timing of alluvial-fan aggradation and elevated sea-surface temperatures in the Gulf of California (a proxy for monsoon activity). Specifically, both records exhibit two peaks from the latest Pleistocene to the present; i.e., one at ca. 15–7 ka cal BP (calibrated years, before present) and another at ca. 6–3 ka cal BP. The dual-pulsed nature of alluvial-fan aggradation from the latest Pleistocene to present was also emphasized by Bull (1991), who noted "the aggradation event that was associated with the Pleistocene-Holocene climate change consisted of two main pulses (Q3a and Q3b) with an intervening period of stream-channel incision" (p. 114). Miller et al. (2010) considered the dual-pulsed nature of aggradation to be inconsistent with PVCH. However, two pulses of aggradation separated by approximately 4–8 ka and resulting from a single pulse of sediment yield is precisely what the PVCH predicts (documented in Sect. 2). A sediment pulse from upstream catchments can overwhelm the ability of a downstream valley-floor channel or alluvial fan to convey that increase in sediment, leading to a "primary" phase of aggradation. When sediment supply declines during the waning phase of the sediment pulse, fluvial channels incise into the sediments just deposited and an abandoned terrace is formed. Sediments reworked by channel incision, along with sediments still being supplied by the catchment in the waning phase of the sediment pulse, are then deposited in a "secondary" deposit further downstream (Fig. 2a). In this way, a single pulse of sediment can generate two deposits separated by channel incision, as Bull (1991) stated and as Schumm's (1973) concept of complex response formalized.

The PVCH has also been criticized on the basis of process. The conceptual process model that underpins the PVCH as

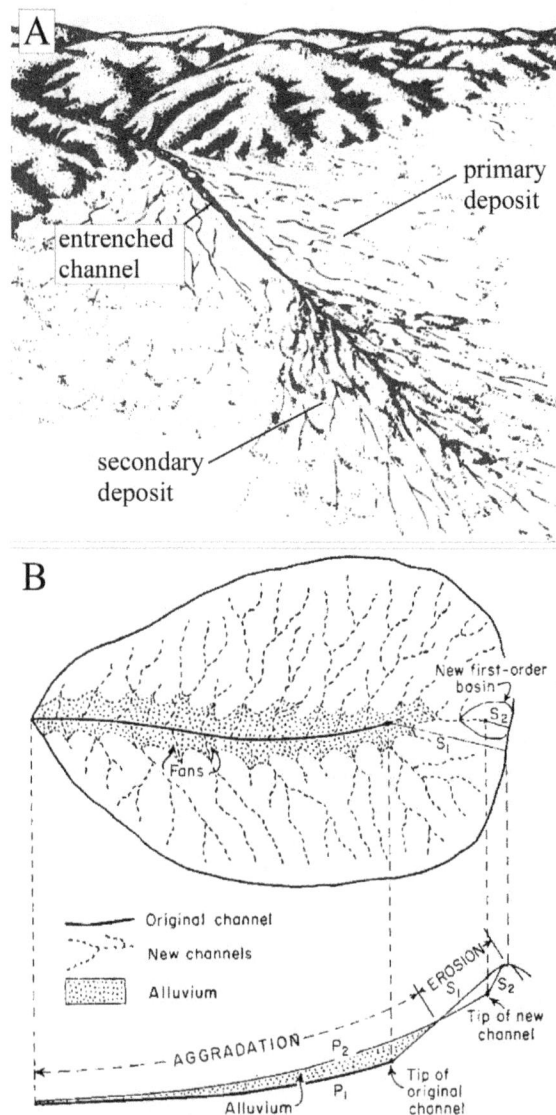

**Figure 2.** Schematic diagrams drawn from the literature illustrating (a) fan and (b) source-catchment responses to semiarid-to-arid transitions. (a) Illustration of paired deposits (i.e., a secondary deposit downstream and inset into a primary deposit) resulting from a single pulse of sediment (from Bull, 1968). In (a), a pulse of sediment from upstream catchments leads to a "primary" phase of aggradation. When sediment supply declines during the waning phase of the sediment pulse, fluvial channels incise into the sediments just deposited and an abandoned fan terrace is formed. Sediments reworked by channel incision, along with sediments still being supplied by the catchment in the waning phase of the sediment pulse, are then deposited in a "secondary" deposit farther downslope. (b) Schematic diagram illustrating the response of a catchment to an increase in erosivity (from Strahler, 1968). Low-order drainages grow headward, converting hillslopes, and into low-order fluvial channels capable of rapidly transporting material stored as colluvium during the previous period. When channels cease growing headward (i.e., when the catchment stabilizes to a new, higher drainage density), the pulse of sediment wanes and channel incision/secondary aggradation is triggered (not shown).

proposed by Bull (1991) requires that hillslopes be stripped nearly to bedrock in some locations, increasing the area of outcrops in the catchment and hence reducing the availability of soil for transport, triggering channel incision and terrace abandonment downstream (Bull, 1991). McDonald et al. (2003) correctly argued that Bull's (1991) conceptual model is inconsistent with the observed abundance of soils on, and the high-infiltration capacity of, many Holocene desert hillslopes in the Mojave Desert. Modeling studies (discussed in Sect. 3), however, together with the correlation of drainage density with both aridity and percent bare area in catchments of the southwestern US (Melton, 1957), suggest that the principal effect of a reduction in vegetation cover is not an increase in sediment yield from hillslopes but rather an increase in drainage density as hillslopes are converted into low-order channels capable of rapidly entraining colluvium into the fluvial system (Fig. 2b; Strahler, 1958; Tucker and Slingerland, 1997; Pelletier et al., 2011). As such, some modification of the process model underlying the PVCH is necessary even if the PVCH can be shown to exhibit timing that is consistent with available data. More recently, Antinao and McDonald (2013a) challenged the PVCH by noting that Pierson et al. (2007) documented a minimal increase in erosion following the manual removal of *Juniperus* from hillslopes. The Pierson et al. (2007) study is not a proper analog for the woodland-to-desert-scrub transition from the latest Pleistocene to the present in the Mojave Desert; however, because that study was performed in Oregon, erosion was measured in plots less than 10 m wide, and juniper root systems and fallen trees were left intact at the site after cutting (i.e., vegetation cover near the ground surface was higher after cutting than before). In contrast, vegetation cover would be lower and percent bare area would be higher following a woodland-to-desert-scrub transition that occurred over timescales of millennia.

In this paper I first quantify the timing of the retreat of the woodland-to-desert-scrub transition to higher elevations from the latest Pleistocene to the present within the Mojave Desert using all available data from the North American Midden Database of Strickland et al. (2005). Second, I report on the results of a geographic-information-system (GIS)-based analysis that accepts, as input, the timing of woodland-to-desert-scrub transition and a digital elevation model (DEM), and predicts the timing of both the initiation of primary aggradation and channel incision/secondary aggradation as predicted by the PVCH. Thirdly, I compare the predictions of the model to available geochronologic data from the Mojave Desert. My analysis demonstrates that the timing of fluvial-system aggradation from the latest Pleistocene to the present in the Mojave Desert is consistent with the PVCH in eight out of nine sites in the Mojave Desert where the timing is reasonably well constrained. Fourthly, I demonstrate consistency in the relationship between modern vegetation cover and elevation as a function of latitude within the Mojave Desert in order to provide a basis for assuming that the timing of veg-

etation changes as a function of elevation are relatively uniform throughout the study region. Finally, I present an improved process model of how the PVCH works, which includes transient changes in drainage density associated with semiarid-to-arid climate transitions.

## 2  A test of the PVCH for the Mojave Desert

In this section I reexamine the conclusions of Antinao and McDonald (2013a) using the same data but a different methodology. Rather than grouping fan-aggradation and paleovegetation sites according to spatial proximity and elevation ranges that are similar but nevertheless are associated with differences in the timing of assumed versus actual paleovegetation changes of up to 6 ka, my methodology honors the dominant role of elevation in controlling plant distributions in the Mojave Desert by first quantifying the relationship between the elevational lower limit of woodland plants versus time in the Mojave Desert and then applying that relationship to a GIS analysis that predicts the timing of both primary aggradation and channel-incision/secondary-aggradation based on the timing of the woodland-to-desert-scrub transition in the source catchments upstream from every point on the landscape. In both Antinao and McDonald (2013a) and this study, the commonness/abundance of *Juniperus* is used as a proxy for woodland versus desert scrub (i.e., *Juniperus* rare/absent) vegetation types. Where biomass data are available in the southwestern US, the woodland-to-desert-scrub transition is associated with a step change in biomass (e.g., Whittaker and Niering, 1975). Since woody biomass "reduces runoff and overland-flow erosion by improving water infiltration, reducing impacts by water droplets, intercepting rain and snow, and physically stabilizing soil by their roots and leaf litter" (Kort et al., 1998), a transition from commonness/abundance to rarity/absence of *Juniperus* is likely to also be associated with a step-change increase in sediment supply to fluvial systems downstream.

I analyzed the North American Midden Database (Strickland et al., 2005) to identify the elevational lower limit of *Juniperus* as a function of time from the latest Pleistocene to the present in the Mojave Desert (Fig. 3). The solid curve in Fig. 3, which correctly differentiates all 87 (within $2\sigma$ uncertainty; one rare/absent *Juniperus* data point lies to the left of the solid curve but the solid curve is consistent with that point within the 95 % confidence interval) available localities in terms of common/abundant versus rare/absent *Juniperus*, illustrates the dominant role of elevation in controlling woodland versus desert scrub vegetation types from 17–0 ka cal BP. The consistency of the data – i.e., within age uncertainty all common/abundant records reside on one side of the solid curve in Fig. 3 and all rare/absent records reside on the other side – provides confidence in using the curve to predict the timing of the woodland-to-desert-scrub transition

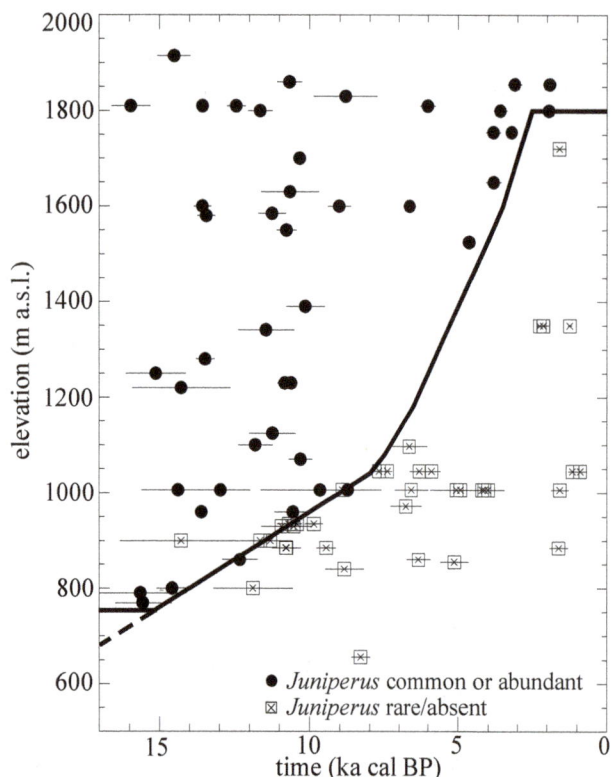

**Figure 3.** Plot of every midden in the North American Midden Database (Strickland et al., 2005) between 34 and 37° N latitude, 117.5 and 115° W longitude, and 500 and 2000 m a.s.l., and 1 and 17 ka in age (cal BP; $^{14}$C years were converted using IntCal09 of Reimer et al. (2009); 95 % confidence intervals are shown) with *Juniperus* common/abundant (closed circles; total of 46 samples) and *Juniperus* rare/absent (open squares; total of 41 samples). In the latter cases, *Juniperus* is specifically noted as absent or a complete taxa list is available that does not include *Juniperus*. The solid curve is the lower elevational limit of woodlands adopted in the model. The dashed line indicates that the lower limit of common/abundant *Juniperus* is poorly constrained from 17 to 15 ka cal BP.

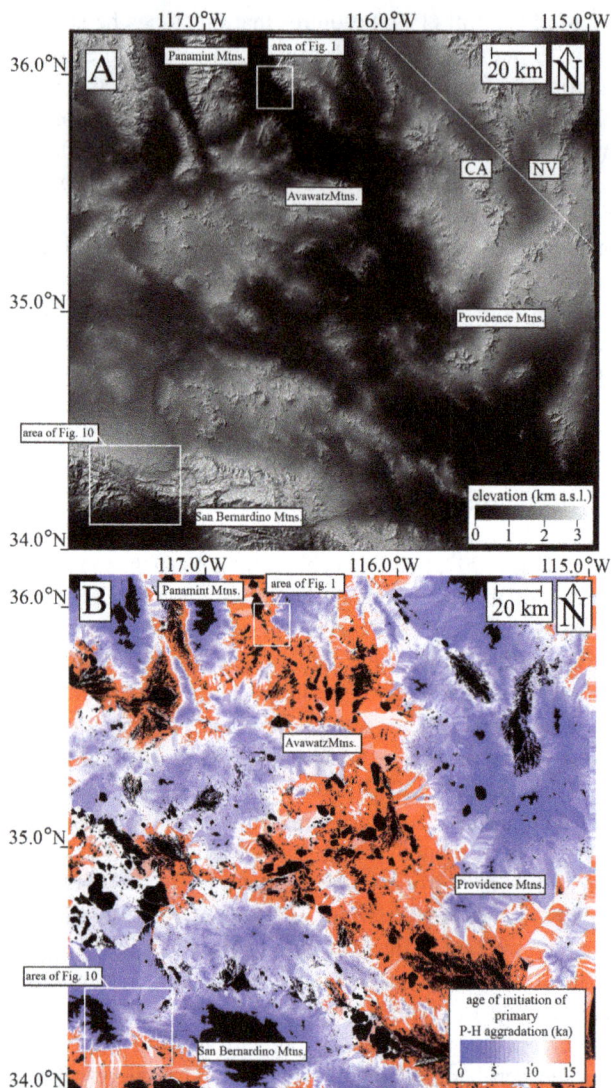

**Figure 4. (a)** Shaded-relief map of the study area. **(b)** Color map of the model-predicted age of initiation of primary aggradation (i.e., unit Q3a of Bull, 1991) where deposits of such ages are present. Reddish areas indicate Pleistocene-aged Q3a deposits while bluish areas indicate Holocene-aged (0–11.7 ka cal BP) Q3a deposits (if Q3a deposits are present). Black areas are locations not downstream (based on flow pathways defined by the modern DEM) from areas that have undergone P–H (Pleistocene–Holocene) changes in *Juniperus* cover.

within source catchments upstream from aggradation sites in the Mojave Desert using elevation data. The elevation–age relationship (i.e., solid curve in Fig. 3) for the lower limit of *Juniperus* is well constrained in the 15–10 ka cal BP interval but significantly less well constrained in the 10–3 ka cal BP interval. However, this uncertainty has little practical effect on the comparison of the model predictions to data because the predicted age of initiation of primary aggradation is between 15 and 10 ka cal BP for all of the sites except two (Johnson Valley and Grassy Valley; Table 1).

Primary aggradation is assumed to be initiated downstream when a small but significant portion (5 % is used here) of the source catchment area (defined as upstream areas with slopes > 20 %) changes from commonness/abundance to rarity/absence of *Juniperus* (Figs. 4–6). Primary aggradation is assumed to terminate when 50 % of the source catch-

ment area has transitioned from commonness/abundance to rarity/absence of *Juniperus*, at which point the source area for sediment derived from woodlands-to-desert-scrub transition is assumed to decline over time, triggering channel incision, terrace abandonment, and secondary aggradation further downstream.

The GIS analysis that predicts the timing of the onset of primary aggradation and the timing of channel incision/onset of secondary aggradation begins by associating each pixel

**Table 1.** Comparison of available data and model predictions for the timing of fluvial-system aggradation and incision from the latest Pleistocene to the present. Sites are listed in order of increasing elevation to highlight the relationship between elevation and the age of initiation of primary aggradation in the data and model (UTM – Universal Transverse Mercator; V. – Valley).

| Locality | UTM Easting (m)[5] | UTM Northing (m) | Elevation[6] (m a.s.l.) | Predicted initiation primary aggrad.[3] (ka BP) | Predicted incision[3] (ka BP) | Measured initiation primary aggrad.[2] (ka BP) | Measured incision[1] (ka BP) |
|---|---|---|---|---|---|---|---|
| Southern DV[7] | 525 510[6] | 3 978 760[6] | 0 | 15 ± 2 | 8 ± 2 | 17.4 ± 2.4 | 10.7 ± 1.54 |
| Chambless[8] | 639 525 | 3 818 044 | 250 | 15 ± 2 | 8 ± 2 | < 9.4 ± 0.67[4] | N/A |
| Kelso Wash[8] | 594 631 | 3 876 604 | 360 | 13 ± 2 | 5 ± 2 | < 13.3 ± 0.7 | N/A |
| Red Pass[9] | 564 267 | 3 909 254 | 500 | 13 ± 2 | 6 ± 2 | 12.1 ± 0.6 <br> 12.8 ± 0.5 <br> 11.5 ± 0.2 | N/A |
| Fenner Wash[8] | 658 533 | 3 843 945 | 520 | 12 ± 2 | 6 ± 2 | 12.1 ± 0.77 <br> 11.1 ± 0.3 <br> 10.6 ± 0.26 | N/A |
| Coyote Wash[8] | 522 313 | 3 875 008 | 540 | 11 ± 2 | 9 ± 2 | < 13.6 ± 0.3 | N/A |
| Sheep Creek[8] | 451 338 | 3 829 441 | 870 | 10 ± 2 | 5 ± 2 | 10.6 ± 0.56 <br> 11.8 ± 0.56 | 7.74 ± 0.54 <br> 8.31 ± 0.51 <br> 9.44 ± 0.49 |
| Johnson V.[10] | 550 100[6] | 3 796 840[6] | 980 | 8 ± 2 | 6 ± 2 | < 12.3 ± 0.9 <br> 10.3 ± 1.1 | N/A |
| Grassy V.[8] | 477 828 | 3 902 734 | 1040 | 6 ± 2 | 5 ± 2 | < 10.6 ± 0.6 | N/A |

[1] Channel-incision/terrace-abandonment ages are given in the two cases where the report makes clear that the sample came from near the top of the deposit. [2] Ages indicated by < are maximum values for the initiation of alluvial aggradation because the underlying unit (groundwater discharge or fluvial) was dated or because partial bleaching of TL samples was noted. [3] Uncertainty associated with model predictions is estimated to be 2 ka based on limitations of the paleovegetation record and the magnitude of typical variations in the model predictions with reasonable variations in the model. [4] Aggradation is initiated much later than predicted at this site. Deposit could be associated with secondary aggradation. [5] UTM coordinates are in NAD83 datum, zone 11; [6] Estimated from aerial photographs and digital elevation models. [7] Sohn et al. (2007); [8] Miller et al. (2010); [9] Mahan et al. (2007); [10] Rockwell et al. (2000).

in the source catchments with the age of the woodland-to-desert scrub transition using the solid curve in Fig. 3. Then, for each pixel in the study site, the percent of the upland source catchment that has undergone a change from commonness/abundance to rarity/absence of *Juniperus* is computed in 1 ka intervals starting at 15 ka cal BP. For example, at 15 ka only the lowest elevations (i.e., those less than or equal to 750 m a.s.l.) of source catchments have undergone a change from commonness/abundance to rarity/absence of *Juniperus*; hence, pixels downstream of such basins will record relatively low percentages for the percentage of the upstream source catchment basin that has undergone such a change. As time passes and woodlands retreat to higher elevations, the GIS analysis computes a progressively larger percentage of each source catchment that has transitioned from commonness/abundance to rarity/absence of *Juniperus*. The GIS analysis requires a map of the source catchment domain for each downstream pixel. The mapping of contributing area is performed in this study using the multiple-flow-direction (MFD) algorithm of Freeman (1991). I used the MFD method of Freeman (1991) to delineate source areas because it more

faithfully represents flow routing pathways in distributary channel systems (which are common in the study area) compared with alternatives such as the $D\infty$ method (Pelletier, 2008).

When 5 % of the source catchment area has undergone a change from commonness/abundance to rarity/absence of *Juniperus*, primary aggradation is assumed to be initiated. Similarly, when 50 % of the source catchment area has undergone a change from commonness/abundance to rarity/absence of *Juniperus*, channel incision/secondary aggradation is assumed to be initiated. The 5 and 50 % threshold values are not unique, but the results are not highly sensitive to these values within reasonable ranges; i.e., the predicted ages of aggradation and incision differ by 2 ka or less for 96.5 and 98.6 % of the study area where aggradation is predicted to occur as these values are varied over reasonable ranges from 5 to 15 % and 40 to 60 %, respectively. The systematic uncertainty of the GIS analysis is taken to be approximately 2 ka (Table 1) based on the uncertainty associated with the 5 and 50 % thresholds as well as the uncertainty of the timing

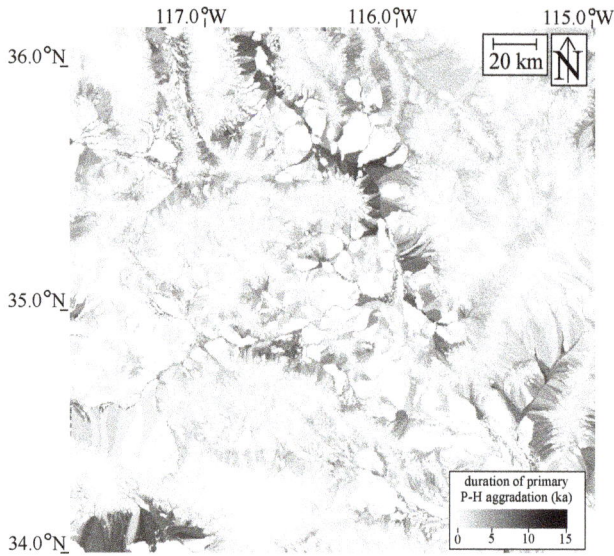

**Figure 5.** Grayscale map of the duration of primary aggradation associated with vegetation changes of the latest Pleistocene to mid-Holocene in the Mojave Desert (study area is the same as that shown in Fig. 4). Values range from 4–8 ka cal at alluvial sites fed by catchments with a wide range of elevations in the 750–1800 m range (typically lower-elevation sites) to 1–4 ka cal at sites fed by catchments with a smaller range of such elevations (typically higher-elevation sites).

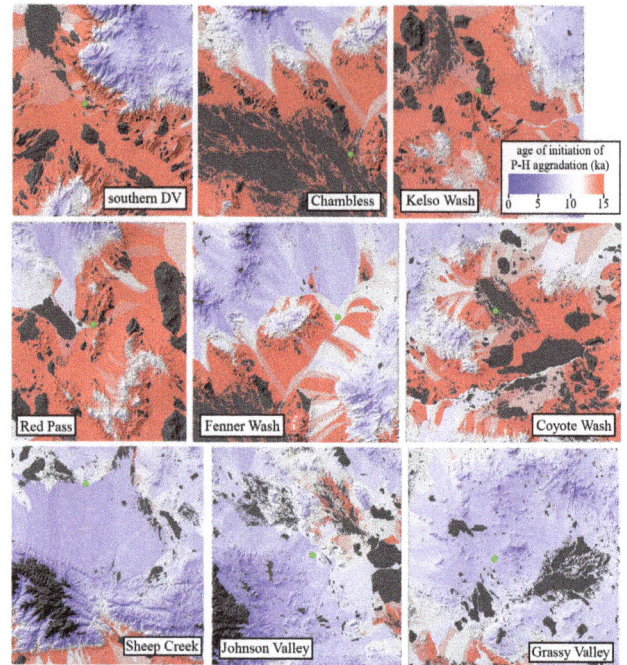

**Figure 6.** Subsets of Fig. 4 shown in greater detail and overlain on a shaded-relief image for each of the nine areas where the timing of fan aggradation is reasonably well constrained. Study areas are shown as lowest to highest elevation (of the dated sample locations) from upper left to lower right. The predicted age of initiation occurs systematically later at higher elevations, as also documented in Table 1 and Fig. 7.

of the woodland-to-desert scrub transition at each elevation (i.e., the solid curve in Fig. 3).

The results of the GIS analysis indicate that the PVCH predicts that the initiation of primary aggradation associated with vegetation changes from the latest Pleistocene to the present was highly variable in space and occurred over a wide range of time from ca. 15–6 ka cal BP (but possibly earlier given uncertainty in the timing of woodland-to-desert-scrub transition at the lowest elevations ca. 17–15 ka cal BP; see dashed line in Fig. 3). According to the PVCH, aggradation began earlier at sites fed by catchments of lower elevation and later at sites fed by catchments of higher elevation (Figs. 4, 6, 7; Table 1). The predictions of the PVCH are consistent with eight out of nine fan-aggradation sites in the Mojave Desert (Table 1). The PVCH predicts that the duration of primary aggradation (which is also equivalent to the time lag between the initiation of primary aggradation and channel incision/secondary aggradation) was also highly variable, ranging from 4–8 ka at sites fed by catchments with a wide range of elevations in the 750–1800 m range to 1–4 ka at sites fed by catchments with a smaller range of such elevations (Fig. 5, Table 1).

It should be noted that the GIS analysis predicts the timing of aggradation in some upland catchments where fill terraces may not form or may not be preserved. It is nevertheless appropriate to include such areas in the analysis because aggradation can occur locally in such areas. Fill terraces

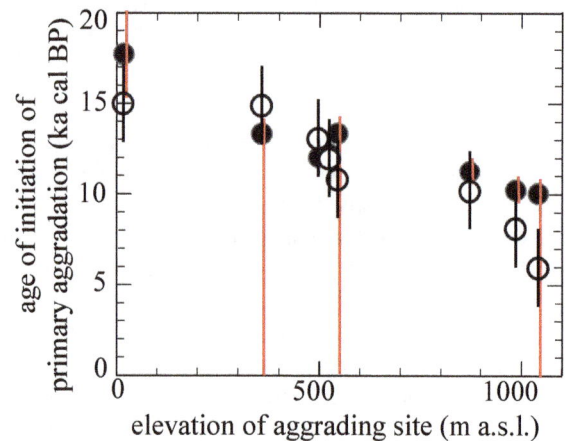

**Figure 7.** Plot of measured (closed circles) and predicted (open circles) ages of the initiation of primary aggradation according to the PVCH. All data from Table 1 are plotted except for Chambless site, which was not included because of ambiguity in whether the site represents primary or secondary aggradation (see Sect. 4 for discussion). Uncertainties of the measured ages are shown in red. Ages increase with decreasing elevation of the site of aggradation (which correlates with the lowest elevations of the source catchments). Uncertainty values are provided and explained in Table 1.

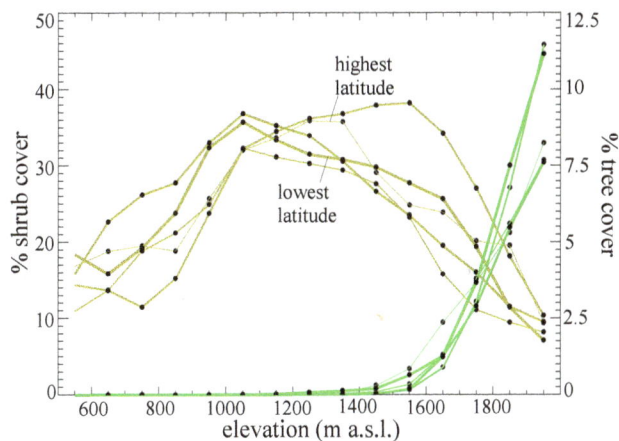

**Figure 8.** Plot of the average percent shrub (in brown) and tree (in green) cover in the Mojave Desert as a function of elevation for 5 bins of latitude (thinner lines indicate higher latitudes) using the US Geological Survey's (2013) LANDFIRE database. The data exhibit some spatial variation in the vegetation-cover-versus-elevation relationship, but there is no systematic difference across latitudes.

are preserved, for example, in many high-elevation, rapidly eroding catchments of the Transverse Ranges (located in the SW corner of the study area) (Bull, 1991). A more sophisticated analysis would predict the timing of both aggradation/incision and the likelihood of preservation of a fill deposit, but such an approach would introduce new parameters into the analysis and is unnecessary for addressing the key questions of this paper (i.e., the timing of aggradation and incision). Also, flow-routing pathways are identified using the modern DEM (the US National Elevation Dataset), yet channels have been modified by late-Quaternary fluvial-system aggradation and incision. As such, some portions of alluvial fans and lowland basins comprised of late-Pleistocene and/or Holocene deposits are now disconnected from source catchments (via channel incision and terrace abandonment) and hence appear as black in Figs. 4–6. In such areas one can estimate the predictions of the PVCH approximately using the predicted ages from nearby areas of similar elevation.

The study region considered in this paper is large; i.e., the entire Mojave Desert and some portions of adjacent regions. Adopting such a large study area has the advantage that a lot of data can be brought to bear in order to precisely constrain the relationship between vegetation changes, elevation, and time/age (Fig. 3). One could argue, however, that the study area (Fig. 4) is too large for a single paleovegetation change vs. age relationship (i.e., Fig. 3) to be applicable throughout the region (as assumed in the above analysis). We can address this issue, at least for modern vegetation (Fig. 8). I used gridded 30 m/pixel estimates of shrub and tree cover from the US Geological Survey's (2013) LANDFIRE database to quantify the relationship between percent shrub/tree cover and elevation using 100 m bins of elevation from 500 to 2000 m a.s.l. LANDFIRE is a US Geological Survey mapping program

that produces high-resolution geospatial data for vegetation and fire regimes in the US. I divided the study region illustrated in Fig. 4 (i.e., 34–36.2° N, 115–117.8° W) into five equally spaced bins of latitude and computed the relationship between average shrub/tree cover and elevation for each bin separately (Fig. 8). The results demonstrate that there is no systematic change in the relationship between modern vegetation cover and elevation with latitude from 34 to 36.2° N in the Mojave Desert. That is not to say that elevation is a perfect predictor of vegetation cover. Clearly, slope gradient and aspect, spatial variations in soil properties, etc. influence the elevational zonation of plants in the southwestern US locally. The scatter of the data plotted in Fig. 8 reflects the significant spatial variations in plant cover that exist within areas of equal elevation. Rather, this analysis demonstrates that any systematic differences in average vegetation cover across latitudes is much smaller than the differences across elevation.

## 3 An improved process model for PVCH

Melton (1957) documented a strong positive correlation between drainage density and both aridity and percent bare area in the southwestern US. Melton's findings provide a basis for predicting an increase in drainage density during semiarid-to-arid transitions. Similarly, Pelletier et al. (2013) quantified drainage density across the elevation/precipitation/vegetation gradient of the Santa Catalina Mountains in Arizona (from Sonoran desert scrub at low elevations that have a mean annual precipitation of $0.2\,\mathrm{m\,a^{-1}}$ through pinyon/juniper woodland to mixed conifer forest at high elevations that have a mean annual precipitation of $0.8\,\mathrm{m\,a^{-1}}$) and obtained a similar inverse relationship between drainage density and water-availability/vegetation cover. Unfortunately, we lack studies that demonstrate an increase in sediment yield from hillslopes and/or drainage density resulting from a transition from woodland to desert scrub vegetation in the Mojave Desert specifically. However, anthropogenic disturbances in the Mojave Desert that are associated with reductions in vegetation cover consistently result in higher erosion rates from hillslopes and low-order fluvial valleys (e.g., Lovich and Bainbridge, 1999; Iverson, 1980, and references therein).

Theoretical models for the controls on drainage density further suggest that vegetation cover is the most important climatically related variable controlling drainage density. Perron et al. (2008, 2009), for example, demonstrated that the spacing of first-order valleys (closely related to drainage density) is a function of the relative rates of colluvial sediment transport (dominant on hillslopes) and slope-wash/fluvial erosion (dominant in valley bottoms). A reduction in vegetation cover decreases the rate of colluvial sediment transport because fewer plants are present to drive bioturbation, while simultaneously increasing slope-wash/fluvial-erosion rates via a reduction in protective cover. Both of these effects combine to promote higher drainage density. Multiple lines

68 Earth Surface Science: Properties, Processes and Interactions

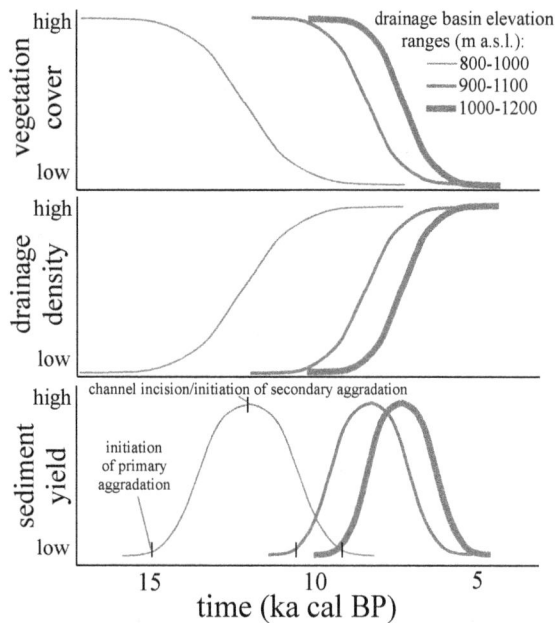

**Figure 9.** Conceptual model illustrating relationships among vegetation cover, drainage density, and sediment yield from source catchments for three hypothetical catchments (each with 200 m of relief) of different mean elevation.

of evidence also indicate that sediment yield is far more sensitive to changes in vegetation cover (via a threshold shear stress for detachment or entrainment) than changes in runoff intensity (Tucker and Slingerland, 1997). The critical shear stress criterion for detachment is (Tucker and Slingerland, 1997)

$$P_s A S^2 > \tau_c^3, \tag{1}$$

where $P_s$ is runoff ($L\,T^{-1}$), $A$ is contributing area ($L^2$), $S$ is slope (L/L), and $\tau_c$ is the shear stress threshold that depends on vegetation cover ($L\,T^{-1/3}$). Equation (1) shows that a twofold decrease in $\tau_c$ is equivalent to an eightfold increase in in runoff intensity. I cannot quantify $\tau_c$ for desert scrub versus woodland vegetation types, but Prosser and Dietrich (1995), working in central coastal California, documented a strong sensitivity of $\tau_c$ to percent bare area, and Melton (1957) documented a strong correlation between drainage density and both aridity and percent bare area in the southwestern US. The results of Prosser and Dietrich (1995) were obtained in the relatively humid climate of coastal California, but Al-Hamdan et al. (2013) developed a theoretical model that suggests the sensitivity of erosion to bare area is a general phenomenon. According to Eq. (1), a twofold decrease in $\tau_c$ can increase the drainage density (which goes as the square root of contributing area) by almost threefold.

Figure 9 presents a schematic diagram of an alternative conceptual model for the PVCH that includes elevation and is qualitatively based on the results of numerical modeling studies (e.g., Tucker and Slingerland, 1997). This fig-

ure shows variations in sediment yield (bottom panel) resulting from hypothetical changes in mean vegetation cover (top panel) and drainage density (middle panel) for three hypothetical drainage basins with elevations of 800–1000, 900–1100, and 1000–1200 m a.s.l. The lowest drainage basin (800–1000 m a.s.l.) experiences the initiation of aggradation first (ca. 15 ka cal BP) because the loss of *Juniperus* is time-transgressive, proceeding from low to high elevations through time. Numerical models predict that sediment pulses produced via an increase in drainage density are temporary (i.e., they are sediment *pulses* with both waxing and waning phases) because first-order valley heads eventually stabilize (following a transition from lower to higher drainage density) as they become adjusted to a smaller contributing area associated with the lower vegetation cover and hence lower shear stress threshold for detachment associated with arid climates. As such, the accelerating increase in drainage density that occurs in the 15–12 ka time interval for the lowest drainage basin is accompanied by an increase in sediment yield to a peak value of ca. 12 ka (Fig. 9). Numerical models demonstrate that the stabilization of valley heads following drainage-network expansion causes a reduction in sediment supply that triggers channel incision, terrace abandonment, and secondary aggradation further downstream (Tucker and Slingerland, 1997) (ca. 12 ka for the lowest drainage basin). The response of the higher-elevation drainage basins is similar except that aggradation is delayed relative to lower-elevation drainage basins and the duration of primary aggradation is predicted to be somewhat shorter as a result of the increase in the rate of the recession of *Juniperus* at higher elevations (1100–1800 m a.s.l.) relative to lower elevations.

## 4 Discussion

Aggradation begins at a time consistent with the predictions of the PVCH in all cases except one (Chambless) in which the measured initiation of aggradation occurs significantly later than the prediction (Figs. 4–7; Table 1). There is no discrepancy, however, if the dated deposit at Chambless corresponds to a site of secondary aggradation (i.e., if sediments with a depositional age of ca. 15 ka cal BP are located upstream) because the predicted age of channel incision/secondary aggradation according to the GIS analysis is consistent with the measured age (Table 1). The Chambless site is not included in Fig. 7 because of this ambiguity. The likelihood that two pulses of aggradation will occur from one pulse of sediment yield complicates testing of any model for late-Quaternary fluvial-system aggradation, but this complexity can be addressed in future studies with detailed geologic mapping and simultaneous dating of both primary and nearby/inset secondary late-Quaternary deposits.

In the two sites where the timing of incision is constrained (southern Death Valley and Sheep Creek), the model underpredicts the age of incision by approximately 3 ka. It is difficult to draw conclusions from a sample size of two, but the discrepancy between the predicted and measured incision ages could be due to the relatively large uncertainty of the timing of paleovegetation changes within the 10–3 ka interval and/or the relatively large uncertainty associated with ages of incision measured (as done here) using the highest stratigraphic age (which necessarily overestimates the age of incision).

The results of this paper are consistent with the time-transgressive nature of aggradation and incision with elevation documented by Weldon (1986) in his study of the late-Quaternary history of Cajon Pass (located near the southwestern corner of the study area shown in Fig. 4) despite the more Mediterranean climate of the Transverse Ranges and the associated differences in vegetation types compared with the Mojave Desert (Fig. 10). Weldon (1986) documented a wave of aggradation followed by incision that moved up Cajon Pass from elevations of 500 (i.e., the Freeway Crossing) to 1500 m a.s.l. (the summit) during the time interval from 15 to 6 ka cal BP (Fig. 10a). Weldon's (1986) study is particularly valuable because it demonstrates the time-transgressive nature of aggradation with elevation at a single site rather than by combining many study sites within a large region. Weldon (1986) also demonstrated that the hillslope sediment yield from 15 to 6 ka cal BP was an order of magnitude higher than sediment yields during either the late Pleistocene or the mid–late Holocene. The PVCH accurately predicts the time-transgressive wave of aggradation documented by Weldon (1986) (Fig. 10c).

The PVCH purportedly fails in two sites considered by Antinao and McDonald (2013a), i.e., the western side of the Providence Mountains (dated by Clarke, 1994) and the Sierra El Mayor piedmont of Baja California deposits (dated by Spelz et al., 2008; Armstrong et al., 2010). The Baja California sites are not part of the Mojave Desert but are important to address here because, according to Antinao and McDonald (2013a), they provide a basis for a reduced relevance of PVCH in the late-Quaternary evolution of the southwestern US. Fan aggradation is not always triggered by climatic variations, however, so it is important to consider other possible triggers for aggradation when evaluating the PVCH in specific cases. The Kelso Dunes have migrated across the distal portion of the western piedmont of the Providence Mtns. in late-Quaternary time. The site location map in Fig. 2 of Clarke (1994) and the stratigraphic columns in Fig. 3 of Clarke (1994) clearly show that the deposits dated by Clarke (1994) include aeolian sand deposits at least 1 m thick at each sample locality and grade to aeolian sand deposits at least 8 m thick at the distal end of the fan. Seven out of the eight sediment samples dated by Clarke (1994) were aeolian sediments and the sole fluvial sediment sample dated was located directly atop aeolian sediments. Therefore, despite the common interpretation that the Clarke (1994) ages record climatically driven fluvial-system aggradation, it is at least possible that the ages measured by Clarke (1994) more

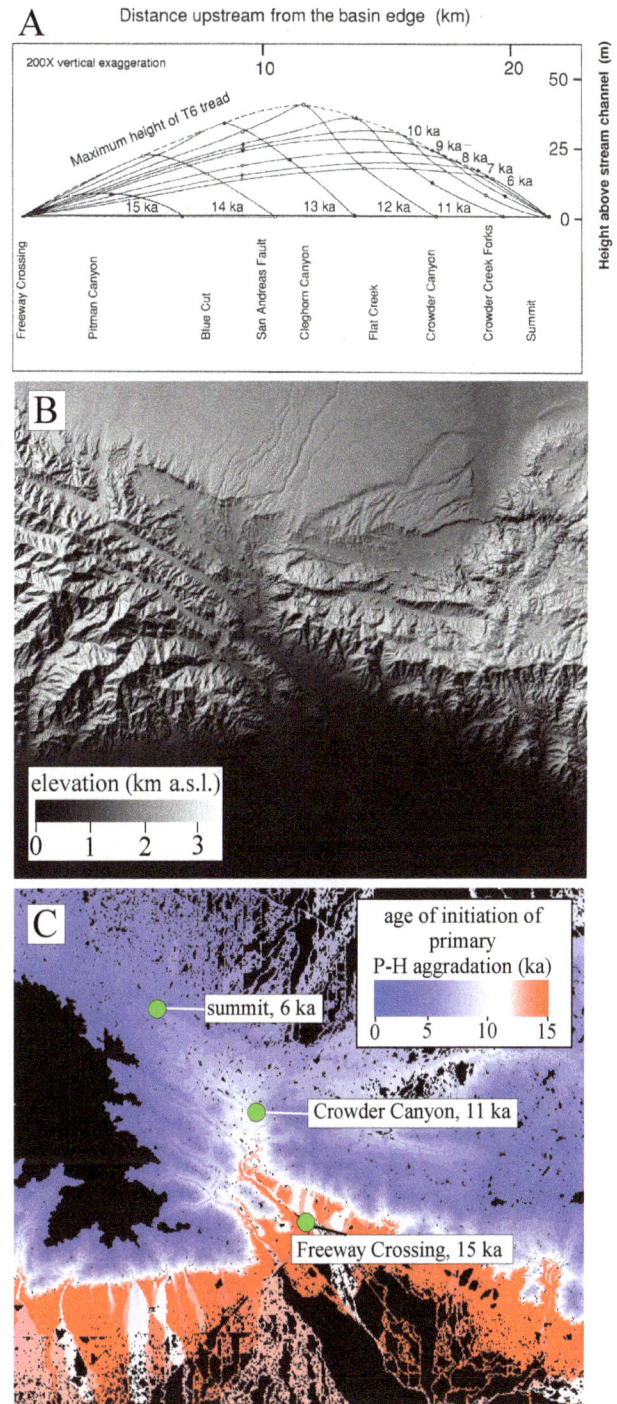

**Figure 10.** Comparison of (**a**) measured (from Bull, 1991 after Weldon, 1986) and (**c**) predicted age of initiation of primary aggradation in Cajon Pass, California, according to the PVCH. Location of (**c**) shown in Fig. 4. Shaded-relief map of the area shown in (**b**).

strongly reflect the history of the Kelso Dunes in addition to fluvial aggradation triggered by the local base-level rise associated with the migration of the Kelso Dunes across the distal portion of the fan. Given this possibility, the Clarke (1994) data may not be the most reliable data to use when testing alternative models for the climatic triggering of fluvial-system aggradation.

Similarly, fan deposits on the eastern piedmont of Sierra El Mayor dated by Armstrong et al. (2010) are potentially problematic for the purposes of testing the PVCH because they were deposited atop, and shortly following deposition of, fluvial and deltaic sediments of the Colorado River (Armstrong et al., 2010), which today is located less than 1 km from the dated fan deposits. As such, aggradation of the Colorado River may have resulted in a local increase in base level, triggering aggradation of fan sediments without a climatically driven change in upstream sediment supply. Although the response of fluvial channels to base-level rise depends on a number of factors including the magnitude of the base-level change and the slope of the channel affected by base-level changes (Schumm, 1993), Leopold and Bull (1979) presented one specific example in which a modest increase in base level (i.e., $\sim 3$ m) triggered aggradation at a distance of $\sim 1$ km. As such, it is possible that deposition on the western piedmont of Sierra El Mayor is principally influenced by the base-level control exerted by the Colorado River rather than by a climatically driven increase in sediment supply.

Spelz et al. (2008) dated fan aggradation on the western piedmont of the Sierra El Mayor that is likely unaffected by the Colorado River. However, the cosmogenic ages of Spelz et al. (2008) on boulders of the terrace associated with latest Pleistocene aggradation varied by 300 % (700 % if all of the data are considered – one date of 76 ka cal BP was excluded in the average). As such, the age control at this site is not ideal. Spelz et al. (2008) used a weighted average of boulder ages to obtain an estimated surface age of $15.5 \pm 2.2$ ka cal BP (i.e., well before the retreat of late-Pleistocene plants at $10.7 \pm 0.5$ ka cal BP as constrained by the presence of boojum trees (*Fouquieria columnaris*) (Anderson and Van Devender, 1995)). However, a more accurate surface-exposure age is, in many cases, obtained by using the youngest boulder age that is not an obvious outlier, especially when inheritance is a potentially important factor (e.g., Applegate et al., 2010; Heyman et al., 2011). In this case that age would be $11 \pm 1.3$ ka cal BP; i.e., broadly consistent with the timing of paleovegetation changes.

Many authors, including Bull (1991), Harvey et al. (1999), McDonald et al. (2003), Miller et al. (2010), and Antinao and McDonald (2013a, b) have invoked changes in the frequency and/or intensity of extreme storms to drive fluvial-system aggradation in the southwestern US. Miller et al. (2010), for example, documented a correlation between sea-surface temperatures in the Gulf of California (a proxy for monsoon activity) and alluvial-fan aggradation in the Mojave Desert. The role of monsoon thunderstorms specifically in driving aggradation in the Mojave Desert is uncertain given the limited impact of the North American Monsoon (NAM) system on the Mojave Desert (Higgins et al., 1997). Recent investigations of the spatial extent of NAM show no significant impact on the Mojave Desert (e.g., Dominguez et al., 2009). The correlation between sea-surface temperatures in the Gulf of California (which drive NAM storms) and alluvial-fan aggradation in the Mojave Desert is excellent for the 6–3 ka cal BP time period but significantly poorer for the latest-Pleistocene-to-early-Holocene time period; i.e., primary aggradation occurred ca. 14–7 ka cal BP while elevated sea-surface temperatures occurred 15–11 ka cal BP.

Antinao and McDonald (2013b) argued, as an alternative to the NAM-driven extreme-storm hypothesis of Miller et al. (2010), that more frequent and/or intense El-Niño-like conditions in the tropical Pacific increased moisture delivery to the southwestern US ca. 14.5–8 ka cal BP, triggering fan aggradation. It is clear that dissipating tropical storms are responsible for generating most of the extreme floods in the historical record in drainage basins larger than $\sim 10^3$ km$^2$ in the Mojave and Sonoran deserts, including tropical storms Norma (1970), Kathleen (1976), Octave (1983), and Nora (1997). As such, the importance of tropical storms in triggering modern regional flooding in the southwestern US is undeniable. The relative importance of monsoon versus tropical storms is largely a function of drainage-basin area, with tropical storms dominating the largest floods for many drainage basins greater than $\sim 10^3$ km$^2$ and NAM-driven convective storms dominating the extreme-storm record for many smaller drainage basins (e.g., Hirschboeck, 1988).

It is unclear, however, whether either of these mechanisms (increased monsoon activity or increased El-Niño-like conditions driving more frequent or more intense tropical storm activity) is consistent with the time-transgressive nature of fluvial-system aggradation in the Mojave Desert with elevation, in which the initiation of primary aggradation occurred earlier (i.e., ca. 15–12 ka cal BP) at lower-elevation sites, 400–900 m a.s.l., and later (i.e., ca. 12–6 ka cal BP) at higher-elevation sites, 900–1500 m. More fundamentally, it is equally unclear whether an increase in the frequency or intensity of extreme storms is more likely to trigger fan aggradation or incision. Fluvial-system aggradation requires an increase in sediment yield from hillslopes *without a comparable increase in the ability of fluvial channels to transport that increase in sediment* (aggradation can also be triggered by a decrease in the ability of fluvial channels to transport sediment without any change in sediment supply from hillslopes, but this scenario is not relevant to the extreme-storm hypothesis since more intense storms are unlikely to be less effective at transporting the sediment delivered to them compared with less intense storms). That is, any climate change that simultaneously increases hillslope sediment yield *and* the ability of fluvial channels to convey that increase in sediment yield might not result in aggradation except in the lowest elevations of closed basins (e.g., playas). In contrast, the conceptual model for the PVCH presented here predicts valley-floor and alluvial-fan aggradation specifically because the sediment pulse from hillslopes was not accompanied by a significant increase in the ability of fluvial channels to convey that increase in sediment yield. It is certainly possible that periodic episodes of increasing tropical-storm activity accompanied vegetation changes to generate the late-Quaternary alluvial record of the Mojave Desert. As such, more research is needed to distinguish among these hypotheses and/or demonstrate how they act in concert.

The PVCH may be applicable to other sites worldwide that have experienced a transition from semiarid to arid climates during the late Quaternary. Due to the fact that constraints on paleovegetation and the timing of fluvial-system aggradation are rarely present in the same location, however, the PVCH has rarely been tested outside of the deserts of North America (where packrat middens are available and have been studied for decades, for example). An exception is the Nahal Yael, a drainage basin in southern Israel where Bull and Schick (1979) applied an early version of the Bull (1991) model. Enzel et al. (2012) recently showed that the Nahal Yael did not experience a semiarid-to-arid climatic transition, however. As such, the PVCH does not apply to that site. The case of the Nahal Yael underscores the importance of having reliable local paleoclimate/paleovegetation data when attempting to apply or test the PVCH.

## 5  Conclusions

This paper builds upon previous studies (e.g., Weldon, 1986) as well as comprehensive new geochronologic studies (e.g., Miller et al., 2010) to further demonstrate that fluvial-system aggradation from the latest Pleistocene to the present in the southwestern US was time-transgressive with elevation. As such, the timing of the initiation of primary fluvial-system aggradation and channel incision/secondary aggradation exhibit spatial variability that depends sensitively on the elevation ranges of source catchments. In order to predict these timings for specific locations using the PVCH, I tightly constrained the relationship between paleovegetation changes and elevation and then used a GIS analysis that incorporated the elevation ranges of source catchments. The PVCH predicts the correct timing of the initiation of aggradation in eight out of nine cases in the Mojave Desert where reasonable age control exists. It is likely that changes in the frequency or intensity of extreme storms have contributed to cycles of fluvial-system aggradation and incision in the southwestern US, but the results of this paper suggest that the recent trend of discounting the importance of hillslope vegetation changes deserves reexamination.

**Acknowledgements.** This work was partially supported by NSF award no. 0309518. I wish to thank Phil Pearthree and Vic Baker for comments on an early draft that helped me to improve the paper. P. Kyle House, Arnaud Temme, and two anonymous reviewers provided helpful reviews during the review process that led to a significantly improved paper.

Edited by: A. Temme
Reviewed by: P. K. House and two anonymous referees

## References

Al-Hamdan, O. Z., Pierson, F. B., Nearing, M. A., Williams, C. J., Stone, J. J., Kormos, P. R., Boll, J., and Weltz, M. A.: Risk assessment of erosion from concentrated flow on rangelands using overland flow distribution and shear stress partitioning, Trans. Am. Soc. Agri. Biol. Eng., 56, 539–548, doi:10.13031/2013.42684, 2013.

Anders, M. D., Pederson, J. L., Rittenour, T. M., Sharp, W. D., Gosse, J. C., Karlstrom, K. E., Crossey, L. J., Goble, R. J., Stockli, L., and Yang, G.: Pleistocene geomorphology and geochronology of eastern Grand Canyon: linkages of landscape components during climate changes, Quaternary Sci. Rev., 24, 2428–2448, doi:10.1016/j.quascirev.2005.03.015, 2005.

Anderson, R. S. and Van Devender, T. R.: Vegetation history and paleoclimates of the coastal lowlands of Sonora, Mexico – pollen records from packrat middens, J. Arid Environ., 30, 295–306, doi:10.1016/S0140-1963(05)80004-7, 1995.

Antinao, J.-L. and McDonald, E.: A reduced relevance of vegetation change for alluvial aggradation in arid zones, Geology, 41, 11–14, doi:10.1130/G33623.1, 2013a.

Antinao, J.-L. and McDonald, E.: An enhanced role for the Tropical Pacific on the humid Pleistocene-Holocene transition in southwestern North America, Quaternary Sci. Rev., 78, 319–341, doi:10.1016/j.quascirev.2013.03.019, 2013b.

Applegate, P. J., Urban, N. M., Laabs, B. J. C., Keller, K., and Alley, R. B.: Modeling the statistical distributions of cosmogenic exposure dates from moraines, Geosci. Model Devel., 3, 293–307, doi:10.5194/gmd-3-293-2010, 2010.

Armstrong, P., Perez, R., Owen, L. A., and Finkel, R. C.: Timing and controls on late Quaternary landscape development along the eastern Sierra el Mayor, northern Baja California, Mexico, Geomorphology, 114, 415–430, doi:10.1016/j.geomorph.2009.08.005, 2010.

Bull, W. B.: The alluvial fan environment, Progr. Phys. Geog., 1, 222–270, doi:10.1177/030913337700100202, 1968.

Bull, W. B.: Geomorphic responses to climatic change, Oxford, UK, Oxford University Press, p. 326, doi:10.1002/gea.3340080106, 1991.

Christensen, G. E. and Purcell, C.: Correlation and age of Quaternary alluvial-fan sequences, Basin and Range province, southwestern Unites States, in: Soils and Quaternary Geology of the Southwestern United States, edited by: Weide, D. L., Geological Society of America Special Paper 203, 115–122, 1985.

Clarke, M. L.: Infrared stimulated luminescence ages from aeolian sand and alluvial fan deposits from the eastern Mojave Desert, California, Quaternary Geochronology, Quaternary Sci. Rev., 13, 533–538, doi:10.1016/0277-3791(94)90073-6, 1994.

Dominguez, F., Villegas, J. C., and Breshears, D. D.: Spatial extent of the North American Monsoon: Increased cross-regional linkages via atmospheric pathways, Geophys. Res. Lett., 36, L07401, doi:10.1029/2008GL037012, 2009.

Enzel, Y., Amit, R., Grodek, T., Ayalon, A., Lekach, J., Porat, N., Bierman, P., Blum, J. D., and Erel, Y.: Late Quaternary weathering, erosion, and deposition in Nahal Yael, Israel: An "impact of climatic change on an arid watershed"?, Geol. Soc. Am. Bull., 124, 705–722, doi:10.1130/B30538.1, 2012.

Freeman, G. T.: Calculating catchment area with divergent flow based on a rectangular grid, Comput. Geosci., 17, 413–422, 1991.

Harvey, A. M., Wigand, P. E., and Wells, S. G.: Response of alluvial fan systems to the late Pleistocene to Holocene climatic transition, Contrasts between the margins of pluvial Lakes Lahontan and Mojave, Nevada and California, USA, Catena, 36, 255–281, 1999.

Held, I. M. and Soden, B. J.: Robust responses of the hydrological cycle to global warming, J. Climate, 19, 5686–5699, doi:10.1175/JCLI3990.1, 2006.

Heyman, J., Stroeven, A. P., Harbor, J. M., and Caffee, M. W.: Too young or too old: Evaluating cosmogenic exposure dating based on an analysis of compiled boulder exposure ages, Earth Planet Sci. Lett., 302, 71–80, doi:10.1016/j.epsl.2010.11.040, 2011.

Higgins, R. W., Yao, Y., and Wang, X. L.: Influence of the North American Monsoon System on the U.S. Summer Precipitation Regime, J. Climate, 10, 2600–2622, doi:10.1175/1520-0442(1997)010<2600:IOTNAM>2.0.CO;2, 1997.

Hirschboeck, K. K.: Flood hydroclimatology, in Flood Geomorphology, edited by: Baker, V. R., Kochel, R. C., and Patton, P. C., John Wiley & Sons, 189–205, 1988.

Iverson, R. M.: Processes of accelerated pluvial erosion on desert hillslopes modified by vehicular traffic, Earth Surf. Proc. Landf., 5, 369–388, doi:10.1002/esp.3760050407, 1980.

Kort, J., Collins, M., and Ditsch, D.: A review of soil erosion potential associated with biomass crops, Biomass Bioenerg., 14, 351–359, doi:10.1016/S0961-9534(97)10071-X, 1998.

Langbein, W. B. and Schumm, S. A.: Yield of sediment in relation to mean annual precipitation, Am. Geophys. Union Trans., 39, 1076–1084, doi:10.1029/TR039i006p01076, 1958.

Lau, W. K.-M., Wu, H.-T., and Kim, K.-M.: A canonical response of precipitation characteristics to global warming from CMIP5 models, Geophys. Res. Lett., 40, 3163–3169, doi:10.1002/grl.50420, 2013.

Leopold, L. B. and Bull, W. B.: Base level, aggradation, and grade, Proc. Am. Phil. Soc., 123, 168–202, 1979.

Lovich, J. E. and Bainbridge, D.: Anthropogenic Degradation of the Southern California Desert Ecosystem and Prospects for Natural Recovery and Restoration, Environ. Manag., 24, 309–326, doi:10.1007/s002679900235, 1999.

Mahan, S. A., Miller, D. M., Menges, C. M., and Yount, J. C.: Late Quaternary stratigraphy and luminescence geochronology of the northeastern Mojave Desert, Quaternary Int., 166, 61–78, 2007.

McDonald, E. V., McFadden, L. D., and Wells, S. G.: Regional response of alluvial fans to the Pleistocene–Holocene climatic transition, Mojave Desert, California, in: Paleoenvironments and paleohydrology of the Mojave and southern Great Basin deserts, edited by: Enzel, Y., Wells, S. G., and Lancaster, N., Geological Society of America Special Paper 368, 189–205, 2003.

Melton, M. A.: An analysis of the relations among elements of climate, surface properties, and geomorphologym Tech. Rept. 11, Project NR 389-042, Office of Naval Research, Department of Geology, Columbia University, New York, 1957.

Menges, C. M., Taylor, E. M., Workman, J. B., and Jayko, A. S.: Regional surficial deposit mapping in the Death Valley area of California and Nevada in support of ground- water modeling, in: Quaternary and Late Pliocene Geology of the Death Valley Region – Recent Observations on Tectonics, Stratigraphy, and Lake Cycles, edited by: Machette, M. N., Johnson, M. L., and Slate, J. L., Pacific Cell Friends of the Pleistocene Field Trip, February 17–19, 2001, H151–H166, US Geological Survey Open-File Report 01-51, 2001.

Miller, D. M., Schmidt, K. M., Mahan, S. A., McGeehin, J. P., and Owen, L. A.: Holocene landscape response to seasonality of storms in the Mojave Desert, Quaternary Int., 215, 45–61, doi:10.1016/j.quaint.2009.10.001, 2010.

Murray, A. B., Lazarus, E., Ashton, A., Baas, A., Coco, G., Coulthard, T., Fonstad, M., Haff, P., McNamara, D., Paola, C., Pelletier, J., and Reinhardt, L.: Geomorphology, complexity, and the emerging science of the Earth's surface, Geomorphology, 103, 496–505, doi:10.1016/j.geomorph.2008.08.013, 2009.

Pelletier, J. D.: Quantitative Modeling of Earth Surface Processes, Cambridge University Press, New York, doi:10.1017/CBO9780511813849, 2008.

Pelletier, J. D., Quade, J., Goble, R. J., and Aldenderfer, M. S.: Widespread hillslope gullying on the southeastern Tibetan Plateau: Human or climate-change induced?, Geol. Soc. Am. Bull., 123, 1926–1938, doi:10.1130/B30266.1, 2011.

Pelletier, J. D., Barron-Gafford, G. A., Breshears, D. D., Brooks, P. D., Chorover, J., Durcik, M., Harman, C. J., Huxman, T. E., Lohse, K. A., Lybrand, R., Meixner, T., McIntosh, J. C., Papuga, S. A., Rasmussen, C., Schaap, M., Swetnam, T. L., and Troch, P. A.: Coevolution of nonlinear trends in vegetation, soils, and topography with elevation and slope aspect: A case study in the sky islands of southern Arizona, J. Geophys. Res. Earth Surf., 118, 741–758, doi:10.1002/jgrf.20046, 2013.

Perron, J. T., Dietrich, W. E., and Kirchner, J. W.: Controls on the spacing of first-order valleys, J. Geophys. Res., 113, F04016, doi:10.1029/2007JF000977, 2008.

Perron, J. T., Kirchner, J. W., and Dietrich, W. E.: Formation of evenly spaced ridges and valleys, Nature, 460, 502–505, doi:10.1038/nature08174, 2009.

Pierson, F. B., Bates, J. D., Svejcar, T. J., and Hardegree, S. P.: Runoff and erosion after cutting western juniper, Rangeland Ecol. Manag., 60, 285–292, 2007.

Prosser, I. P. and Dietrich, W. E.: Field experiments on erosion by overland flow and their implication for a digital terrain model of channel initiation, Water Resour. Res., 31, 2867–2876, doi:10.1029/95WR02218, 1995.

Reimer, P. J., Baillie, M. G. L., Bard, E., Bayliss, A., Beck, J. W., Blackwell, P. G., Bronk Ramsey, C., Buck, C. E., Burr, G. S., Edwards, R. L., Friedrich, M., Grootes, P. M., Guilderson, T. P., Hajdas, I., Heaton, T. J., Hogg, A. G., Hughen, K. A., Kaiser, K. F., Kromer, B., McCormac, F. G., Manning, S. W., Reimer, R. W., Richards, D. A., Southon, J. R., Talamo, S., Turney, C. S. M., van der Plicht, J., and Weyhenmeyer, C. E.: IntCal09 and Marine09 radiocarbon age calibration curves, 0–50,000 years cal BP, Radiocarbon, 51, 1111–1150, 2009.

Reinhardt, L., Jerolmack, D. J., Cardinale, B., Vanacker, V., and Wright, J.: Dynamic interactions of life and its landscape: feedbacks at the interface of geomorphology and ecology: Earth Surf. Process. Landf., 35, 78–101, doi:10.1002/esp.1912, 2010.

Rockwell, T. K., Lindvall, S., Herzberg, M., Murbach, D., Dawson, T., and Berger, G.: Paleoseismology of the Johnson Valley, Kickapoo, and Homestead Valley faults: Clustering of earthquakes in the eastern California shear zone, Bull. Seis. Soc. Am., 90, 1200–1236, doi:10.1785/0119990023, 2000.

Schumm, S. A.: Geomorphic thresholds and complex response of drainage systems, in: Fluvial Geomorphology: SUNY Binghamton Publication in Geomorphology, edited by: Morisawa, M., 299–310, 1973.

Schumm, S. A.: River response to baselevel change: Implications for sequence stratigraphy, J. Geol., 101, 279–294, doi:10.1086/648221, 1993.

Sohn, M. F., Mahan, S. A., Knott, J. R., and Bowman, D. D.: Luminescence ages for alluvial fan deposits in Southern Death Valley: Implications for climate-driven sedimentation along a tectonically active mountain front, Quaternary Int., 166, 49–60, 2007.

Spelz, R. M., Fletcher, J. M., Owen, L. A., and Caffee, M. W.: Quaternary alluvial-fan development, climate and morphologic dating of fault scarps in Laguna Salada, Baja California, Mexico, Geomorphology, 102, 578–594, doi:10.1016/j.geomorph.2008.06.001, 2008.

Strahler, A. N.: Dimensional analysis applied to fluvially eroded landforms: Geol. Soc. Am. Bull., 69, 279–300, doi:10.1130/0016-7606(1958)69[279:DAATFE]2.0.CO;2, 1958.

Strickland, L. E., Thompson, R. S., Anderson, K. H., and Pelltier, R. T.: Late Quaternary biogeographic and climatic changes in western North America: Evidence from mapped arrays of packrat midden data, American Geophysical Union, Fall Meeting 2005, abstract PP11B–1470, 2005.

Tucker, G. E. and Slingerland, R.: Drainage basin response to climate change, Water Resour. Res., 33, 2031–2047, doi:10.1029/97WR00409, 1997.

US Geological Survey: LANDFIRE Existing Vegetation Type layer (2013, June – last update), US Department of Interior, Geological Survey, Digital data available at: http://landfire.cr.usgs.gov/viewer/, 2013.

Weldon, R. J.: The late Cenozoic geology of Cajon Pass: implications for tectonics and sedimentation along the San Andreas fault, Ph.D. Dissertation, California Institute of Technology, 1986.

Whittaker, R. H. and Niering, W. A.: Vegetation of the Santa Catalina Mountains, Arizona. V. Biomass, Production, and Diversity along the Elevation Gradient, Ecology, 56, 771–790, doi:10.2307/1936291, 1975.

# 7

# Ancient pre-glacial erosion surfaces preserved beneath the West Antarctic Ice Sheet

K. C. Rose[1], N. Ross[2], T. A. Jordan[3], R. G. Bingham[4], H. F. J. Corr[3], F. Ferraccioli[3], A. M. Le Brocq[5], D. M. Rippin[6], and M. J. Siegert[7]

[1]Bristol Glaciology Centre, School of Geographical Sciences, University of Bristol, Bristol BS8 1SS, UK
[2]School of Geography, Politics & Sociology, Newcastle University, Newcastle upon Tyne NE1 7RU, UK
[3]British Antarctic Survey, High Cross, Madingley Road, Cambridge CB3 0ET, UK
[4]School of GeoSciences, University of Edinburgh, Edinburgh EH8 9XP, UK
[5]School of Geography, University of Exeter, Exeter EX4 4RJ, UK
[6]Environment Department, University of York, York YO10 5DD, UK
[7]Grantham Institute and Department of Earth Science and Engineering, Imperial College London, London SW7 2AZ, UK

*Correspondence to:* K. C. Rose (kathrynrose100@hotmail.com) and M. J. Siegert (m.siegert@imperial.ac.uk)

**Abstract.** We present ice-penetrating radar evidence for ancient (pre-glacial) and extensive erosion surfaces preserved beneath the upstream Institute and Möller ice streams, West Antarctica. Radar data reveal a smooth, laterally continuous, gently sloping topographic block, comprising two surfaces separated by a distinct break in slope. The erosion surfaces are preserved in this location due to the collective action of the Pirrit and Martin–Nash hills on ice sheet flow, resulting in a region of slow flowing, cold-based ice downstream of these major topographic barriers. Our analysis reveals that smooth, flat subglacial topography does not always correspond to regions of either present or former fast ice flow, as has previously been assumed. We discuss the potential origins of the erosion surfaces. Erosion rates across the surfaces are currently low, precluding formation via present-day glacial erosion. We suggest that fluvial or marine processes are most likely to have resulted in the formation of these surfaces, but we acknowledge that distinguishing between these processes with certainty requires further data.

## 1 Introduction

The Institute and Möller ice streams (IMIS) drain around 20 % of the area of the West Antarctic Ice Sheet (WAIS) (Fig. 1). Despite their significance as fast-flowing outlet glaciers within an ice sheet that is regarded as potentially unstable (Joughin et al., 2014), until recently relatively little was known about the glacial history of this region (Bingham and Siegert, 2007). In order to address this issue, an aerogeophysical survey was undertaken across the ice streams and surrounding locations (Ross et al., 2012). Mapping ice sheet boundary conditions in these little-explored regions is of great importance, as subglacial topography can exert a strong control on ice dynamics (Joughin et al., 2009) and may retain a long-term record of geomorphic processes and ice sheet evolution (Young et al., 2011). Analysis of bed topography can therefore be used to make inferences about the nature and evolution of palaeo-landscapes in Antarctica (Rose et al., 2013).

The IMIS survey has provided a wealth of new information, elucidating the tectonic, topographic and hydrological settings of this region (Ross et al., 2012; Jordan et al., 2013; Le Brocq et al., 2013; Siegert et al., 2014). These, in turn, have contributed to our understanding of early ice inception (Ross et al., 2014) and ice sheet sensitivity in the Weddell Sea sector (Siegert et al., 2013; Wright et al., 2014). The macro-scale geomorphology of the IMIS sector, however, has yet to be considered. In particular, the region between

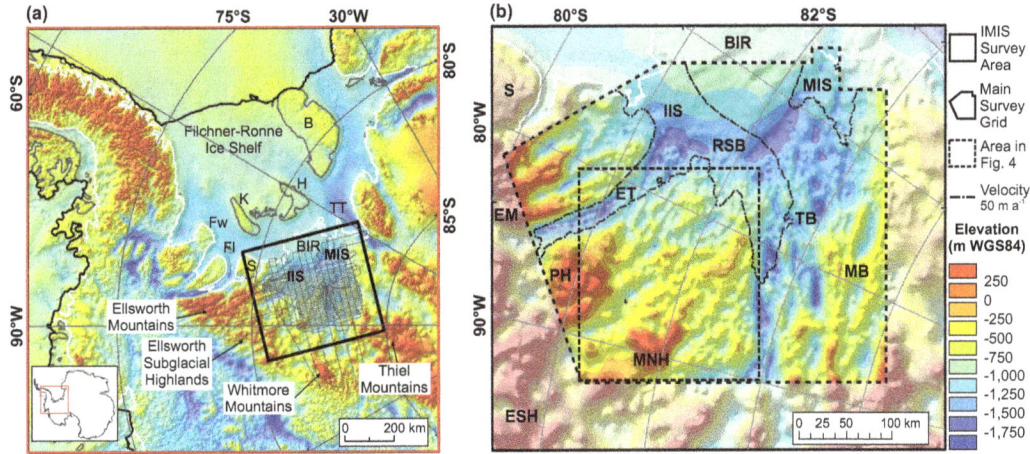

**Figure 1.** (a) Subglacial topography of the Weddell Sea sector, West Antarctica (Fretwell et al., 2013). The white line marks the grounding line, black line marks the outer ice shelf edge, solid grey lines mark islands, (derived from MODIS MOA imagery; Haran et al., 2005, updated 2013), and light-grey gridded lines mark flight lines from a high-resolution modern airborne survey (Ross et al., 2012). The inset with the red box shows the location of (a) in West Antarctica. (b) Subglacial topography of the Institute and Möller ice streams, derived from IMIS survey (Ross et al., 2012), and overlain on semi-transparent Bedmap2 topography (Fretwell et al., 2013). The $50\,\mathrm{m\,a^{-1}}$ ice sheet surface velocity contour is also shown (Rignot et al., 2011). Annotations: B – Berkner Island; BIR – Bungenstock Ice Rise; EM – Ellsworth Mountains; ESH – Ellsworth Subglacial Highlands; ET – Ellsworth Trough; Fl – Fletcher Promontory; Fw – Fowler Peninsula; H – Henry Ice Rise; IIS – Institute Ice Stream; K – Korff Ice Rise; MB – Marginal Basins; MIS – Möller Ice Stream; MNH – Martin–Nash Hills; PH – Pirrit Hills; RSB – Robin Subglacial Basin; S – Skytrain Ice Rise; TB – Transitional Basins; TT – Thiel Trough (southerly margin).

the Robin Subglacial Basin and the mountain ranges of the Ellsworth–Whitmore Mountains block has received little attention (Fig. 1). Here, we focus on this striking region, where a zone of apparently flat and smooth topography, located between the deeper elongate troughs and basins that underlie the fast-flowing tributaries of the Institute Ice Stream (IIS), is imaged and mapped. We inspect and analyse the morphology of this sector in order to interpret this feature. We discuss different erosion regimes associated with erosion surfaces and favour a fluvial or marine mechanism of formation. Differentiating between these two processes would require further investigation and additional data (e.g. apatite fission track analysis, offshore seismic and sedimentary records, regional rock outcrop analysis, glacial isostatic adjustment reconstruction). Our findings do, however, provide the foundation from which the evolution of this landscape and its relationship with long-term glacial history may be deciphered.

## 2 Regional topographic and geological setting

The Weddell Sea sector of the WAIS (Fig. 1), in which the IMIS are contained, is characterised by a broad, marine (below sea level) embayment (Ross et al., 2012). The IMIS drain into the Filchner–Ronne Ice Shelf and their lower trunks (0–130 km upstream of the grounding line) are separated by the Bungenstock Ice Rise (Fig. 1b). The area immediately inland of the Bungenstock Ice Rise is dominated by the $\sim 1.5$ km deep Robin Subglacial Basin. Two primary tributaries feed the fast ice flow ($> 50\,\mathrm{m\,a^{-1}}$) of the IIS (Fig. 1b). One is un-

derlain by the linear Ellsworth Trough and the other by the Transitional Basins – a series of major topographic depressions (40–80 km long, 10–20 km wide, $\sim 1.8$–1.9 km deep) which likely have a tectonic origin and are dispersed in a more complex right-stepping, en échelon pattern (Jordan et al., 2013). In contrast, the onset of fast flow of the Möller Ice Stream (MIS) initiates closer to the grounding line, in proximity to the Robin Subglacial Basin and the northern margin of the Transitional Basins (Fig. 1b). Along the eastern edge of the IMIS survey area, and south-east of the Transitional Basins, lies a series of elongate Marginal Basins (Jordan et al., 2013; Rose et al., 2014). The IMIS are bordered by numerous uplands, including the Ellsworth, Whitmore and Thiel mountains and the Ellsworth Subglacial Highlands (Fig. 1a). The main focus of this paper is the inter-tributary area of the IIS, between the Robin Subglacial Basin and the Pirrit and Martin–Nash hills further inland.

In terms of their geological setting, the upper parts of the IMIS flow over the Ellsworth–Whitmore Mountains block (Fig. 2a), one of four distinct tectonic blocks or micro-plates that make up the West Antarctic crust (Dalziel and Elliot, 1982). The other tectonic blocks include the Antarctic Peninsula, Thurston Island and Marie Byrd Land (Fig. 2a). To the north of the IMIS, the Filchner–Ronne Ice Shelf is underlain by the Weddell Sea Embayment, which separates the Antarctic Peninsula and Ellsworth–Whitmore Mountains blocks from the more tectonically stable East Antarctic Craton (Studinger and Miller, 1999). The Weddell Sea Embayment lies in a key position between West and East Antarctica

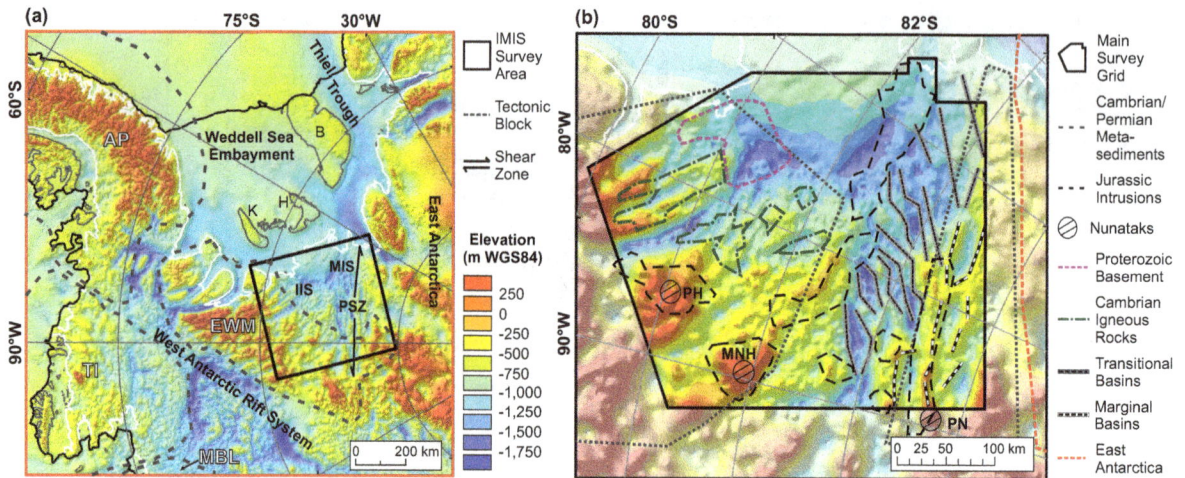

**Figure 2. (a)** Subglacial topography (Fretwell et al., 2013) overlain by the main tectonic features (Jordan et al., 2013) inferred for the Weddell Sea sector, West Antarctica. **(b)** Subglacial topography of the Institute and Möller ice streams overlain by inferred geological features and sediments. The lines denoting the Transitional and Marginal Basins mark their structural boundaries. The line between West and East Antarctica is inferred (Jordan et al., 2013). Annotations: AP – Antarctic Peninsula block; B – Berkner Island; EWM – Ellsworth–Whitmore Mountains block; H – Henry Ice Rise; K – Korff Ice Rise; MBL – Marie Byrd Land block; MIS – Möller Ice Stream; MNH – Martin–Nash Hills; PH – Pirrit Hills; PN – Pagano Nunatak; PSZ – Pagano Shear Zone; TI – Thurston Island block.

in the region where Gondwana break-up initiated $\sim 180$ Myr ago (Storey et al., 2001; Dalziel et al., 2013). It is characterised by a region of thinned crust which extends $\sim 1200$ km south from the foot of the continental slope to the Ellsworth–Whitmore Mountains block (Studinger and Miller, 1999; Jordan et al., 2013). In addition, a major strike-slip boundary (the Pagano Shear Zone) is inferred between the southeastern edge of the Ellsworth–Whitmore Mountains block and East Antarctica, in the region of the Transitional and Marginal Basins (Fig. 2). This boundary parallels (and likely dictates) the eastern shear margin of the MIS, and may extend along the Thiel Trough, defining the south-eastern edge of the Weddell Sea Embayment (Jordan et al., 2013). The timing of movement along the Pagano Shear Zone is uncertain, but plate reconstructions based on palaeomagnetic data suggest that the Ellsworth–Whitmore Mountains block reached its current position, relative to the Antarctic Peninsula and East Antarctica, by $\sim 175$ Ma (Dalziel et al., 2013).

Geological and geophysical surveys suggest that the bedrock underlying the IMIS is predominantly folded Palaeozoic metasediments and volcanic rocks (512–250 Ma), overlying more deeply buried Mesoproterozoic basement (> 1000 Ma) (Garrett et al., 1987; Storey and Dalziel, 1987; Curtis, 2001; Jordan et al., 2013) (Fig. 2b). The Palaeozoic bedrock has a dominant NW–SE structural trend, approximately parallel to the Ellsworth Mountains and adjacent Ellsworth Trough (Storey and Dalziel, 1987; Jordan et al., 2013), which developed during the $\sim 250$ Ma Gondwanide orogeny (Curtis, 2001). Several large Jurassic ($\sim 175$ Ma) granite intrusions crosscut older sediments forming isolated nunataks, such as the Pirrit Hills and the Martin–Nash Hills

(Fig. 2b), which protrude above the ice surface within the IMIS catchment (Storey et al., 1988). Geophysical data show the subglacial extent and shape of these intrusions. The Pirrit and Martin–Nash hills, for example, appear roughly circular in shape, whilst other structurally controlled intrusions that flank the Transitional Basins and underlying Pagano Shear Zone are more linear in form (Fig. 2b). The latter runs at an approximately 30° angle to the structural trend of the Palaeozoic metasediments (Garrett et al., 1988; Jordan et al., 2013). In this region, the Pagano intrusion also forms the northern flank of the elongate Marginal Basins (Fig. 2b), which lie at the boundary between West and East Antarctica at the edge of the IMIS survey area (Jordan et al., 2013).

## 3 Approach

### 3.1 Data collection

In the austral summer of 2010/2011 an airborne geophysical survey was carried out across the IMIS. A central survey grid, with flight line spacing of 7.5 km and tie lines of 25 km, was established over the ice streams, covering an area of $\sim 350 \times 400$ km (Fig. 1a, light-grey gridded lines). Additional exploratory lines, with 50 km spacing, were also flown to link with previous regional surveys (Ross et al., 2012). Approximately 25 000 line kilometres of radio-echo sounding (RES), gravity and magnetic data were collected (Jordan et al., 2013). Here, we discuss only the RES data. These were acquired using a coherent system with a 12 MHz bandwidth and 150 MHz carrier frequency (Corr et al., 2007), providing an approximate 10 m along-track sampling interval. Differential GPS, with a horizontal accuracy of $\sim 5$ cm, was used for positioning. Doppler processing was applied to migrate radar-scattering hyperbolae in the along-track direction (Hélière et al., 2007). The seismic processing software PROMAX was then used to carry out a semi-automated picking sequence that identifies the onset of the received bed echo.

Ice thickness was determined from the two-way travel time of the bed pick using a velocity of $0.168 \, \mathrm{m \, ns^{-1}}$ coupled with a firn layer correction of 10 m (Ross et al., 2012). Bed elevations were then calculated by subtracting ice thickness measurements from ice surface elevations. The latter were determined from measurements of terrain clearance derived from radar/laser altimeter measurements, relative to the WGS84 ellipsoid. Cross-over rms errors are $\sim 18$ m (Ross et al., 2012). These data have since contributed to the Bedmap2 depiction of Antarctic subglacial bed elevation (Fretwell et al., 2013).

## 3.2　DEM, radar echograms and satellite imagery

A digital elevation model (DEM) of subglacial topography was produced by rendering RES-derived bed elevations onto a 1 km grid mesh, using the "Topo to Raster" iterative finite difference interpolation function in ArcGIS. This algorithm employs a nested grid strategy to calculate successively finer grids until the user-specified resolution is obtained (Hutchinson, 1988, 1989), and it has been shown to be particularly effective in rendering glacial terrain (Fretwell et al., 2013). In addition, an isostatic correction was applied to the DEM to account for the removal of the modern ice sheet load. This comprised a simple Airy-type compensation, with an ice density of $915 \, \mathrm{kg \, m^{-3}}$ and a mantle density of $3330 \, \mathrm{kg \, m^{-3}}$. Whilst we acknowledge that this approach does not take into account the full complexity associated with glacio-tectonic interactions, it does provide a general indication of pre-glacial elevations across the region, offering insight into the landscape setting prior to glaciation. Radar echograms were also studied in order to assess detailed (along-track) basal topography and englacial structure. These images offer an immediate regional-scale perspective on the nature of the sub-ice landscape and add specific context to the DEM. Furthermore, the Moderate Resolution Imaging Spectroradiometer (MODIS) Mosaic of Antarctica (MOA) (Haran et al., 2005, updated 2013) was used to understand the morphology of the ice sheet surface (Scambos et al., 2007), as this can provide detailed insight into the nature of the underlying subglacial terrain (Ross et al., 2014).

## 3.3　Geomorphometry

The DEM was analysed in order to identify, map and quantify subglacial morphology and the spatial variability of geomorphic features. In particular, bed slope and hypsometry (area–elevation distribution) were quantified, as outlined below. Geomorphic features were also compared with maps of spectral bed roughness, derived from the IMIS survey radar transects (from Rippin et al., 2014). From the DEM, bed slope represents the gradient (rate of maximum change in $z$ value) from each grid cell to its neighbours and is measured in degrees (Burrough and McDonell, 1998). Abrupt changes in slope can reveal distinct changes in landscapes and specific topographic features, as well as providing an overview of general surface texture.

Hypsometric analysis quantifies the distribution of land surface area with altitude (Strahler, 1952). It is commonly used to understand the relationship between local and regional tectonics and the spatial variability in fluvial and glacial surface processes (Montgomery et al., 2001; Pedersen et al., 2010). Hypsometry identifies the dominant signal of landscape erosion, but is generally scale-dependent (Brocklehurst and Whipple, 2004). In order to assess hypsometry, therefore, the topographic drainage basin encompassing the IMIS was delineated. While this represents an artificial drainage divide boundary, as topography is predominantly below sea level within the survey area, it is a well-defined region that reflects the upslope area that contributes flow (normally water) to a common outlet (topographic low point along the basin boundary). Although it does not necessarily relate to a viable hydrological system (because rivers grade to sea level), the drainage region provides a discrete scale at which to calculate hypsometry (i.e. at basin scale).

Basal roughness is a measure of vertical variation with horizontal distance (e.g. Taylor et al., 2004; Li et al., 2010; Rippin et al., 2011). Low roughness records little vertical variation with horizontal distance, resulting in a surface with a smooth texture, whilst high roughness reflects significant vertical variation with horizontal distance, resulting in a surface with a rough texture. As the nature of basal topography exerts a control on ice flow, quantifying roughness at the ice–bed interface can be used to make inferences about ice-dynamic regimes (Hubbard et al., 2000; Siegert et al., 2005; Bingham et al., 2007; Bingham and Siegert, 2007, 2009). Previously, "smooth" (low-roughness) regions have been associated with fast-flowing ice, the presence of glacial sediments and/or significant glacial landscape modification, whilst "rough" topography is often associated with slow-flowing ice and minimal glacial modification (Bingham and Siegert, 2007). Rippin et al. (2014) used a two-parameter fast Fourier transform to determine total roughness and the standard deviation of along-track bed topography to assess roughness directionality across the IMIS. Total basal roughness represents the amplitude (vertical irregularity) of bed roughness at all wavelengths and can perhaps be viewed as the "net" roughness determined for a given location. Directional roughness takes into account erosion relative to the ice-flow field by determining roughness variations both orthogonal and parallel to ice flow (Rippin et al., 2014; see detailed methods therein). Directional roughness may therefore highlight regions where substantial erosion parallel to ice flow has resulted in the formation of streamlined landforms, such as mega-scale glacial lineations (e.g. Graham et

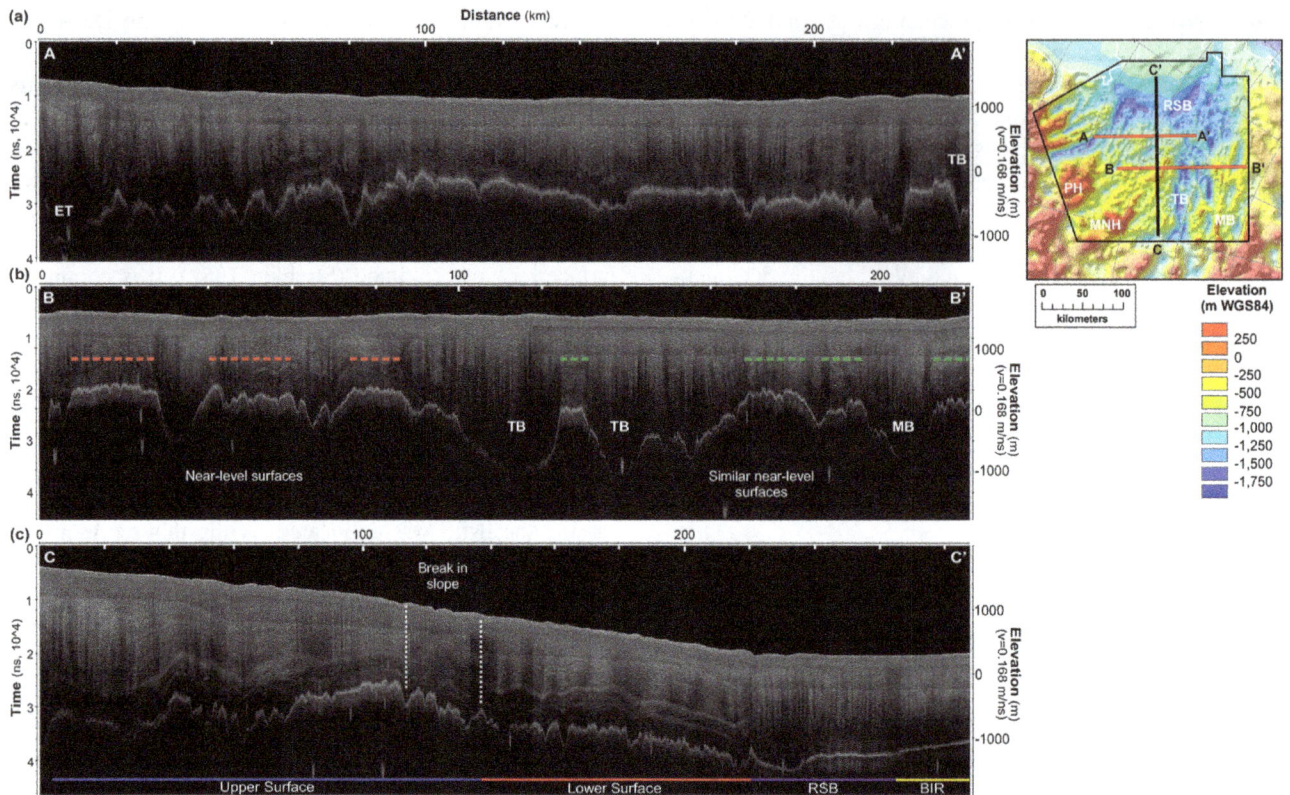

**Figure 3.** Radar echograms showing the gently sloping surfaces identified in subglacial topography. Inset shows the location of transects overlain on subglacial topography. **(a)** Cross-profile transect A–A', located proximal to the Robin Subglacial Basin, between the main tributaries of the Institute Ice Stream. **(b)** Cross-profile transect B–B', located further inland, extending between the Pirrit Hills (panel left) and the Marginal Basins (panel right), crossing the easternmost tributary of the Institute Ice Stream, which is underlain by the Transitional Basins. Cross-profiles show ice flow going into the page. Dashed lines (red and green) highlight regions of consistent summit elevation. **(c)** Long-profile transect C–C', running from the inland edge of the survey grid (panel left) to southern edge of the Bungenstock Ice Rise (panel right). Ice flow is roughly from left to right, but changes orientation across the Robin Subglacial Basin. Elevations are relative; topography in all transects is located below sea level (see inset). Annotations: BIR – Bungenstock Ice Rise; ET – Ellsworth Trough; MB – Marginal Basin; MNH – Martin–Nash Hills; PH – Pirrit Hills; RSB – Robin Subglacial Basin; TB – Transitional Basins.

al., 2007; King et al., 2009). Here, we further develop the study by Rippin et al. (2014) by comparing regional-scale roughness characteristics and landscape geometry.

## 4 Results

### 4.1 Subglacial topography

#### 4.1.1 Radar echograms

Figure 3 displays a series of radar echograms representing two cross-profiles and a single long-profile across the central IMIS survey area. The cross-profile radar echograms illustrate a smooth, gently sloping bed in the region between the Robin Subglacial Basin and the Pirrit and Martin–Nash hills (Fig. 3a and b). Proximal to the Robin Subglacial Basin, the lateral extent of this smooth bed is truncated on either side by the Ellsworth Trough and Transitional Basins (Fig. 3a). Further inland, the smooth bed is still evident but is less ob-

viously continuous, being dissected by a number of typically U-shaped valleys (Fig. 3b, red dashed lines; lines show regions of summit elevation consistency). This radar echogram also shows evidence of similar (but more dissected) near-level surfaces in the region of the Marginal Basins, east of the Transitional Basins (Fig. 3b, green dashed lines). The long-profile radar echogram (Fig. 3c) shows that the surface is divided into two discrete topographic elements, which form a lower and an upper surface. These are separated by two pronounced breaks in slope, approximately 80 and 110 km inland from the edge of the Robin Subglacial Basin (Fig. 3c, vertical white dashed lines). The lower surface appears to dip towards the Robin Subglacial Basin, whilst the upper surface shows a distinct tilt inland.

### 4.1.2 DEM

The DEM shows the full extent of the surfaces identified in the radar echograms, revealing a smooth, gently slop-

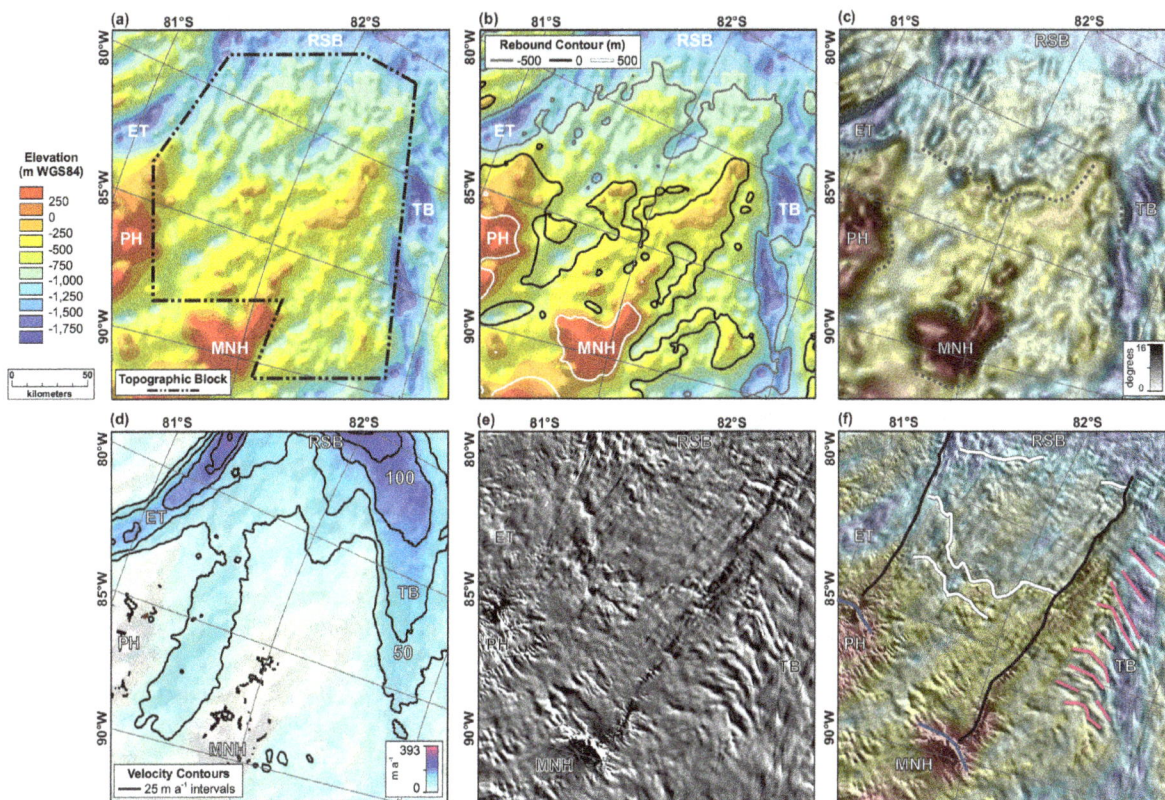

Figure 4. (a) Subglacial topography between the main Institute Ice Stream tributaries. Colour and distance scales for subglacial topography remain consistent for each panel. (b) Present-day bed elevations overlain with elevation contours (500 m intervals) derived from the isostatically adjusted topography. (c) Surface slope (semi-transparent) overlain on subglacial topography. Grey dotted lines mark regions of steep slopes ($> 7°$). (d) Ice sheet surface velocity (semi-transparent) and velocity contours (Rignot et al., 2011) overlain on subglacial topography. (e) MODIS MOA imagery of the ice sheet surface. (f) MODIS MOA imagery (semi-transparent) overlain on subglacial topography and annotated to show the dominant morphological features observed (see Sect. 4.2). Extent of panels shown by rectangular dashed black box in Fig. 1b. Annotations: ET – Ellsworth Trough; MNH – Martin–Nash Hills; PH – Pirrit Hills; RSB – Robin Subglacial Basin; TB – Transitional Basins.

ing, roughly rectangular topographic block, located in the region between the Robin Subglacial Basin and the Pirrit and Martin–Nash hills (Fig. 1b, dashed black box, and Fig. 4a, black dot-dash line). The main block is $\sim 200 \times 150$ km in size and is located between the Ellsworth Trough and the Transitional Basins. Slope analysis highlights five geomorphic features in the topography (Fig. 4c). First, steep slopes ($> 7°$) are associated with the flanks of the Pirrit and Martin–Nash hills. They are also found at the lateral boundaries of the block, along the margins of the Ellsworth Trough and Transitional Basins. Second, these troughs and basins are characterised by steep-sided valley sides and flat-bottomed floors, indicative of U-shaped cross-profiles. Third, the block itself has generally low slope gradients ($< 4°$), reflecting a gently dipping surface profile. Fourth, the middle of the block is dissected by a linear zone of higher slope values ($6–8°$), transverse to ice flow, which divides it into two sections (Fig. 4c). This break in slope is in keeping with observations from radar echograms at the ice surface and bed (Fig. 3c), and

we note that it is also visible in DEMs of the ice sheet surface (e.g. Bamber et al., 2009; Fretwell et al., 2013). Fifth, under present-day ice cover, the elevation of the topographic block is predominantly below sea level (within a range of 200 to $-1500$ m, mean of $-656$ m). When a generalised isostatic correction is applied (Fig. 4b, contours), the elevation of the block rises (within a range of 600 to $-1100$ m), but mean elevation remains below sea level (mean of approximately $-270$ m). It is also noted that the estimated sea-level contour (0 m) (Fig. 4b) corresponds to the break in slope ($\sim 80$ km from the Robin Subglacial Basin) identified across the centre of the block (Fig. 3c, white dashed line, and Fig. 4c).

### 4.1.3 Hypsometry

The hypsometry of the IMIS drainage basin (Fig. 5) displays two area–elevation maxima that are skewed downward ($H_{max}$ values of $-900$ and $-400$ m), demonstrating that the majority of the basin lies below sea level (within a range of 1591 to $-1971$ m, mean of $-643$ m; Fig. 5b). The style of

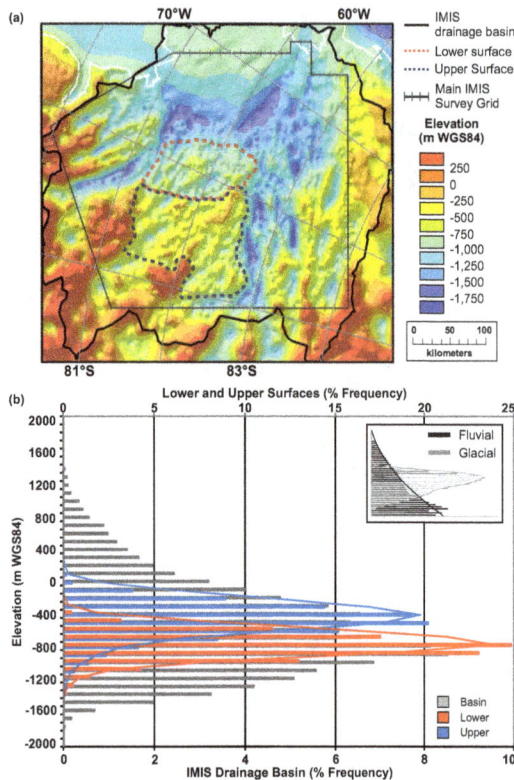

**Figure 5.** (**a**) Drainage basin determined for subglacial topography in the region of the Institute and Möller ice streams (IMIS). (**b**) Hypsometry (area–elevation distribution) determined for the IMIS drainage basin, in comparison with the lower (red) and upper (blue) erosion surfaces. The inset shows characteristic hypsometric distributions for fluvial (black) and glacial (grey) landscapes (Egholm et al., 2009).

hypsometry is consistent with glacial, rather than fluvial, environments (Egholm et al., 2009; Fig. 5b, inset). The area–elevation distribution of the gently sloping topographic block (Fig. 4a, black dot-dash line) was also determined in order to assess its setting within the context of the IMIS basin hypsometry (Fig. 5a). Because the radar data (Fig. 3c), in conjunction with slope analyses (Fig. 4c), reveal the block is divided into two surfaces, its hypsometry was also sub-divided into a lower surface proximal to the Robin Subglacial Basin (Fig. 5a, red dashed line) and an upper surface located further inland (Fig. 5a, blue dashed line).

The hypsometric distributions of the two surfaces fall within the central elevation range of the IMIS drainage basin bed (within a range of $-100$ to $-1100$ m) (Fig. 5b). Both the lower and upper surfaces show hypsometric maxima that are in line with the lower and upper $H_{max}$ peaks recorded in the IMIS basin (lower $H_{max} = -800$ m, upper $H_{max} = -500$ m; Fig. 5b). This highlights that the majority (almost 60 %) of the drainage basin corresponds to the gently sloping block. The two surfaces that comprise the block lie almost completely below sea level (< 1 % area above 0 m), in keeping

with their broad sea embayment setting. For each surface, a large proportion of the area lies within an elevation range of 200 m, highlighting the low relief of this block. Specifically, 50 % of the lower surface area lies between elevations of $-800$ and $-1000$ m and 40 % of the upper surface area lies between elevations of $-400$ and $-600$ m.

### 4.1.4 Roughness

A thorough analysis of spectral basal roughness across the IMIS was carried out by Rippin et al. (2014). Here, we discuss the total basal roughness and roughness directionality derived from that work (Fig. 6), and its relation to the morphological analyses presented above. Generally the block is characterised by low roughness ($\leq 0.1$) at all wavelengths (total roughness) (Fig. 6b). A band of slightly higher total roughness (0.1–0.2) is also evident in the centre of this region, corresponding with the break in slope in the topography (Figs. 3c and 4c). Low total roughness values are also found across the sediment-filled Robin Subglacial Basin and the Bungenstock Ice Rise. Higher total roughness values (> 0.2) are typically associated with the surrounding Ellsworth Mountains and Pirrit and Martin–Nash hills (Fig. 6b). Similarly, when examining the directionality of roughness relative to present-day ice flow, the lowest values are associated with the Robin Subglacial Basin and Bungenstock Ice Rise, whilst the highest values generally correlate with subglacial mountains. In other regions, such as the Transitional Basins, we find that roughness orthogonal to flow is typically higher than that parallel to flow (Fig. 6c and d). In contrast, however, the block shows a different pattern, whereby roughness is lower (0.24) orthogonal to ice flow (Fig. 6c) but higher (0.35) parallel to flow (Fig. 6d). These patterns reveal that the block has generally low basal roughness, dominated by shorter wavelength variations (typically parallel to ice flow), which gives it a distinct roughness character within the IMIS basin (Rippin et al., 2014). Similarly, the Robin Subglacial Basin is also defined by its own distinct (extremely smooth) roughness character that is particularly evident in the patterns of roughness directionality (Fig. 6c and d).

### 4.2 Ice sheet surface imagery

MODIS MOA imagery highlights four key morphological features in the ice sheet surface that reflect distinct changes in the nature (e.g. roughness, elevation) of the underlying subglacial topography (Fig. 4e). First, the isolated granite intrusions of the Pirrit and Martin–Nash hills are characterised by a rough surface texture (Fig. 4f, blue lines), forming clustered "corrugated" features, orientated transverse to ice flow. Second, downstream of each hill is a linear feature, aligned with ice flow in the direction of the grounding line (Fig. 4f, black lines). Third, located between these are a series of linear to curved features that lie transverse to ice flow (Fig. 4f, white

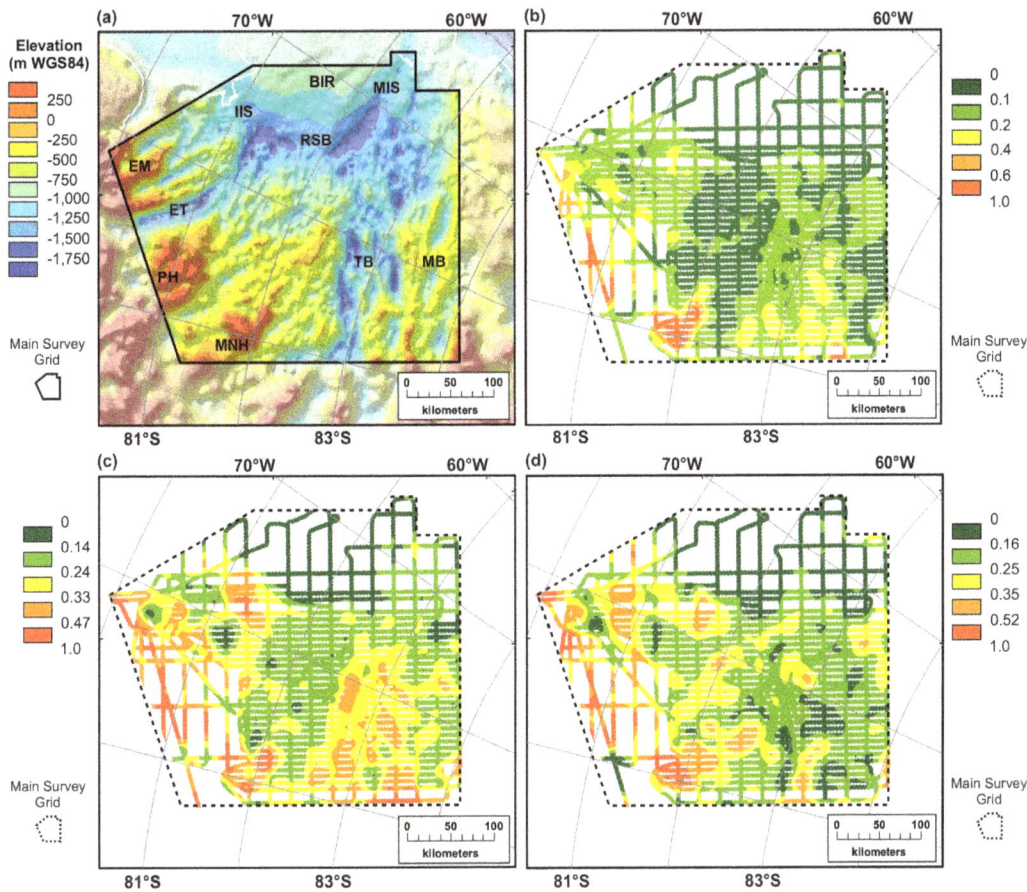

**Figure 6.** (a) Subglacial topography of the Institute and Möller ice streams overlain on semi-transparent Bedmap2 topography. Annotations: BIR – Bungenstock Ice Rise; EM – Ellsworth Mountains; ET – Ellsworth Trough; IIS – Institute Ice Stream; MB – Marginal Basins; MIS – Möller Ice Stream; MNH – Martin–Nash Hills; PH – Pirrit Hills; RSB – Robin Subglacial Basin; TB – Transitional Basins. (b) Total basal roughness determined from bed elevation data along IMIS survey flight lines. (c) Basal roughness determined using the standard deviation of along-track bed topography orthogonal to ice flow. (d) Basal roughness determined using the standard deviation of along-track bed topography parallel to ice to flow. Adapted from Rippin et al. (2014).

lines). The most distinct and laterally continuous curved surface feature runs across the middle of this region. These features, respectively, correspond to the lateral boundaries of the block and the break in slope identified (80 km inland from the Robin Subglacial Basin) in radar echograms (Fig. 3c) and the DEM (Fig. 4c). Fourth, we also note a series of surface features perpendicular to ice flow in the region of the most easterly main tributary of the IIS (Fig. 4f, pink lines). These are clearly associated with ice flow across the basal topography of the Transitional Basins.

## 5  Interpretation

### 5.1  Erosion surfaces

Geomorphic analyses of the DEM and supporting evidence from radar echograms and satellite imagery have enabled us to identify a gently sloping, low-relief topographic block, with associated low basal roughness, in the region between

the Robin Subglacial Basin and the Pirrit and Martin–Nash hills, located further inland. We interpret this region as two large-scale erosion surfaces separated by a marked break in slope (Fig. 3c, white dashed line).

We consider that these surfaces represent erosional rather than depositional topographic features for three reasons. First, they are located inland of the grounding line where the movement of grounded ice and subglacial water is more likely to result in erosion. In particular, in the past (e.g. the LGM), when the ice sheet margin was located at the continental shelf, the erosion surfaces would have been located even further into the ice sheet interior, demonstrating that this is not an obvious position for major glacial–sedimentary deposition. Second, landscape relief points to the region having been subject to erosional processes (Fig. 1). For example, Jurassic granites that comprise the Pirrit and Martin–Nash hills, which protrude above the present-day ice sheet surface, were emplaced between ∼ 175 and 165 Ma (Storey et al., 1988; Lee et al., 2012) and have since been exposed by erosion. The relief today between the mountain peaks and the upper erosion surface demonstrates that a significant amount of rock (at least 2 km) has been removed since the Jurassic. Furthermore, the presence of troughs and basins below sea level, particularly the Ellsworth Trough, indicates significant removal of material by glacial processes. Third, radar echograms show that the gently sloping erosion surfaces display small-scale undulations and intermittent valleys (Fig. 3a), giving this region a different (short-wavelength) roughness (Fig. 6a) that distinguishes it from the likely sediment-filled Robin Subglacial Basin (Rippin et al., 2014). Although we cannot rule out that there may be pockets of sediments or even a thin sedimentary drape in places across the surface, there is no evidence from radar or gravity data for a saturated or thick sedimentary deposit in this region. The surfaces are, therefore, more in keeping with an extensive erosional bedrock landscape setting.

## 5.2 Preservation

Glaciological conditions now and in the past (i.e. since ice sheet glaciation of West Antarctica) are conducive to landscape preservation between the IIS tributaries. MODIS MOA imagery reveals two linear surface features in the lee of the Pirrit and Martin–Nash hills (Fig. 4e and f), which correspond to the lateral boundaries of the block (Fig. 4a). Clearly, these mountains significantly modulate regional ice flow, reducing ice velocities immediately downstream and focusing flow along the adjacent troughs and basins that now host the fast-flowing ($> 50\,\mathrm{m\,a^{-1}}$) tributaries of the IIS (Fig. 4d).

The region was also unlikely to have been a fast flow zone during full glacial conditions during the Quaternary, as there would be slow, interior ice-sheet-type flow (i.e. by internal ice deformation only) across this region when ice was at the shelf edge. In addition, the erosion surfaces display their own distinct roughness signal, with comparatively high values parallel to ice flow, versus those orthogonal to flow (Fig. 6c and d). This indicates that the landscape has not experienced significant streamlining by ice but rather retains a signal of pre-glacial geomorphic processes (Rippin et al., 2014).

## 5.3 Selective glacial erosion

Radar echograms reveal that the block has been subject to a degree of selective glacial erosion following formation. A few valleys (often U-shaped) are visible in cross-profile A, particularly in proximity to the Ellsworth Trough (Fig. 3a), whilst further inland cross-profile B has been more significantly dissected by broader U-shaped valleys (Fig. 3b). The scale and style of these intermittent, U-shaped valleys are suggestive of selective linear erosion by small- to regional-scale, warm-based ice masses (Sugden and John, 1976; Hirano and Aniya, 1988). Given the broad preservation of the topographic block, the ice sheet likely comprised polyther-

mal basal conditions, with spatially restricted zones of warm-based ice focused along valleys to form deep troughs, and cold-based ice associated with low erosion rates located in the lee of the Jurassic intrusions. The locations of the valleys are likely to reflect lines of structural tectonic weaknesses (Jordan et al., 2013) or pre-existing fluvial networks that have been exploited (e.g. Baroni et al., 2005; Rose et al., 2013; Ross et al., 2014).

We propose that incision of the erosion surfaces occurred during periods of intermediate ice sheet cover (cf. Young et al., 2011) when corridors of warm-based ice, subject to topographic steering, flowed around the Pirrit and Martin–Nash hills. Ross et al. (2012) use the location of troughs and elevated bars to infer the position of a former grounding line upstream of the Robin Subglacial Basin. Their findings indicate that, even under a smaller-scale, restricted ice sheet configuration, the majority of ice drainage would likely follow the linear topographic valleys that flank this block (cf. Rose et al., 2014). This would minimise the degree of glacial incision between the valleys (i.e. where there may have been cold-based ice), helping to retain pre-glacial landscape signals across the erosion surfaces. Subsequently, once the ice sheet expanded to a continental scale, buttressing of ice sheet flow by the Pirrit and Martin–Nash hills would also restrict extensive landscape modification by glacial erosion downstream of these major mountain massifs. The existence of an extensive, although incised, erosion surface demonstrates that it is possible to preserve ancient surfaces beneath an ice sheet at low elevations, and not just at high elevations associated with thin, cold-based ice (Rose et al., 2013; Ross et al., 2014). It is also interesting to note that whilst regions of smooth and flat topography are typically associated with relatively fast ice flow and often inferred marine sediments (Joughin et al., 2006; Peters et al., 2006; Bingham and Siegert, 2007), we do not find that relationship here. Instead, the fastest ice flow exploits the deeply incised troughs and basins that flank the low-roughness erosion surfaces.

## 5.4 Regional extent

We also find additional evidence indicating that the surfaces may have been more extensive in the past. Radar echograms, for example, hint at the existence of a more laterally continuous erosion surface extending across to the Transitional and Marginal Basins (Fig. 3b, green dashed lines). This is indicated by the presence of a few, isolated, flat-topped hills between the basins (Figs. 1b and 3b). These basins are tectonically controlled (Jordan et al., 2013), and it is likely that they developed (overdeepened) due to glacial erosion (down-cutting) exploiting these pre-existing tectonic structures. Low roughness values parallel to ice flow highlight that glacial erosion has been particularly effective along these basins (Fig. 6d). As a result, only remnants of this surface, between basins, have remained (Burbank and Anderson, 2012). Without such structural controls driving erosion in these locations, a greater proportion and a more continuous surface may have been retained. These smaller, localised remnants of the erosion surface are harder to identify from the DEM (because it is derived from an aerogeophysical survey with a line spacing of 7.5 km), but their existence is reflected in the cumulative proportion of area–elevation shown in the hypsometry. We find that 60 % of area–elevation distribution in the IMIS drainage basin corresponds to the elevations of the block, but the block only accounts for 15 % of the area of the IMIS drainage basin (Fig. 5b). This gives support to the idea that the ancient erosion surface once occupied a much larger proportion of the embayment but that it has since been dissected by glacial (and likely other) erosion processes.

Indeed, if we examine the landscape at a regional scale, it is easy to see how a larger-scale, more continuous erosion surface may have existed across the Weddell Sea Embayment in the past. In particular, Berkner Island and the Henry and Korff ice rises stand out as regions of gently dipping, smooth topography now encompassed by the Filchner–Ronne Ice Shelf (Fig. 1a). Closer to the present-day grounding line, the Skytrain Ice Rise, Fletcher Promontory and Fowler Peninsula also provide evidence for low-gradient, low-relief surfaces (Doake et al., 1983). The geometry of these features is comparable to the form of the topographic block we identify between the primary IIS tributaries. From this we may infer that, coupled with the block, they may once have formed part of a more continuous erosion surface that occupied a significant proportion of the embayment. If such a surface existed, then it also seems likely that the same dominant process(es) of landscape development acted at a regional scale across the Weddell Sea Embayment and not just at a local scale in the region of the block.

# 6 Discussion

We have mapped a distinct topographic block within the IMIS survey area and characterised it as two ancient preserved erosion surfaces. It is, however, much more difficult to attribute the mode of formation of the erosion surfaces to a particular mechanism. This is largely because the processes involved in the formation of geometrically flat bedrock surfaces are subject to long-standing debates and uncertainty. Globally, erosion surfaces have been described in a number of settings and, although it may not be possible to determine with certainty the exact origin of the erosion surfaces identified, here we briefly discuss a few of the settings and erosion regimes that are commonly associated with erosion surfaces cut into basement rocks.

We have two favoured models and associated processes responsible for the formation of the erosion surfaces: the first is marine erosion, the second fluvial erosion. Both are consistent with our evidence for a gently sloping, low-relief topo-

graphic block characterised by a low surface roughness and an elevation near to isostatically rebounded sea level.

## 6.1 Marine erosion model

Marine erosion is concentrated at the interface between land and sea through the constant action of waves impacting a shoreline, often during a period of tectonic (and sea level) quiescence (Burbank and Anderson, 2012). A marine origin for large-scale erosion surfaces is possible and large-scale marine erosion surfaces are often linked with polar glacial environments. In the Northern Hemisphere, shore platforms, known as strandflats, are found along almost the whole west coast of Norway (Klemsdal, 1982), as well as in Greenland (Bonow et al., 2007), Alaska and Scotland (Dawson et al., 2013). These examples demonstrate that erosion surfaces often extend for many hundreds of kilometres (around coastlines) and may be tens of kilometres wide. They are thought to result primarily from marine abrasion, with contributions from subsequent weathering and glacial erosion processes (Fredin et al., 2013). In Norway, strandflats are found both above and below sea level. Typically, they lie in front of higher land or coastal mountains and form either level or gently dipping bedrock surfaces from the coast (Klemsdal, 1982; Fredin et al., 2013). We find some analogy between the Norwegian strandflats and the setting of the IMIS drainage basin, where the Ellsworth Subglacial Highlands and other mountains are found inland of the erosion surfaces (Fig. 1a), and a gentle coastal dip is evident in the lower surface (Fig. 3c).

In West Antarctica, a sequence of palaeo-erosion surfaces has also been identified in the Ross Sea sector, located between −100 and −350 m (Wilson and Luyendyk, 2006a, b). These surfaces are laterally continuous, averaging the same elevation over large distances. They are 100s km wide and discontinuous, separated by troughs that are occupied by the ice streams of the Siple Coast. Using additional evidence from gravity and marine seismic data sets, Wilson and Luyendyk (2006a, b) interpreted these surfaces as the remnants of a former, continuous wave-cut shore platform.

If we apply a marine erosion model for the erosion surfaces we identify beneath the IIS, we suggest that the surfaces were cut at, or just below, sea level by wave action prior to Antarctic glaciation or during periods when the ice sheet was significantly smaller than today. Such a model is consistent with the isostatically rebounded average elevation of the surface (−270 m), its geomorphic context (i.e. at the head of the Weddell Sea Embayment) and its proximity to the mountain ranges of the Ellsworth–Whitmore Mountains block, in a setting analogous to that of the strandflat features of the Arctic. Although we accept that the width of the erosion surfaces is unusually large ($\sim$ 150 km in total, $\sim$ 80 km lower surface, $\sim$ 70 km upper surface) for it to have been cut by processes of marine erosion alone, we note that the Norwegian strandflat is 60 km wide in places (Porter, 1989) and that many Antarctic glacial landforms (e.g. glacial troughs, cirques) are often considerably larger in scale than similar features elsewhere on Earth (Haynes, 1998; Young et al., 2011).

## 6.2 Fluvial erosion model

Fluvial erosion processes are also capable of generating large-scale, low-gradient bedrock surfaces. Rivers work to reduce surface elevations to a base level, typically at sea level, through the erosion, transportation and deposition of material (Burbank and Anderson, 2012). This often occurs in passive continental margin settings (Beaumont et al., 2000), where fluvial erosion cuts a new base level following rifting. This results in a coastal plain which is then backed by escarpments, the number of which depends on lithology and geological structure.

An alternative model for the formation of the erosion surfaces beneath the IIS, therefore, is that it was the result of fluvial peneplain erosion during, and after, the rifting and break-up of Gondwanaland around ∼ 180 Ma. A constraint on the maximum age of the surface is provided by the 175 Ma granitic intrusions of the Pirrit and Martin–Nash hills, which demonstrate that the surface cannot be older than this. As a consequence of the rifting of Gondwanaland, fluvial systems on the passive margins of the rifts incised into the landscape in response to the subsequent downward shift in base level (Beaumont et al., 2000). This incisional phase resulted in the formation of the broad extensive coastal surfaces, backed by erosional escarpments and characterised by occasional insel-bergs, mapped in southern Africa (e.g. Namibia) and South America (Brown et al., 2000). The same processes (i.e. fluvial erosional processes operating in a passive continental margin setting) may also have been responsible for the development of the surfaces we have identified. In the case of the Antarctic features, however, the surfaces were later incised and fragmented by tectonic processes (e.g. major strike-slip faulting) (Jordan et al., 2013) and by selective linear glacial erosion (Ross et al., 2012) (Fig. 3b). Despite the action of these surface processes, large parts of the ancient erosion surfaces remain little modified. They have been preserved by their situation in the interior of the ice sheet, where rates of subglacial erosion are low (Jamieson et al., 2010), and because of the long-term protection offered by the Pirrit and Martin–Nash hills, which have buttressed downstream ice, limiting glacial erosional capacity (see Sect. 5.2). Assuming the fluvial model to be correct, one interesting possibility is that the prominent break in slope that we identify (Figs. 3c and 4c, e, f) may be analogous to the escarpments that back the coastal plains of southern Africa (Brown et al., 2000).

One line of evidence that could be seen as inconsistent with the fluvial model proposed is that the mean elevation of the isostatically rebounded surface is −270 m. However, given the potential age of the surfaces (i.e. they may have formed long before extensive ice-sheet glaciation in Antarctica), uncertainties over the tectonic structure and history of this area, and the simplicity of the calculation of the isostatic correction applied here, it may simply be that the isostatically rebounded elevation of the surface is unrepresentative of the elevation at which the surface originally formed. In the Transantarctic Mountains in East Antarctica, ancient relict surfaces are found at elevations far above present-day, or ice-free, sea level conditions (Sugden et al., 1995; Kerr et al., 2000).

If the fluvial model is correct, however, it opens up the possibility of using detailed investigations of Antarctic subglacial topography to improve our understanding of the nature of Gondwanaland rifting and, from this, the formation of the Antarctic continent. We note that several studies (e.g. Sugden et al., 1995; Kerr et al., 2000; Näslund, 2001; Jamieson and Sugden, 2008) have also proposed the preservation of other ancient (e.g. Gondwanaland) surfaces in East Antarctica. It may be that the flat-lying surfaces identified elsewhere in the Weddell Sea sector of West Antarctica, as well as those in the Ross Sea, previously interpreted as wave-cut shore platforms (Wilson and Ludendyk, 2006a, b), may also be remnants of extensive ancient fluvial surfaces.

## 7  Conclusions

A new DEM, built from an extensive and high-resolution modern airborne radar survey, provides a detailed view of subglacial topography across the IMIS, where previously only sparse data were available (Bingham and Siegert, 2007). We have examined radar echograms and MODIS MOA imagery and applied morphometric analyses to the DEM in order to characterise the landscape. In doing so, we have (1) identified a smooth, laterally continuous, gently sloping topographic block in the region between the Robin Subglacial Basin and the Pirrit and Martin–Nash hills; (2) characterised this block as two surfaces separated by a distinct break in slope; (3) shown that erosion rates across the surfaces are currently low, precluding formation via present-day glacial erosion; and (4) interpreted these features as erosion surfaces.

Our findings show that it is possible for ancient erosion surfaces to be preserved at low elevations beneath ice sheets. We have also identified the Pirrit and Martin–Nash hills as having played a key role in the long-term landscape evolution of this region. By buttressing upstream ice and reducing downstream ice velocities, they have limited glacial erosion rates in their lee, thus enabling long-term preservation of the erosion surfaces. By modulating ice dynamics, these mountain massifs have facilitated a region where smooth, low-geometry basal topography does not correspond to the fast ice flow typically associated with ice streams. Whilst we have characterised the topographic block, we have not been able to constrain the dominant mechanism of formation for the erosion surfaces based on the current data available. Instead, we have presented a number of different erosion regimes (namely fluvial and marine) that are associated with the formation of erosion surfaces in order to inspire further investigation in this region so that the origin of these surfaces may be deciphered with greater certainty in the future.

**Acknowledgements.** This project was funded by UK NERC AFI grant NE/G013071/1. Carl Robinson (Airborne Survey engineer), Ian Potten and Doug Cochrane (pilots), and Mark Oostlander (air mechanic) are thanked for their invaluable assistance in the field. We thank Arjen Stroeven and an anonymous reviewer for helpful comments on an earlier draft of this manuscript. We would also like to extend particular thanks to David Sugden, whose helpful, fruitful and generous discussions enabled us to improve the manuscript greatly.

Edited by: J. Willenbring

# References

Bamber, J. L., Gomez-Dans, J. L., and Griggs, J. A.: A new 1 km digital elevation model of the Antarctic derived from combined satellite radar and laser data – Part 1: Data and methods, The Cryosphere, 3, 101–111, doi:10.5194/tc-3-101-2009, 2009.

Baroni, C., Noti, V., Ciccacci, S., Righini, G., and Salvatore, M.-C.: Fluvial origin of the valley system in northern Victoria Land (Antarctica) from quantitative geomorphic analysis, Geol. Soc. Am. Bull., 117, 212–228, doi:10.1130/B25529.1, 2005.

Beaumont, C., Kooi, H., and Willett, S.: Coupled tectonic-surface process models with applications to rifted margins and collision orogens, in: Geomorphology and Global Tectonics, edited by: Summerfield, M. A., John Wiley & Sons, Chichester, 29–55, 2000.

Bingham, R. G. and Siegert, M. J.: Radar-derived bed roughness characterization of Institute and Möller ice streams, West Antarctica, and comparison with Siple Coast ice, Geophys. Res. Lett., 34, L21504, doi:10.1029/2007gl031483, 2007.

Bingham, R. G., and Siegert, M. J.: Quantifying subglacial bed roughness in Antarctica: implications for ice-sheet dynamics and history, Quaternary Sci. Rev., 28, 223–236, 2009.

Bingham, R. G., Siegert, M. J., Young, D. A., and Blankenship, D. D.: Organized flow from the South Pole to the Filchner-Ronne ice shelf: An assessment of balance velocities in interior East Antarctica using radio echo sounding data, J. Geophys. Res.-Earth, 112, F03S26, doi:10.1029/2006jf000556, 2007.

Bonow, J. M., Lidmar-Bergström, K., Japsen, P., Chalmers, J. A., and Green, P. F.: Elevated erosion surfaces in central West Greenland and southern Norway: their significance in integrated studies of passive margin development, Norweg. J. Geol., 87, 197–206, 2007.

Brocklehurst, S. H. and Whipple, K. X.: Hypsometry of glaciated landscapes, Earth Surf. Proc. Land., 29, 907–926, 2004.

Brown, R. W., Gallagher, K., Gleadow, A. J. W., and Summerfield, M. A.: Morphotectonic evolution of the South Atlantic margins of Africa and South America, in: Geomorphology and Global Tectonics, edited by: Summerfield, M. A., John Wiley & Sons, Chichester, 255-281, 2000.

Burbank, D. W. and Anderson, R. S.: Tectonic Geomorphology, 2nd Edn., Wiley-Blackwell, Chichester, UK, 2012.

Burrough, P. A. and McDonell, R. A.: Principles of Geographical Information Systems, Oxford University Press, New York, 1998.

Corr, H., Ferraccioli, F., Frearson, N., Jordan, T. A., Robinson, C., Armadillo, E., Caneva, G., Bozzo, E., and Tabacco, I. E.: Airborne radio-echo sounding of the Wilkes Subglacial Basin, the Transantarctic Mountains, and the Dome C region, in: The Italian-British Antarctic Geophysical and Geological Survey in Northern Victoria Land 2005–06 – Towards the International Polar Year 2007–08, edited by: Bozzo, E. and Ferraccioli, F., Terra Antartica Reports, Terra Antartica Publications, Siena, 13, 55–63, 2007.

Curtis, M. L.: Tectonic history of the Ellsworth Mountains, West Antarctica: Reconciling a Gondwana enigma, Geol. Soc. Am. Bull., 113, 939–958, 2001.

Dalziel, I. W. D. and Elliot, D. H.: West Antarctica: problem child of Gondwanaland, Tectonics, 1, 3–19, 1982.

Dalziel, I. W. D., Lawver, L., Norton, I. O., and Gahagan, L. M.: The Scotia Arc: Genesis, Evolution, Global Significance, Annu. Rev. Earth Planet. Sci., 41, 767–793, 2013.

Dawson, A. G., Dawson, S., Cooper, A. G., Gemmell, A., and Bates, R.: A Pliocene age and origin for the strandflat of the Western Isles of Scotland: a speculative hypothesis, Geolog. Mag., 150, 360–366, 2013.

Doake, C. S. M., Crabtree, R. D., and Dalziel, I. W. D.: Subglacial Morphology between Ellsworth Mountains and Antarctica Peninsula: New Data and Tectonic Significance, Proceedings of the 4th International Symposium on Antarctic Earth Sciences, Cambridge University Press, New York, 270–273, 1983.

Egholm, D. L., Nielsen, S. B., Pedersen, V. K., and Lesemann, J. E.: Glacial effects limiting mountain height, Nature, 460, 884–888, 2009.

Fredin, O., Bergstrom, B., Eilertsen, R., Hansen, L., Longva, O., Nesje, A., and Sveian, H.: Glacial landforms and Quaternary landscape development in Norway, in: Quaternary Geology of Norway, edited by: Olsen, L., Fredin, O., and Olesen, O., Geological Survey of Norway Special Publication, Geological Survey of Norway, Trondheim, 5–25, 2013.

Fretwell, P., Pritchard, H. D., Vaughan, D. G., Bamber, J. L., Barrand, N. E., Bell, R., Bianchi, C., Bingham, R. G., Blankenship, D. D., Casassa, G., Catania, G., Callens, D., Conway, H., Cook, A. J., Corr, H. F. J., Damaske, D., Damm, V., Ferraccioli, F., Forsberg, R., Fujita, S., Gim, Y., Gogineni, P., Griggs, J. A., Hindmarsh, R. C. A., Holmlund, P., Holt, J. W., Jacobel, R. W., Jenkins, A., Jokat, W., Jordan, T., King, E. C., Kohler, J., Krabill, W., Riger-Kusk, M., Langley, K. A., Leitchenkov, G., Leuschen, C., Luyendyk, B. P., Matsuoka, K., Mouginot, J., Nitsche, F. O., Nogi, Y., Nost, O. A., Popov, S. V., Rignot, E., Rippin, D. M., Rivera, A., Roberts, J., Ross, N., Siegert, M. J., Smith, A. M., Steinhage, D., Studinger, M., Sun, B., Tinto, B. K., Welch, B. C., Wilson, D., Young, D. A., Xiangbin, C., and Zirizzotti, A.: Bedmap2: improved ice bed, surface and thickness datasets for Antarctica, The Cryosphere, 7, 375–393, doi:10.5194/tc-7-375-2013, 2013.

Garrett, S. W., Herrod, L. D. B., and Mantripp, D. R.: Crustal structure of the area around Haag Nunataks, West Antarctica: new aeromagnetic and bedrock elevation data, in: Gondwana Six: Structure, Tectonics and Geophysics, edited by: McKenzie, G. D., AGU Geophysical Monograph, Washington, D.C., 109–116, 1987.

Garrett, S. W., Maslanyj, M. P., and Damaske, D.: Interpretation of aeromagnetic data from the Ellsworth Mountains-Thiel Mountains ridge, West Antarctica, J. Geol. Soc. Lond., 145, 1009–1017, 1988.

Graham, A. G. C., Lonergan, L., and Stoker, M. S.: Evidence for Late Pleistocene ice stream activity in the Witch Ground Basin, central North Sea, from 3D seismic reflection data, Quaterny Sci. Rev., 26, 627–643, 2007.

Haran, T., Bohlander, J., Scambos, T., Painter, T., and Fahnestock, M.: MODIS Mosaic of Antarctica (MOA) Image Map, National Snow and Ice Data Centre, Boulder, Colorado, USA, 2005, updated, 2013.

Haynes, V. M.: The morphological development of alpine valley heads in the Antarctic Peninsula, Earth Surf. Proc. Land., 23, 53–67, 1998.

Hélière, F., Lin, C. C., Corr, H., and Vaughan, D.: Radio echo sounding of Pine Island Glacier, West Antarctica: Aperture synthesis processing and analysis of feasibility from space, IEEE T. Geosci. Remote, 45, 2573–2582, 2007.

Hirano, M. and Aniya, M.: A rational explanation of cross-profile morphology for glacial valleys and of glacial valley development, Earth Surf. Proc. Land., 13, 707–716, 1988.

Hubbard, B. P., Siegert, M. J., and McCarroll, D.: Spectral roughness of glaciated bedrock geomorphic surfaces: Implications for glacier sliding, J. Geophys. Res., 105, 21295–21303, 2000.

Hutchinson, M. F.: Calculation of hydrologically sound digital elevation models, in: Third International Symposium on Spatial Data Handling, Columbus, Ohio, International Geographical Union, Sydney, 117–133, 1988.

Hutchinson, M. F.: A New Procedure for gridding elevation and stream line data with automatic removal of spurious pits, J. Hydrol., 106, 211–232, 1989.

Jamieson, S. S. R. and Sudgen, D. E.: Landscape evolution in Antarctica, in: Antarctica: A Keystone in a Changing World, edited by: Cooper, A. K., Barrett, P. J., Stagg, H., Storey, B., Stump, E., Wise, W., and the 10th ISAES editorial team, Proceedings of the 10th International Symposium on Antarctic Earth Sciences, The National Academies Press, Washington, D.C., 39–54, 2008.

Jamieson, S. S. R., Sudgen, D. E., and Hulton, N. R. J.: The evolution of the subglacial landscape of Antarctica, Earth Planet. Sc. Lett., 293, 1–27, 2010.

Jordan, T. A., Ferraccioli, F., Ross, N., Corr, H. F. J., Leat, P. T., Bingham, R. G., Rippin, D. M., le Brocq, A., and Siegert, M. J.: Inland extent of the Weddell Sea Rift imaged by new aerogeophysical data, Tectonophysics, 585, 137–160, 2013.

Joughin, I., Bamber, J. L., Scambos, T., Tulaczyk, S., Fahnestock, M., and MacAyeal, D. R.: Integrating satellite observations with modelling: basal shear stress of the Filchner-Ronne ice streams, Antarctica, Philos. T. Roy. Soc. A, 364, 1795–1814, 2006.

Joughin, I., Tulaczyk, S., Bamber, J. L., Blankenship, D., Holt, J. W., Scambos, T., and Vaughan, D. G.: Basal conditions for Pine Island and Thwaites Glaciers, West Antarctica, determined using satellite and airborne data, J. Glaciol., 55, 245–257, 2009.

Joughin, I., Smith, B. E., and Medley, B.: Marine ice sheet collapse potentially under way for the Thwaites Glacier Basin, West Antarctica, Science, 344, 735–738, 2014.

Kerr, A., Sugden, D. E., and Summerfield, M. A.: Linking tectonics and landscape development in a passive margin setting: the Transantarctic Mountains, in: Geomorphology and Global Tectonics, Summerfield, M. A., John Wiley & Sons, Chichester, 303–319, 2000.

King, E. C., Hindmarsh, R. C. A., and Stokes, C. R.: Formation of mega-scale glacial lineations observed beneath a West Antarctic ice stream, Nat. Geosci., 2, 585–588, 2009.

Klemsdal, T.: Coastal classification and the coast of Norway, Norsk Geografisk Tidsskrift, 36, 129–152, 1982.

Le Brocq, A. M., Ross, N., Griggs, J. A., Bingham, R. G., Corr, H. F. J., Ferraccioli, F., Jenkins, A., Jordan, T. A., Payne, A. J., Rippin, D. M., and Siegert, M. J.: Evidence from ice shelves for channelized meltwater flow beneath the Antarctic Ice Sheet, Nat. Geosci., 6, 945–948, 2013.

Lee, H. M., Lee, J. I., Lee, M. J., Kim, J., and Choi, S. W.: The A-type Pirrit Hills Granite, West Antarctica: an example of magmatism associated with the Mesozoic break-up of the Gondwana supercontinent, Geosci. J., 16, 421–433, 2012.

Li, X., Sun, B., Siegert, M. J., Bingham, R. G., Tang, X. Y., Zhang, D., Cui, X. B., and Zhang, X. P.: Characterization of subglacial landscapes by a two-parameter roughness index, J. Glaciol., 56, 831–836, 2010.

Montgomery, D. R., Balco, G., and Willett, S. D.: Climate, tectonics and the morphology of the Andes, Geology, 29, 579–582, 2001.

Näslund, J.-O.: Landscape development in western and central Dronning Maud Land, East Antarctica, Antarct. Sci., 13, 302–311, 2001.

Pedersen, V. K., Egholm, D. L., and Nielsen, S. B.: Alpine glacial topography and the rate of rock column uplift: a global perspective, Geomorphology, 122, 129–139, 2010.

Peters, L. E., Anandakrishnan, S., Alley, R. B., Winberry, J. P., Voigt, D. E., Smith, A. M., and Morse, D. L.: Subglacial sediments as a control on the onset and location of two Siple Coast ice streams, West Antarctica, J. Geophys. Res.-Solid, 111, B01302, doi:10.1029/2005jb003766, 2006.

Porter, S. C.: Some geological implications of average Quaternary glacial conditions, Quatern. Res., 32, 245–261, 1989.

Rignot, E., Mouginot, J., and Scheuchl, B.: Ice flow of the Antarctic ice sheet, Science, 333, 1427–1430, 2011.

Rippin, D. M., Vaughan, D. G., and Corr, H. F. J.: The basal roughness of Pine Island Glacier, West Antarctica, J. Glaciol., 57, 67–76, 2011.

Rippin, D. M., Bingham, R. G., Jordan, T. A., Wright, A. P., Ross, N., Corr, H. F. J., Ferraccioli, F., Le Brocq, A., Rose, K. C., and Siegert, M. J.: Basal roughness of the Institute and Möller ice streams, West Antarctica: determining past ice dynamical regimes, Geomorphology, 214, 139–147, 2014.

Rose, K. C., Ferraccioli, F., Jamieson, S. S. R., Bell, R. E., Corr, H., Creyts, T. T., Braaten, D., Jordan, T. A., Fretwell, P. T., and Damaske, D.: Early East Antarctic Ice Sheet growth recorded in the landscape of the Gamburtsev Subglacial Mountains, Earth Planet. Sc. Lett., 375, 1–12, 2013.

Rose, K. C., Ross, N., Bingham, R. G., Corr, H. F. J., Ferraccioli, F., Jordan, T. A., Le Brocq, A., Rippin, D. M., and Siegert, M. J.: A temperate former West Antarctic ice sheet suggested by an extensive zone of subglacial meltwater channels, Geology, 42, 971–974, 2014.

Ross, N., Bingham, R. G., Corr, H. F. J., Ferraccioli, F., Jordan, T. A., Le Brocq, A., Rippin, D. M., Young, D., Blankenship, D. D., and Siegert, M. J.: Steep reverse bed slope at the grounding line of the Weddell Sea sector in West Antarctica, Nat Geosci., 5, 393–396, 2012.

Ross, N., Jordan, T. A., Bingham, R. G., Corr, H. F. J., Ferraccioli, F., Le Brocq, A., Rippin, D. M., Wright, A. P., and Siegert, M. J.: The Ellsworth Subglacial Highlands: inception and retreat of the West Antarctic Ice Sheet, Geol. Soc. Am. Bull., 126, 3–15, 2014.

Scambos, T. A., Haran, T. M., Fahnestock, M. A., Painter, T. H., and Bohlander, J.: MODIS-based Mosaic of Antarctica (MOA) data sets: Continent-wide surface morphology and snow grain size, Remote Sens. Environ., 111, 242–257, 2007.

Siegert, M. J., Taylor, J., and Payne, A. J.: Spectral roughness of subglacial topography and implications for former ice-sheet dynamics in East Antarctica, Global Planet. Change, 45, 249–263, 2005.

Siegert, M. J., Ross, N., Corr, H., Kingslake, J., and Hindmarsh, R.: Late Holocene ice-flow reconfiguration in the Weddell Sea sector of West Antarctica, Quaternary Sci. Rev., 78, 98–107, 2013.

Siegert, M. J., Ross, N., Corr, H., Smith, B., Jordan, T., Bingham, R. G., Ferraccioli, F., Rippin, D. M., and Le Brocq, A.: Boundary conditions of an active West Antarctic subglacial lake: implications for storage of water beneath the ice sheet, The Cryosphere, 8, 15–24, doi:10.5194/tc-8-15-2014, 2014.

Storey, B. C. and Dalziel, I. W. D.: Outline of the structural and tectonic history of the Ellsworth Mountains-Thiel Mountains ridge, West Antarctica, in: Gondwana Six; Structure, tectonics, and geophysics, edited by: McKenzie, G. D., AGU Geophysical Monograph, Washington, D.C., 117–128, 1987.

Storey, B. C., Hole, M. J., Pankhurst, R. J., Millar, I. L., and Vennum, W.: Middle Jurassic within-plate granite in West Antarctica and their bearing on the break-up of Gondwanaland, J. Geol. Soc., 145, 999–1007, 1988.

Storey, B. C., Leat, P. T., and Ferris, J. K.: The location of mantle plume centres during the initial stages of Gondwana break-up, Geol. Soc. Am. Special Paper, 352, 71–80, 2001.

Strahler, A. N.: Hypsometric (area-altitude) analysis of erosional topography, Geol. Soc. Am. Bull., 63, 1117–1142, 1952.

Studinger, M. and Miller, H.: Crustal structure of the Filchner-Ronne shelf and Coats Land, Antarctica, from gravity and magnetic data: Implications for the breakup of Gondwana, J. Geophys. Res., 104, 20379–20394, 1999.

Sugden, D. E. and John, B. S.: Glaciers and Landscape, Edward Arnold, London, 1976.

Sugden, D. E., Denton, G. H., and Marchant, D. R.: Landscape evolution of the Dry Valleys, Transantarctic Mountains: tectonic implications, J. Geophys. Res., 100, 9949–9967, 1995.

Taylor, J., Siegert, M. J., Payne, A. J., and Hubbard, B.: Regional-scale bed roughness beneath ice masses: measurement and analysis, Comput. Geosci., 30, 899–908, 2004.

Wilson, D. S. and Luyendyk, B. P.: Bedrock plateaus within the Ross Embayment and beneath the West Antarctic Ice Sheet, formed by marine erosion in Late Tertiary time, in: Antarctica: contribution to global earth sciences, edited by: Fütterer, D. K., Damaske, D., Kleinschmidt, G., Miller, H., and Tessensohn, F., Springer-Verlag, Berlin, 123–128, 2006a.

Wilson, D. S. and Luyendyk, B. P.: Bedrock platforms within the Ross Embayment, West Antarctica: hypotheses for ice sheet history, wave erosion, Cenozoic extension and thermal subsidence, Geochem. Geophy. Geosy., 7, 1–23, 2006b.

Wright, A. P., Le Brocq, A. M., Cornford, S. L., Bingham, R. G., Corr, H. F. J., Ferraccioli, F., Jordan, T. A., Payne, A. J., Rippin, D. M., Ross, N., and Siegert, M. J.: Sensitivity of the Weddell Sea sector ice streams to sub-shelf melting and surface accumulation, The Cryosphere, 8, 2119–2134, doi:10.5194/tc-8-2119-2014, 2014.

Young, D. A., Wright, A. P., Roberts, J. L., Warner, R. C., Young, N. W., Greenbaum, J. S., Schroeder, D. M., Holt, J. W., Sugden, D. E., Blankenship, D. D., van Ommen, T. D., and Siegert, M. J.: A dynamic early East Antarctic Ice Sheet suggested by ice-covered fjord landscapes, Nature, 474, 72–75, 2011.

# 8

# Neotectonics, flooding patterns and landscape evolution in southern Amazonia

**U. Lombardo**

Universität Bern, Geographisches Institut, Hallerstrasse 12, 3012 Bern, Switzerland

*Correspondence to:* U. Lombardo (lombardo@giub.unibe.ch)

**Abstract.** The paper examines the role of neotectonic activity in the evolution of the landscape in southern Amazonia during the Holocene. It uses both new and published data based on the analysis of remote sensing imagery and extensive field work in the Llanos de Moxos, Bolivian Amazon. The study of the region's modern and palaeorivers, ria lakes, palaeosols and topography provides a strong case in favour of the thesis that the northern part of the Llanos de Moxos constitutes the southern margin of the Fitzcarrald Arch and that it has experienced uplift during the Holocene. The paper assesses the extent and timing of the neotectonic activity in light of the new data and reconstructs the evolution of the landscape since the late Pleistocene. The evidence suggests that at least two uplift events took place: a first uplift in the late Pleistocene, which caused the formation of Lake Oceano, and a second uplift during the mid-Holocene, which formed Lake Rogaguado. These two uplifts appear to be linked to the knickpoints observed close to the towns of Guayaramerín and Puerto Siles respectively. The backwater effect due to these uplifts transformed the region's major rivers in seasonal ria lakes, causing the deposition of thick organic clay layers along the Beni, Mamoré and Madre de Dios river banks. I argue that neotectonic episodes could have dramatically changed the drainage of the Llanos de Moxos, determining its flooding regime, soil properties and forest–savannah ecotone. These results stress the need for geomorphologists, palaeo-ecologists and archaeologists to take into account neotectonics when reconstructing the region's past.

## 1 Introduction

Palaeo-ecological reconstructions are a fundamental step in order to understand the origins of modern landscapes and the potential changes they could undergo due to climate change. Understanding how a particular region or ecosystem responded to past human disturbances and changes in past climate can tell us much about its resilience (Tinner and Ammann, 2005; Bush and Silman, 2007; Dearing, 2008). At the same time, understanding the processes behind the formation of modern landscapes and ecosystems can help assess to what extent these are "natural" or anthropogenic, eventually contributing to better informed conservation policies (Angermeier, 2000; Willis and Birks, 2006; Vegas-Vilarrúbia et al., 2011). These considerations are particularly true for the Llanos de Moxos (LM) (Fig. 1), which cover most of the savannah of southern Amazonia. Since the beginning of the Holocene this region has experienced both climate and

human induced environmental changes (Mayle et al., 2000; Whitney et al., 2011; Urrego et al., 2013; Lombardo et al., 2013a, b).

The LM, located between the central Andes and the Brazilian Shield, is one of the largest seasonally flooded savannahs of South America. This region covers most of the southern part of the Amazon Basin and represents about 10 % of the wetlands of South America (Junk, 2013). The forest–savannah ecotone is largely determined by seasonal floods (Mayle et al., 2007). The patchwork of savannahs and forests that covers the LM is a key element for the survival of its rich biodiversity, including rare and threatened species (Herzog et al., 2012; Wallace et al., 2013). The recent designation of three new protected areas has made the LM the world's largest Ramsar site (http://www.worldwildlife.org/press-releases/bolivia-designates-world-s-largest-protected-wetland). In addition to its biodiversity, the LM is also of great palaeo-

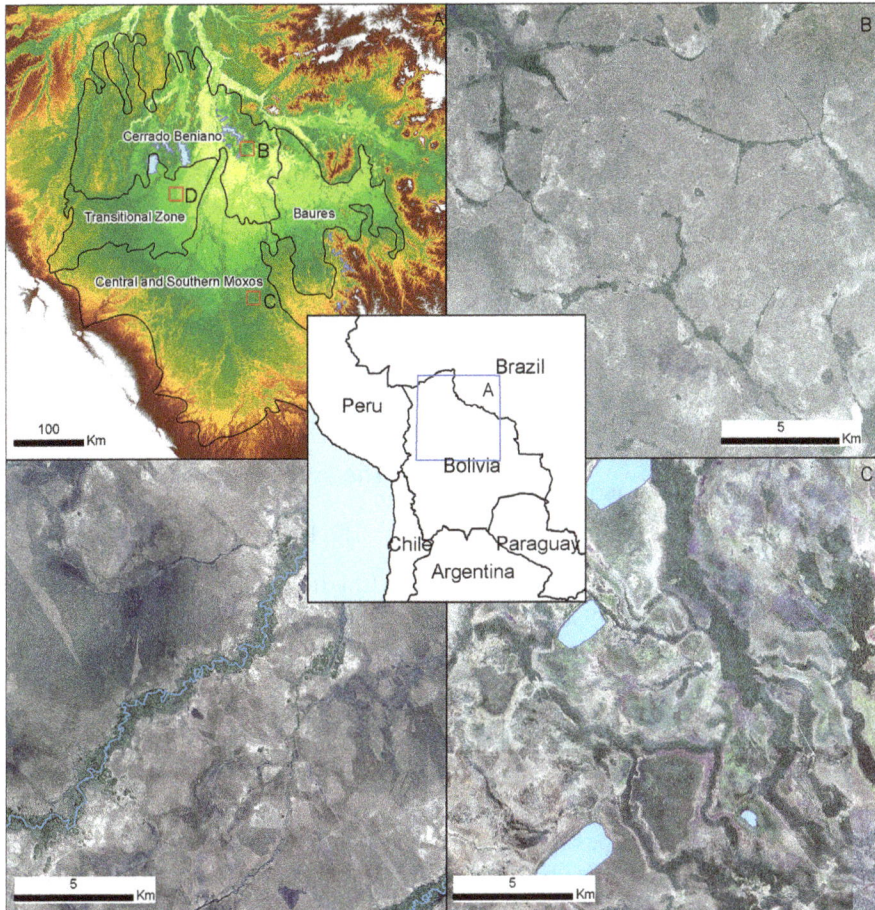

**Figure 1.** Geoecological subregions of the LM. (**a**) Map of the LM's subregions (based on Langstroth, 2011). (**b**) Satellite image of the Cerrado Beniano, a hilly landscape where forests grow in depressions. (**c**) Satellite image of the central and southern Moxos, a flat landscape where forests grow on elevated fluvial levees. (**d**) Satellite image of the transitional zone, where forests are relegated to the floodplains of underfit rivers.

ecological importance, as this region constitutes the southern border of the Amazonian rainforest, hence a preferential area to study past forest–savannah dynamics (Mayle et al., 2000). Inundation patterns, or flood pulses, have important effects on the biogeochemistry and the ecology of floodplain ecosystems, as they determine the occurrence and distribution of plants and animals, they affect primary and secondary production and influence decomposition and nutrient cycles in water and soils (Junk et al., 1989; Junk, 2013). Nowadays, the LM is constituted by different ecoregions which are characterized by specific vegetation assemblages which depend on the interplay between precipitation, drainage and soils (Navarro, 2011).

Recent research shows that the LM was first inhabited by hunter-gatherers at the beginning of the Holocene (Lombardo et al., 2013b). Its landscape has since then been significantly changed during the late Holocene by the so-called "Earth-mover" societies, who built an extensive array of earthworks (Erickson, 2008; Mann, 2008; Lombardo and Prümers, 2010; Lombardo et al., 2011b). We know very little about the way

early hunter-gatherers dealt with the seasonal floods during the early to mid-Holocene, but evidence suggests that they had to abandon the LM around 4000 yr BP (before present) because of a large-scale environmental change, which also caused the burial of the early archaeological sites with fluvial sediments (Lombardo et al., 2013b). Inundation patterns also played an important role in shaping the way in which the late Holocene "Earthmovers" adapted to their local environment and reshaped the landscape, determining the spatial distribution of different earthworks (Lombardo et al., 2011b); especially in order to improve the drainage of agricultural land (Lombardo et al., 2011a, 2012).

Therefore, the reconstruction of past environments, landscapes and human–environment interactions in southern Amazonia throughout the Holocene greatly depends on our ability to reconstruct past inundation patterns.

Up to now, with the exception of a recent study which reconstructs 6 kyr of palaeovegetation in the Baures subregion (Fig. 1) (Carson et al., 2014), palaeo-ecological reconstructions in the Bolivian Amazon have been mostly based on pollen and charcoal archives from lakes located outside the margins of the LM (Mayle et al., 2000; Burbridge et al., 2004; Whitney et al., 2011; Urrego et al., 2013). These archives have been interpreted as the result of changes in climate and pre-Columbian human impacts, implicitly assuming that other factors, such as neotectonics, did not play an important role in the evolution of the landscape during the Holocene. However, several authors have stressed that the LM shows manifold evidence of neotectonic activity during this period (Hanagarth, 1993). In particular, the existence of hundreds of rectangular and oriented lakes have been attributed to the propagation to the surface of bedrock faults (Plafker, 1964, 1974; Allenby, 1988); the formation of ria lakes in the northern part of the LM and the changes in the meandering behaviour of the Mamoré River at Puerto Siles have been associated with uplift and tilting movements in the northern LM (Hanagarth, 1993; Dumont and Fournier, 1994) and the counter clockwise movement of the Beni and Grande rivers has been associated with the tilting of the basement (Dumont and Fournier, 1994; Dumont, 1996). A ria lake forms when the lower part of a tributary is ponded at the margin of the trunk river because of a quick depositional event of the main river, a raise of the base level due to faulting or tilting, or because of regional subsidence (Schumm et al., 2002).

These studies suggest that neotectonics had a significant impact on the formation of the modern landscape of the LM and, hence, on past inundation patters. However, because of the lack of chronological data, digital elevation models (DEMs), remote sensing imagery, and the difficulty of performing fieldwork in the region at the time of these early studies, published work to date does not provide quantitative measures of the neotectonic movements and do not define a chronological framework for the tectonic activity. Furthermore, previous observations based on remote sensing imagery have not been corroborated by field evidence. The extent and the chronology of neotectonic activity are extremely important in order to assess the potential impact that neotectonics could have had on the evolution of southern Amazonia. Neotectonics could have brought about changes in the drainage of the LM during the Holocene which, in turn, could have altered the flooding regime, vegetation assemblages, landscape evolution, and human–environment interactions in the region.

In the present paper, previous studies are revisited in light of new remote sensing imagery, digital elevation models and extensive field work. Based on this evidence, I propose a general framework to explain the contribution of neotectonics to the evolution of the landscape and inundations patterns in southern Amazonia. The paper describes the physical setting of the LM and the geomorphological evidence that supports the existence of neotectonic activity. The paper is organized in three main sections. The first part, Sect. 2, describes the study area, focusing on three aspects: the general landscape, the modern flooding patterns and the geology. The second part, Sect. 4, assesses the evidence of neotectonic activity. This is subdivided into two parts: rivers/palaeorivers and lakes. Here, the first paragraphs review the existing literature; the following paragraphs present new data and the results of the study. The third part, Sect. 5, addresses the following questions:

1. What could be the general mechanism behind neotectonics in southern Amazonia?

2. What was the extent and timing of neotectonic activity?

3. How did neotectonics affect the LM hydrology during the Holocene?

## 2 The study area

### 2.1 The LM and its subregions

The LM, in southern Amazonia, is a large seasonally flooded savannah criss-crossed by strips of forest. However, important differences exist between its subregions (Fig. 1). The northern LM, also called Cerrado Beniano, is covered by reddish, relatively well drained lateritic crusts, which host Cerrado-like savannahs (Langstroth, 2011; Navarro, 2011). Forests grow in the lower part of the landscape, mostly in the small valleys between lateritic hills where sediment accumulation provides enough humidity and room for tree roots to grow (Fig. 1b). Compared to the rest of the LM, which are almost totally flat, this region is hillier and shows noticeable topographic changes. As it can be seen in the DEM in Fig. 1a, most of the Cerrado Beniano is well above the level of the seasonal floods. Hence, here, the forest–savannah ecotone is determined primarily by soil properties and local precipitation, and is relatively independent from the LM's flood regime. Some of the largest lakes in the LM are located here.

In the central and southern LM (Fig. 1c) the topographic control on the savannah–forest assemblage is inverted with respect to the northern LM, in that forested areas tend to be concentrated on high ground rather than valleys. The central and southern parts of the LM are an extremely flat floodplain, mostly covered by seasonally flooded savannahs, which, in general, are fluvial backswamps. Savannahs are interspersed by strips of forests, which mainly grow on the fluvial deposits that remain above the level of the seasonal floods (Mayle et al., 2007). Most of the soils are relatively fertile because they generally form on recently deposited, clay-rich, Andean sediments. The most common soil types are Gleysols in the savannahs, Fluvisols in the recent fluvial levees and Luvisols and Cambisols in the older palaeolevees (Langstroth, 1996; Boixadera et al., 2003; Lombardo et al., 2014). Here soil properties and the forest–savannah ecotone along fluvial splays and levees are primarily determined by the inundation patterns. This is the part of the LM where hundreds of rectangular and oriented lakes formed.

In the western LM, the Cerrado Beniano is separated from the central LM by a transitional zone (Fig. 1d). This encompasses the savannahs north and west of Santa Ana de Yacuma. This area is very flat and crossed by underfit rivers. These rivers flow on ancient courses of the Beni River (Dumont and Fournier, 1994) and do not overflow during the rainy season (Walker, 2004). The savannah is seasonally flooded, but floods are caused by local precipitation alone, with no contribution from river overflow. The lack of fluvial sediment deposition, combined with strong redoximorphic conditions, results in highly hydromorphic soils, very poor in nutrients, often with very low pH values and toxic levels of exchangeable aluminium (Navarro, 2011; Lombardo et al., 2013a).

The northeastern part of the LM, the Baures subregion, is characterized by a very poorly drained savannah dotted with several forested areas. These forests grow on 2–3 m high elevated platforms that constitute outcrops of tertiary rocks (Pitfield and Power, 1987) (Fig. 2b). This is the less accessible and therefore less studied area of the LM.

With an average slope of 0.014 %, the LM is extremely flat and subject to seasonal floods. Seasonal floods and backswamp sedimentation along the basins of the Mamoré and Beni rivers are primarily controlled by El Niño/La Niña cycles (Aalto et al., 2003) and the South Atlantic sea surface temperature anomalies (Ronchail et al., 2005).

Precipitation in the LM range from 1350 to 2450 mm yr$^{-1}$ (yearly average precipitation between 1971 and 2000 at Trinidad; Ronchail and Gallaire, 2006), while the flooded area is in the range of 30 000–92 000 km$^2$ (Hamilton et al., 2004; Melack and Hess, 2011). This shows the high sensitivity of the LM and its flooding regime to relatively small changes in precipitation. Flood dynamics are also influenced by the Mamoré River, which is the base level for most of the rivers in the region (Bourrel and Pouilly, 2004). The duration and extent of the inundations depend on the combination of precipitation in the upper catchment of the Mamoré River, the impeded drainage of local precipitation (Bourrel et al., 2009) and the elevation of the water table (Ronchail et al., 2005; Fan et al., 2013). A lag of 1–3 months is observed between the time of maximum precipitation in the upper catchment of the Mamoré and the peak in floodplain inundation (Hamilton et al., 2004). This results from the movement of river runoff from upland catchments to the floodplains, as well as from the delay in the drainage of inundated areas to the rivers (Hamilton et al., 2004). On an interannual basis, consecutive years with large inundations are caused by groundwater storage (Ronchail et al., 2005).

## 2.2 The geology of the Llanos de Moxos

The Andes–Amazon foreland basin system is not a typical foreland basin system (DeCelles and Giles, 1996); it is divided into two different basins by the uplifting Fitzcarrald Arch, a 600 m high bulge (Espurt et al., 2007, 2010). It has been hypothesized that the Fitzcarrald Arch is the projection to the surface of a buoyant flat slab resulting from the subduction of the Nazca Ridge (Espurt et al., 2008). The formation of the flat slab about 4 Myr ago would have brought about the back-arc uplift of the Fitzcarrald Arch (Espurt et al., 2007, 2008; Regard et al., 2009). The LM constitutes the foredeep of the southern Amazonian foreland basin. It is located between the central Andes, to the W–SW, and the Brazilian Shield, to the E. The Brazilian Shield dips gently towards the Andes underlying unconsolidated sediments (Plafker, 1964; Hanagarth, 1993). Very little is known about the stratigraphy and thickness of these foreland deposits. Geophysical surveys in the southern part of the LM indicate the presence of thick layers of fluvial sediments covering the Brazilian Shield (Plafker, 1964). The sediments overlying the crystalline bedrock thicken gradually from the margin of the Brazilian Shield towards the west. The thickness of the sediments is more than 5500 m near the Andean foothills and approximately 800 m at a distance of 150 km from the Andes, in the village of Perú (Plafker, 1964) (Fig. 2b). In the northern part of the LM, recent surveys have shown the presence of metamorphic rock at 165 m below Santa Ana de Yacuma and at 65 m below the town of Magdalena (UNASBVI-JICA, 2009). A 200 m long and 50 m high rocky exposure, belonging to the Brazilian Shield, outcrops in the location of El Cerro, 60 km north of Santa Ana de Yacuma. Several authors have identified an uplifting region in northern Beni; the so-called Linea Bala–Rogaguado in Fig. 2b (Allenby, 1988; Dumont, 1996; Dumont and Fournier, 1994; Plafker, 1964; Hanagarth, 1993). The Linea Bala–Rogaguado, also called Reyes-Puerto Siles axis (Dumont, 1996), has been interpreted as a SW–NE faulting system which represent the south-western extension of the Tapajó fault in central Brazil (Dumont and Fournier, 1994). More recently, it has been proposed that the Puerto Siles uplift is part of the SE–NW oriented Andean forebulge axis (Aalto et al., 2003; Roddaz et al., 2006) (Fig. 2b).

## 3  Methods

The study examines the evidence of neotectonic activity in southern Amazonia and its impact on flooding patterns and landscape evolution during the Holocene. It focuses on the analysis of modern and palaeorivers, ria lakes, palaeosols and landscape topography. For this it combines the analysis of remote sensing images, extensive fieldwork in the LM and a review of existing studies. ArcGis 10.1 has been used to produce the topographic maps shown in Figs. 1a, 2, 3a, 6a and 9 and the topographic profiles shown in Figs. 5, 6a and 10; these are based on the hole-filled Shuttle Radar Topography Mission (SRTM) imagery, where elevation values for water bodies are interpolated (Jarvis et al., 2008). Images of landscape features used in Figs. 1b–d, 3b–d, 7a and 10b are retrieved from Google Earth. Images used in Figs. 4 and 10a

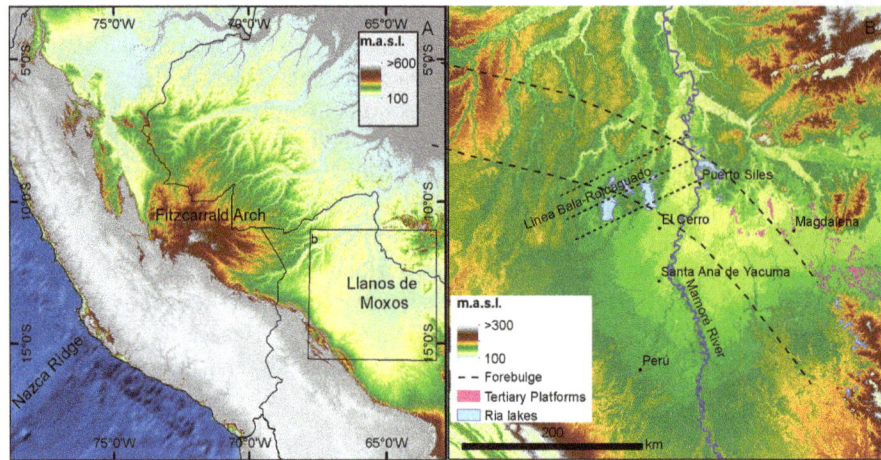

**Figure 2.** Geological setting of the LM. (**a**) Topographic map of central western South America showing how the south-eastern part of the Fitzcarrald Arch constitutes the LM's north-western border. This, together with the Brazilian Shield in the north-east, forms a continuous barrier which impedes the drainage of the LM. (**b**) Topographic map of the LM showing the location of the Linea Bala–Rogaguado (as in Dumont and Fournier, 1994) and the Andean forebulge (as proposed by Aalto et al., 2003).

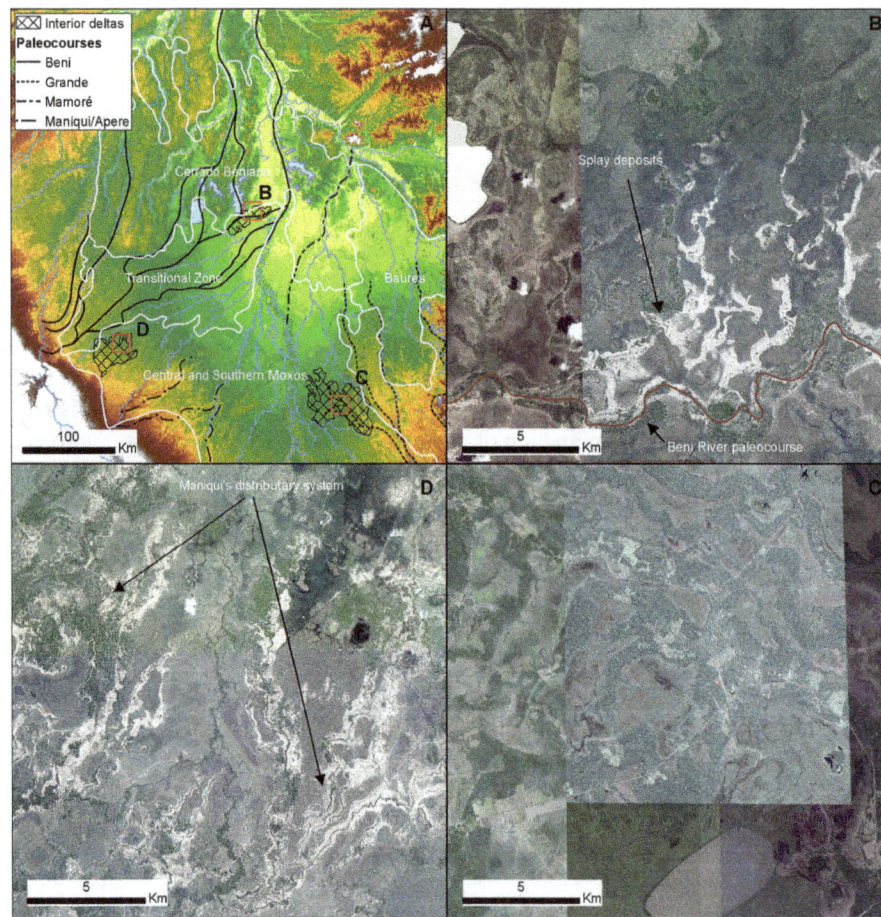

**Figure 3.** Map of rivers and palaeorivers in the LM. (**b**) An example of crevasse splays at the distal part of a palaeocourse of the Beni River. (**c, d**) Interior deltas formed by repeated avulsions of the Grande and Maniqui rivers, respectively.

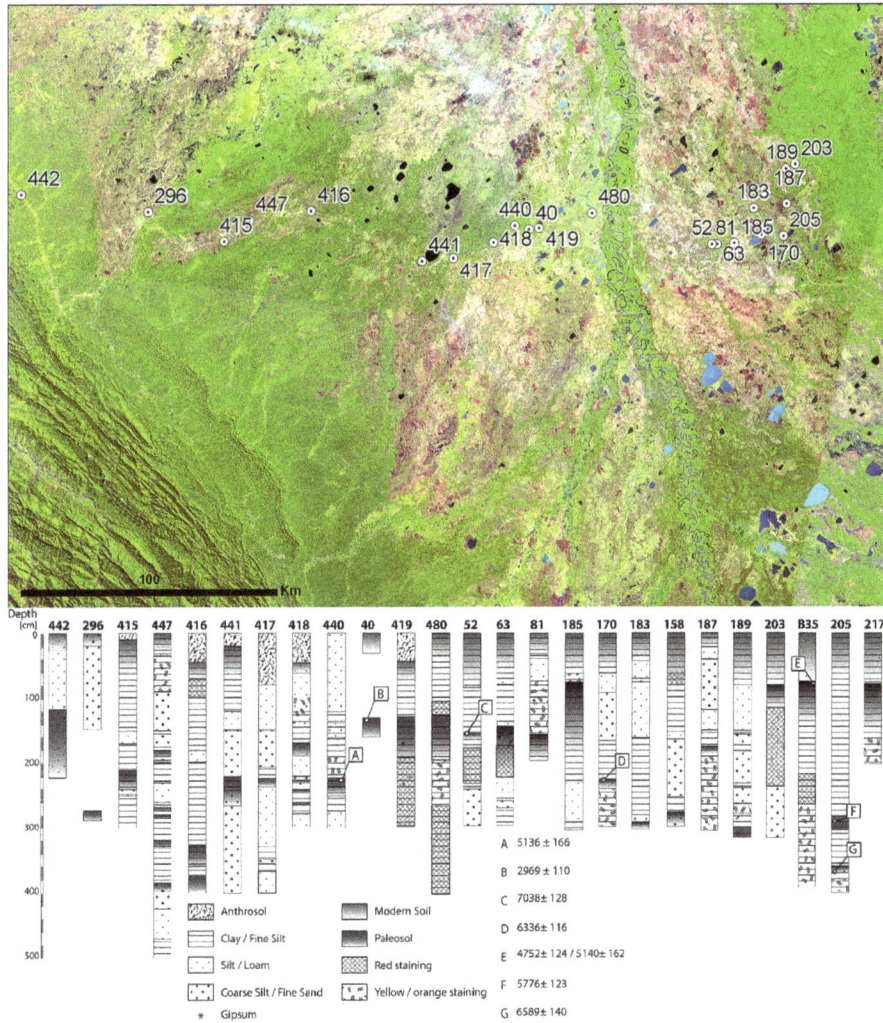

**Figure 4.** Cored profiles along a 300 km long transect crossing the central and southern LM. Palaeosols are always found below alluvia. Sequences of several organic horizons separated by alluvial deposits (for example core 447) indicate frequent avulsions. Radiocarbon ages indicate that the central and southern LM were covered with fluvial sediments during the mid- to late Holocene. AMS (applied mass spectrometry) radiocarbon ages of palaeosols in profile 40 (uncalibrated age 2900 ± 35 yr BP) and 440 (uncalibrated age 4520 ± 40 yr BP) were measured at the Poznan Radiocarbon Laboratory. The rest of the radiocarbon ages have been taken from Lombardo et al. (2012).

are Landsat ETM (Enhanced Thematic Mapper) and have been downloaded from GLOVIS (http://glovis.usgs.gov/). The Mamoré River has been manually digitalized over the hole-filled SRTM. Elevation points within the Mamoré River channel have been extracted from the hole-filled SRTM at each vertex of the digitalized line and plotted against distance downstream (Fig. 5b) and against their latitude (Fig. 5a, c). The latter facilitates the direct comparison between the slope and the morphology of the river. River reaches have been identified on the basis of the remote sensing imagery and after a visual analysis of the river geometry and slope. The average slope of each reach is represented as a line (Fig. 5a, c). Sinuosity was measured for each river reach (1–9 in Fig. 5) as the river length divided by the valley length. In order to investigate the origin and time of formation of

Lake Rogaguado, the lake has been cored in four places; however, only the data from two of the cores is presented in this paper, cores 398 and 397, as the other two cores did not reach the bottom of the lacustrine sediments. Lake cores have been taken with a gravimetric corer. In order to reach the sediments below the lacustrine phase, location 398 has been cored deeper using a Livingstone corer. The radiocarbon age from the base of the lacustrine phase of core 378 comes from a fragment of wood; it has been calibrated with Calib 7.0 (http://calib.qub.ac.uk/calib/calib.html) using the SHcal13 calibration curve (Hogg et al., 2013). In order to assess the sedimentary regime across the central and southern LM, a stratigraphic transect has been made from the Andes to the eastern LM (Fig. 4). Stratigraphic cores along this transect have been taken with a Waker motor vibracorer and described in the field. The description and radiocarbon ages of the cores taken east of the Mamoré River are based on Lombardo et al. (2012).

## 4  The evidence of neotectonic activity in southern Amazonia

### 4.1  Rivers and palaeorivers

Tropical alluvial rivers can change in response to different factors, notably changes in climate, neotectonics or their intrinsic evolution over time (Schumm et al., 2002; Bridge, 2009; Burbank and Anderson, 2011; Gupta, 2011). In particular, river avulsions take place when the main channel becomes infilled with sediments, triggering the diversion of the river into a new course (Slingerland and Smith, 2004; Buehler et al., 2011). River avulsions are the main mechanism behind the formation of distributary fluvial systems (DFS), which fill alluvial basins (Hajek and Edmonds, 2014; Weissmann et al., 2010; Hartley et al., 2010). High aggradation rates in the main river channel can result from a variety of causes such as (i) an increase in the sedimentary load of the river; (ii) an increase in the peak water discharge of the river or (iii) a decrease in the channel gradient, which can be caused by a rise of the base level or downstream tectonic uplift (Slingerland and Smith, 2004). Therefore, the interpretation of environmental and palaeo-environmental fluvial forms and palaeoforms is not always straightforward. Avulsive processes and the formation of DFS within the LM have been described in the case of the Beni River (Plafker, 1964, 1974; Hanagarth, 1993; Dumont, 1996; Hartley et al., 2010) the Grande River (Hanagarth, 1993; Dumont, 1996; Lombardo et al., 2012; Plotzki, 2013) and the Mamoré River (Plotzki et al., 2011, 2013) (Fig. 3a) but the mechanisms behind these avulsions are controversial. The northward migration of the Beni River deflection point has been associated with a fault connected to the Andean foothill margin (Dumont, 1996; Plafker, 1964; Hanagarth, 1993; Allenby, 1988). Based on the assumption that the Grande River changed its course following a counter clockwise movement, the tilting of the basement, and the consequent uplift of its southeastern margin, has been considered responsible for its avulsions (Hanagarth, 1993; Dumont, 1996). More recently, however, it has been suggested that the megafan of Grande River did not follow a counter clockwise movement (Plotzki, 2013). The study of an interior delta/sedimentary lobe at the distal part of one of the Grande River palaeocourses, seems to indicate that the Grande River avulsive phase may have been triggered by an increase in river discharge or sedimentary load that could have been brought about by a mid- to late Holocene climate shift towards wetter conditions (Lombardo et al., 2012; Plotzki et al., 2013). Similarly, the avulsion of the Mamoré seems to be related to an increase in discharge probably caused by a change in climate (Plotzki et al., 2013). The interior delta formed by the Maniqui River (Fig. 3d) has

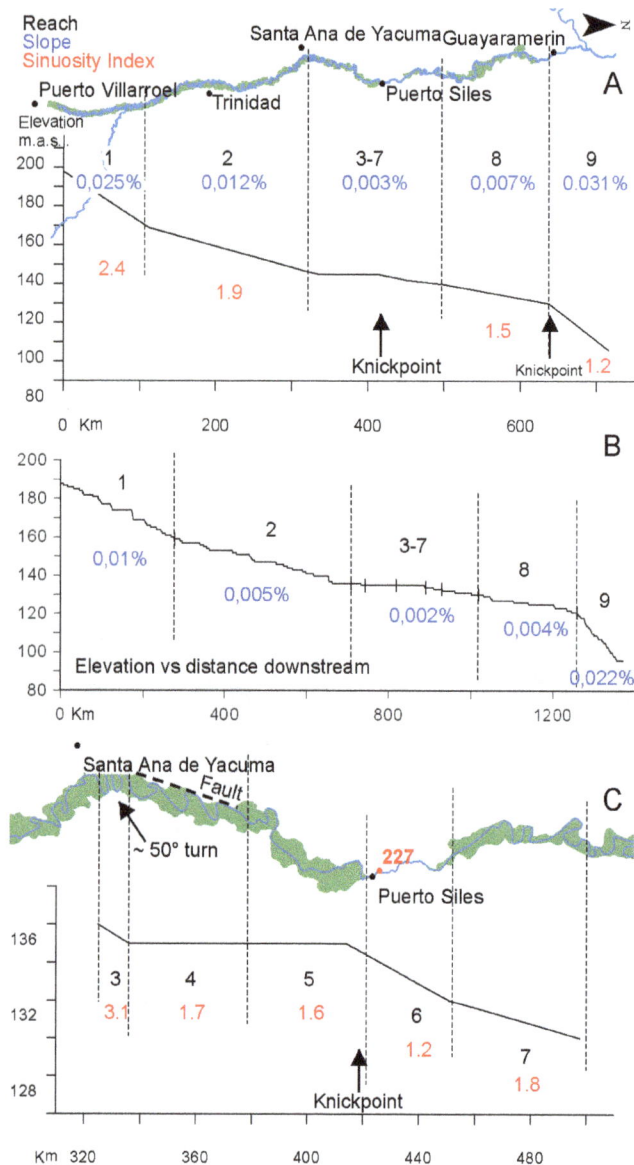

**Figure 5.** Geometric characteristics of the Mamoré River. (**a**) Valley river slope and sinuosity index (SI) of the Mamoré between Puerto Villarroel and Guayaramerín are plotted against each other. Here, the valley slope changes twice, indicating the presence of knickpoints: one in Puerto Siles and the other one close to Guayaramerín. (**b**) The slope is plotted against river length, showing the same pattern as in (**a**). In (**c**), a more detailed map of the section between Santa Ana de Yacuma and 100 km downstream of Puerto Siles shows a series of compressed meanders at Santa Ana, followed by a 50° turn and three asymmetric meanders. After Puerto Siles, the SI declines abruptly and the gallery forest disappears. Point 227 is where photos (**c**) and (**b**) in Fig. 7 were taken.

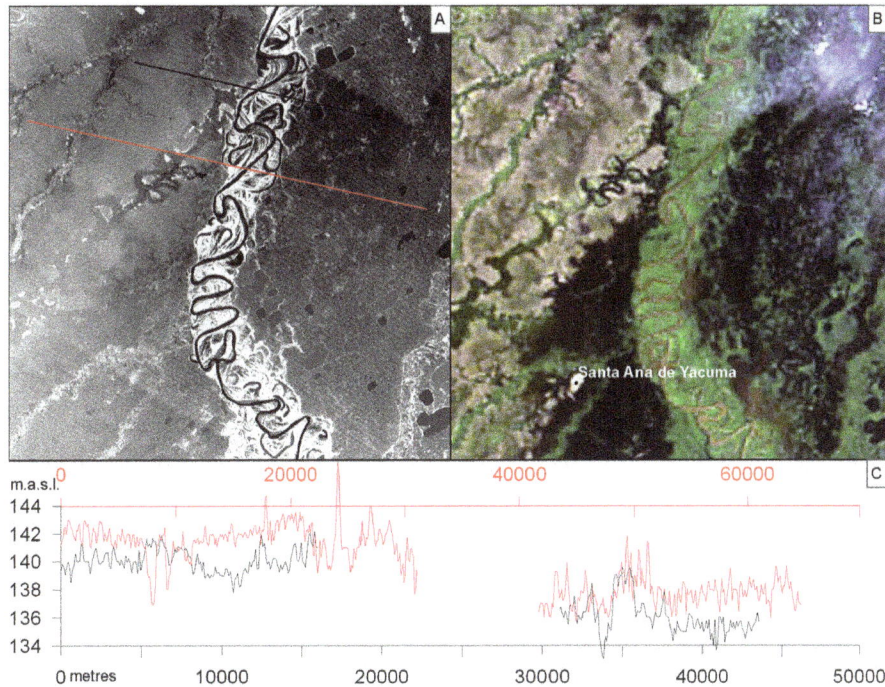

**Figure 6. (b)** Location of the topographic profiles shown in **(c)**. **(b)** MODIS image of the flooded areas around Santa Ana de Yacuma in 2007 (black areas). **(c)** Topographic profiles across the asymmetric meanders of reach 4 (Fig. 5c), indicating that the western side of the river (the dry area shown in **b**) is about 5 m above the level of the eastern side of the Mamoré. This further supports the hypothesis of the presence of an active fault as indicated in Fig. 5c.

been described by Hanagart and Sarmiento (1990); they observed that the formation of the Maniqui megafan was caused by a reduction in the slope, which induced frequent avulsions.

A situation similar to that of the Maniqui and Grande river deltas (Fig. 3c) is found in the distal part of a palaeocourse of the Beni River (Fig. 3b), where numerous crevasse splays formed. Although these fluvial forms have received little attention in the literature, their location on the southern border of the Cerrado Beniano (the Beni crevasse splays) and the transitional zone (the Maniqui River megafan), suggests that their formation could respond to a change in slope, as suggested by Hanagart and Sarmiento (1990) for the Maniqui, which was probably caused by a downstream uplift.

The central and southern Llanos de Moxos are covered by many palaeolevees belonging to DFS. This is characteristic of actively subsiding basins (Hartley et al., 2010; Latrubesse et al., 2010; Weissmann et al., 2010, 2013; Rossetti et al., 2012). In order to study these DFS, 25 stratigraphic cores were retrieved along a 300 km line crossing the central LM (Fig. 4). All the profiles reveal intercalations of clays, loams, silts, fine sands and organic-rich palaeosols. The association of levee sands and silts, splay sands, backwater loams and (organic) clays is characteristic of an avulsive fluvial setting, which is consistent with the formation of megafans in subsiding basins. Radiocarbon dating of palaeosols found below these alluvia indicate depo-

sition during the mid- to late Holocene (Fig. 4). Data from drilling reported in Plafker (1964) reveal that this assemblage of Holocene fluvial sediments is similar to the rest of Quaternary and possibly Tertiary sediments constituting the 800 m of basinal infilling at the site of Perú. Although the evolution of specific rivers at any given time could have been influenced by different factors, the general landscape of the central and southern Llanos de Moxos is that of a typical subsiding basin.

However, there are characteristic features in river patterns that can be associated with neotectonic activity more clearly than avulsive processes. The existence of underlying active tectonic processes can be noted from the analysis of features such as changes in the meandering pattern and the presence of knickpoints (Schumm et al., 2002; Burbank and Anderson, 2011). Evidence of neotectonic activity in the LM can be inferred from the analysis of the Mamoré River. The sharp reduction in the sinuosity of the Mamoré River north of Puerto Siles (Fig. 5) has been considered direct evidence of uplifting along the Linea Bala–Rogaguado (Dumont and Fournier, 1994). The existence of this uplift is consistent with the presence of several rapids located downstream of Puerto Siles (Dumont and Fournier, 1994). In the lower part of the Mamoré and the Beni rivers, rapids are caused by the outcropping of pre-Cambrian rocks belonging to the Brazilian Shield (Hanagarth, 1993).

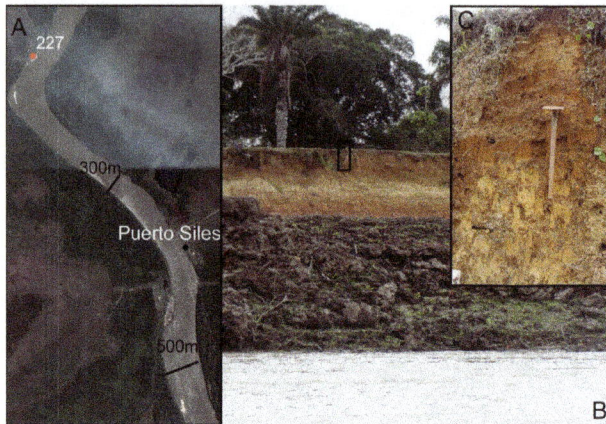

**Figure 7.** Changes in the geomorphic features of the Mamoré River at Puerto Siles. **(a)** Google Earth image showing the change in Mamoré's width and the outcropping of lateritic crusts; **(b)** The Mamoré River bank at point 227: a lateritic crust covered by old fluvial sediments; **(c)** Photograph of a well developed Luvisol at the top of the fluvial levee at location 227.

**Figure 8.** An erosional feature incising the savannahs of the transitional zone along the road from Santa Ana de Yacuma to lake Rogaguado (the dry area shown in Fig. 6b), indicating that these savannahs are established in uplands subject to erosion.

The presence of the Puerto Siles uplift is corroborated in this study by the analysis of the SRTM imagery (Fig. 5). The analysis of the topographic profile of the Mamoré River shows several changes along its course. In its first 400 km (reaches 1–5 in Fig. 5) the Mamoré River behaves as a free meandering river, with its slope and sinuosity decreasing downstream. This behaviour is interrupted at Puerto Siles, where the lowest sinuosity value is associated with a clear increase in slope and to the appearance of lateritic crusts within the river channel, causing the formation of rapids. The change experienced by the Mamoré River at Puerto Siles is also noted in the river's topography, which goes from concave to convex at this point. This change in Puerto Siles is preceded by three compressed meanders followed by three asymmetric meanders (reaches 3 and 4 in Fig. 5). A clear change in the direction of the river, with a clockwise turn of about 50°, is associated with the compressed meanders. Compressed meanders indicate a downstream increase of the river bank's resistance to erosion, often the result of the appearance of bedrock in the channel following localized uplift (Holbrook and Schumm, 1999). This situation is similar to what can be seen in the rest of the Amazon Basin, where discharge characteristics and bank stability play an important role in channel pattern variations (Baker, 1977). The topographic profiles across reach 4 of the Mamoré River show that the western side of the river is about 5 m higher than the eastern side (Fig. 6a). This difference in elevation between the two sides dramatically affects the flooding regime; the savannah in the north of Santa Ana de Yacuma (the transitional zone) is hardly ever flooded by the Mamoré outflow, whilst the savannah in the eastern side is severely flooded every year (Fig. 6b). Therefore, the asymmetric meanders and the 50° change in direction can be interpreted as the result

of a normal fault with its footwall on the western side of the river.

In addition, the presence of underfit rivers in the transitional zone suggests that this subregion is uplifting. This uplift is evident in the formation of a new drainage network constituted by small creeks, which are a common element of the landscape easily seen in the field. Figure 8 shows one of these erosive features, which are formed when rain water travels from the waterlogged savannah to the underfit rivers.

The topographic profile of the Mamoré shows how, at Puerto Siles, the meander belt that borders the river for the first 400 km is replaced by two irregular forested levees. No oxbows are present here. A 40 % decrease in the channel's width is also observed here. The changes at Puerto Siles suggest that the Mamoré experiments acceleration and deepening of the channel. These changes increase the capacity of the river to transport sediment and, in turn, this changes its behaviour: from depositing sediments it now begins to erode its bed. A further confirmation of this change is that downstream from Puerto Siles the Mamoré's river banks are made by thick lateritic crusts covered by old fluvial levees, with no sign of recent fluvial deposition (Fig. 7b, c). Here, well-developed Luvisols are found on top of the levees. About 200 km downstream, close to Guayaramerín, a second knickpoint appears. After this second knickpoint, the Mamoré reaches its steepest slope (reach 9 in Fig. 5).

The analysis of the SRTM indicates that the general downslope topography of the LM is interrupted at Puerto Siles where the terrain becomes hillier. Despite the fact that these hills significantly change the slope of the LM, the river network seems to be unaffected by them. In fact, as we can see in the SRTM images shown in Fig. 9, all the rivers maintain their northward convergent path, cutting through these

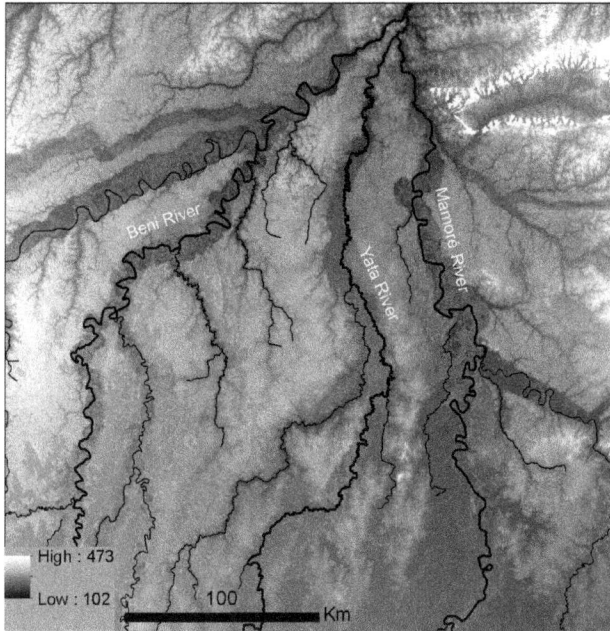

**Figure 9.** SRTM image showing an example of a river network in the Cerrado Beniano. All rivers flow north, against the slope of the general terrain, suggesting that the development of the drainage network preceded the uplift that crated the hilly landscape.

hills; no rivers flow south. This setting suggests that the courses of these rivers were established before the onset of the uplift, and that the rivers have overcome this obstacle by incising the uplifting rocks and sediments. This is consistent with the hypothesis that these hills constitute the southern margin of the Fitzcarrald Arch, which started uplifting about 4 Myr ago (Espurt et al., 2007). Not all the valleys crossing the uplifted area have been active at the same time, as shown by the palaeocourses of the Beni River: the Beni occupied at least two other valleys (now occupied by the Yata and Tapado rivers) before establishing its modern course (Dumont and Fournier, 1994).

## 4.2 Lakes

There are more than 800 lakes in the LM, most of which are rectangular and oriented. Until recently, the most accepted hypothesis behind the formation of these rectangular and oriented lakes of the LM has been tectonic (Plafker, 1964; Allenby, 1988). According to Plafker, the lakes' rectangular shapes results from the propagation of bedrock fractures through unconsolidated sediments and the sinking of basement blocks. However, the coring of three of these lakes has shown that tectonics cannot explain their formation, as a horizontal palaeosol is found at the same depth both outside and inside the lakes (Lombardo and Veit, 2014). Given the available data, the most likely mechanism behind the formation of the rectangular and oriented lakes seems to be the erosive action of waves on the lakes' shores due to constant winds

(Lombardo and Veit, 2014). Apart from the rectangular and oriented lakes, a second group of lakes, found mainly on the eastern and northern borders of the LM, are the ria lakes. Ria lakes are a valuable source of information to reconstruct past environments, sedimentary histories and tectonic processes (Behling et al., 2001; Irion et al., 2006, 2011). They resemble water reservoirs resulting from a dam, where a pre-existing valley is flooded and transformed into a lake. In the Bolivian Amazon, ria lakes have been reported in the southeastern and in the northern parts of the LM. In the southeastern side they have been formed by the rapid deposition of sediments by the Grande River, which dammed fluvial valleys previously eroded in the Brazilian Shield (Dumont and Fournier, 1994). Ria lakes that have been described in the northern LM are connected to the Linea Bala-Rogaguado (Hanagarth, 1993; Dumont and Fournier, 1994). Here there are two types of lakes. The first type, which includes Lago Largo and Agua Clara (Fig. 10), is the typical ria lake which results from a valley flooded by damming; hence, similar to those formed by the former Grande River. The second type, such as Rogaguado and Ginebra, is constituted by lakes that have formed within a valley that became flooded because of tilting (Dumont and Guyot, 1993; Dumont and Fournier, 1994). Ria lakes of tectonic origin are not limited to the LM; they have also been described in the Marañon and Ucayali basins (Räsänen et al., 1987; Dumont, 1993). Many other ria lakes are found along the Negro, Xingú and Tapajó rivers in the Brazilian Amazonia. These seem to have formed because of the increased river incision caused by the drop of the ocean water level during the Last Glacial Maximum (Irion et al., 2011), but it has been suggested that tectonic movements could have also played a role. (Latrubesse and Franzinelli, 2002).

The counter clockwise shift of the Beni River (Dumont, 1996) has left several palaeofloodplains, one of which is nowadays occupied by Lake Rogaguado. If we assume that, before the formation of the lake, when the Beni River flowed through the Rogaguado Valley its slope was relatively constant, the northern uplift or tilting that generated the lake, would have, therefore, also changed the slope of the original valley. The measurement of this change could provide constraints on the extent of the tectonic movement. Figure 10 shows the the original slope of Rogaguado's floodplain and the amount of vertical movement which created the lake. In order to confirm the mechanism at the origin of these lakes and constrain the timing of their formation, Lake Rogaguado was cored at several locations. Figure 11 shows the two cores retrieved from Lake Rogaguado that reached the bottom of the lacustrine sediments. The bottom part of core 398 includes the terrestrial phase that preceded the formation of the lake. This is a palaeosol characterized by abundant red oxides at its base, indicating recurrent wet and dry periods. The amount of organic matter increases towards the top of the terrestrial phase, which probably corresponds to a fossil Ah horizon. This terrestrial phase seems to correspond

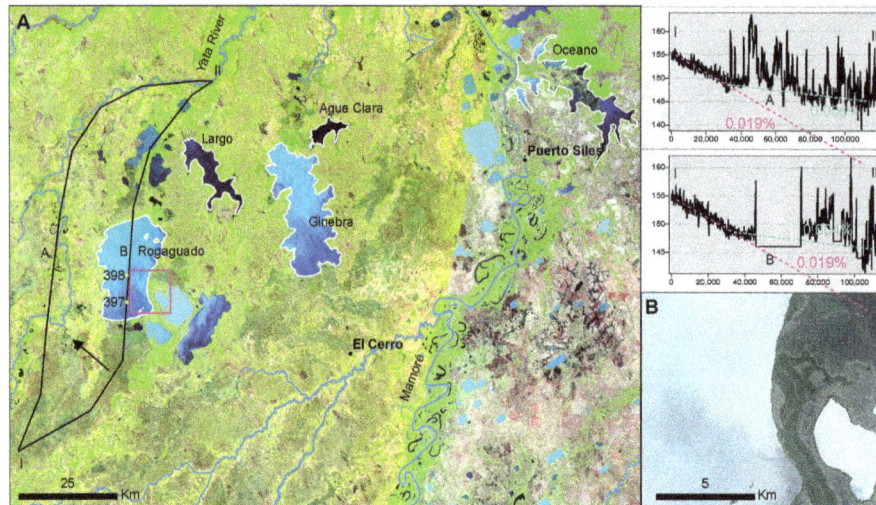

**Figure 10.** Landsat image of ria lakes in the Cerrado Beniano. (**a**) Yellow points in Lake Rogaguado indicate the location of the cores shown in Fig. 11. Topographic profiles from I to II show a reduction in the slope, coinciding with two 90° turns of the Yata River (profile A) and the southern shore of Lake Rogaguado (profile B). The topographic profile B also shows an inversion of slope north of Lake Rogaguado, which embanked the valley and formed the lake. (**b**) A palaeocourse of the Beni River which has been cut by the lake. Topographic profiles are based on the Hole-filled SRTM version 4 (Jarvis et al., 2008). The image of the palaeo-Beni River in (**b**) is from Google Earth.

to the seasonally flooded backswamps of the ancient Beni River. The remains of the palaeochannel and levees of the ancient Beni River are still visible in the south-eastern margin of the lake (Fig. 10b). The stratigraphic units (as inferred from changes in colour and material) of core 398 correlate with those of core 397, which is located about 6.6 km further south. Figure 11 shows the two cores in their relative position. At the time of the coring, the depth of the water at 397 was 2.4 m. Assuming that the clay layer separating the terrestrial phase from the lacustrine one has, in this core, a similar thickness to that in core 398, the depth of the palaeosol can be estimated as being 3 m below the water level. The slope of the palaeosol below the lake, 0.0045 % (30 cm in 6.6 km), is consistent with the average slope of the valley northward from the lake. This suggests that the lake was formed by an uplift on its northern side, causing the tilting that dammed the valley. However, the depth of the palaeosol with respect to the savannah south of the lake indicates also the presence of a normal fault, with a southern footwall and a throw between 2 and 3 m, which probably defined the lake's southern border. The point where the valley's slope changes is about 8 km south of the lake's shore, coinciding with the sharp change in the direction of the Yata River (black arrow in Fig. 10). This data suggests that the formation of Lake Rogaguado, and probably of Lake Ginebra too, is the product of a combination of tilting and faulting, maybe resulting from an extensional setting related to an upward pulling north of the lake. In core 398, radiocarbon dating of a wood fragment from the bottom of the lacustrine phase yields the age of 5791 ± 155 cal yr BP, while the palaeosol that corresponds to the top of the terrestrial phase

has been dated 10 910 ± 124 cal yr BP. The radiocarbon age of the palaeosol is based only on the soluble fraction, because no humins could be retrieved. Therefore, it should be interpreted as the average age of the palaeosol and not as the age of its burial. The palaeosol is separated from the organic sediments of the lacustrine phase by a 6 cm thick layer of clays and silts. This layer of sediments contains no organic matter, suggesting it was deposited soon after formation of the lake in a quick event. The sediments could have come from the erosion of the recently formed lake shores. This preliminary data suggests that Lake Rogaguado was formed in the mid-Holocene. There is a group of ria lakes that have not yet been described. They are located on the eastern margin of the Mamoré River, north of Puerto Siles (Fig. 10). Lake Oceano is the largest. The damming of the valleys that formed these lakes was likely caused by the infilling of the valleys' bottom with sediments from the Mamoré river. Lake Oceano offers a modern analogue of the formation of ria lakes in the southeastern LM by Grande River.

## 5 Discussion

In the present paper, geomorphological evidence of neotectonics in southern Amazonia is reviewed and corroborated with new data obtained from the analysis of remote sensing imagery and field work. A first attempt to establish the extent and timing of the neotectonic activity is made. Here, the three research questions formulated in the introduction are addressed in light of the new evidence presented.

## 5.1 What could be the general mechanism behind neotectonics in southern Amazonia?

None of the two mechanisms proposed in previous studies as being behind neotectonic activity in the LM, the reactivation of the Tapajó fault in central Brazil (Dumont and Fournier, 1994) and the uplift of the Andean forebulge (Aalto et al., 2003), seem to be able to explain all the geomorphological evidence of neotectonics reviewed here. The reactivation of a supposed south-western extension of the Tapajó fault in central Brazil could explain the existence of a fault system along the Linea Bala–Rogaguado. However, it cannot explain the meandering pattern of the Mamoré River, which responds to uplifts occurring about 100 kilometres south (Fig. 6). Nor can it explain the closeness of the Brazilian Shield to the surface, and the outcropping of it in several locations in the south and south-east of the Linea Bala–Rogaguado. Moreover, the neotectonic setting in the Tapajó region is constituted by two grabens trending NW–SE and NNE–SSW and a set of ENE–WSW strike-slip faults (Costa et al., 2001). It is unclear how these extensional structures in central Brazil could propagate into the normal faulting/tilting system associated with the Bala–Rogaguado region in northern Beni. Likewise, there is little evidence of uplift along the forebulge axis as identified by Aalto et al. (2003) (Fig. 1b). The presence of the Andean forebulge can explain the uplift at Puerto Siles and the consequent change in the behaviour of the Mamoré River. However, it is not compatible with the region's overall fluvial pattern and the topography of the LM. The uplift of the forebulge is very small (a few hundred metres over a distance of several hundred kilometres) and an inherited (preflexural) topographic relief can be more important than the forebulge in determining the drainage divide (Garcia-Castellanos and Cloetingh, 2011). Nevertheless, in models of continental and exorheic basins, the effect of the forebulge on the drainage network is generally evident (Garcia-Castellanos, 2002). In an extremely flat landscape, such as that of the LM, an uplift along the forebulge axes which is able to create the Puerto Siles knickpoint would be visible in the subsequent formation of a divergent drainage network. Yet, along the alleged forebulge axes, rivers, and even recent small drainage creeks, seem to be unaffected by the forebulge. In addition (i) some of the lowest-lying areas of the LM are found right along the forebulge, which is very unusual for an uplifting region and (ii) the alleged forebulge would be orthogonal to the Linea Bala–Rogaguado, therefore unable to explain the formation of the ria lakes.

This study suggests that the uplifting Fitzcarrald Arch is the most likely mechanism behind the observed uplift along the Linea Bala–Rogaguado, the general topography of the LM, the outcropping of the basement in the northern LM, the presence of underfit rivers crossing the savannahs in the transitional zone, the meandering pattern of the Mamoré River and the general shape of the fluvial network. The fault that caused the change in the direction of the

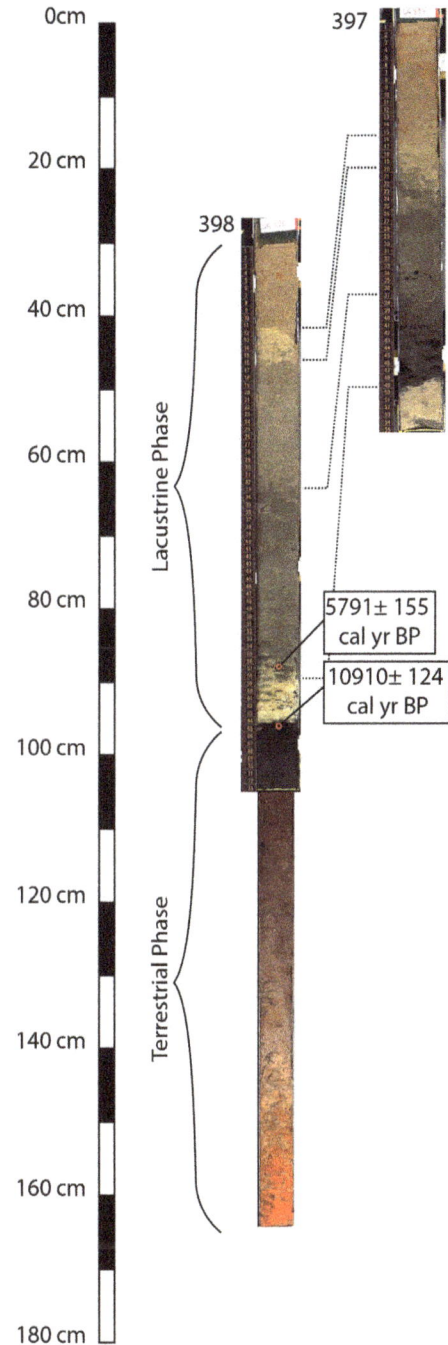

**Figure 11.** Cores from Lake Rogaguado. Core 398 includes a top part, which is the lacustrine phase that followed the formation of the lake, and a bottom part, which is the terrestrial phase which preceded the formation of the Lake. The red mottling at the base of the core indicates alternation of wet and dry periods, suggesting that savannah-like conditions prevailed in the backswamps of the palaeo-Beni River. The age of the beginning of the lacustrine phase is based on an AMS radiocarbon analysis of a wood sample performed at LARA AMS Laboratory in Bern (uncalibrated age $5104 \pm 76$ yr BP). The AMS radiocarbon age of the fossil Ah was measured by DirectAMS using the humate fraction (uncalibrated age $9685 \pm 40$ yr BP).

Mamoré River (Fig. 5, reach 4) and the direction of the Linea Bala–Rogaguado, is consistent with the orientation of the Nazca Ridge, suggesting that the Nazca Ridge subduction is the most likely mechanism behind the tilting/uplift along the Linea Bala–Rogaguado. The backward sedimentation induced by the rapid uplift of the Fitzcarrald Arch could have buried and hidden the Andean forebulge.

In more general terms, the Bolivian Amazon can be divided into two areas under different depositional regimes: the uplifting north, crossed by incising rivers, which includes Pando and the northern LM; and the subsiding central and southern LM which are crossed by free meandering rivers. These two settings are the expression in the surface of the flat slab subduction in the north and normal subduction in the south. These two areas are separated by a transitional zone, which is flat, crossed by underfit rivers and also uplifting.

## 5.2 What is the extent and timing of the neotectonic activity?

It is unlikely that the different neotectonic elements of the landscape that have been reviewed here (the formation of ria lakes, the uplifted transitional zone, the avulsions of the Beni River, the formation of interior deltas and the changes in the meandering pattern of the Mamoré) River all happened at the same time and are the result of the same tectonic event. The data here presented suggest that the most likely scenario is that several tectonic events created and modified lakes and rivers at different times. The radiocarbon ages from the bottom of the lacustrine phase of Lake Rogaguado (which indicate that the lake formed about 6000 BP) and the radiocarbon ages of the palaeosols underlying fluvial deposits in central LM (which point to a mid- to late Holocene burial of the palaeosol) are, for now, the only fixed elements around which a relative chronology can be built. The formation of Lake Rogaguado is here interpreted as the last step in a series of correlated events.

### 5.2.1 Pleistocene

At the end of the Pleistocene, the Beni River flowed from east to west, up to a few kilometres north of Santa Ana de Yacuma, and then turned north in what is the modern Mamoré (Fig. 12a). The sediments that constitute the fluvial levees north of Puerto Siles (Fig. 7c) were probably deposited by the Beni River at this time. The Mamoré River used to flow further east, following the course now occupied by the far smaller Machupo River. Grande River was not a tributary of the Mamoré, but travelled north, discharging into the Iténez River, as well as the Mamoré (Fig. 12a).

### 5.2.2 Early Holocene

Probably during the late Pleistocene or at the beginning of the Holocene, and because of a fault along the Andean pied-

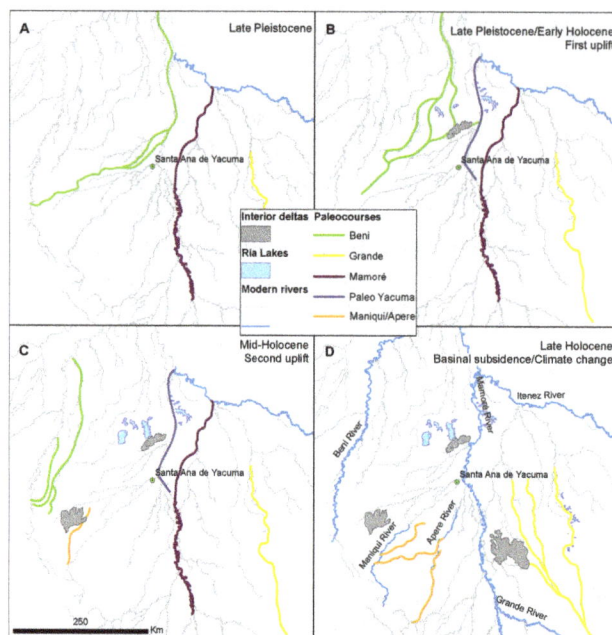

**Figure 12.** Reconstruction of the evolution of the landscape in the LM since the late Pleistocene. (**a**) shows the situation during the late Pleistocene with the Beni River flowing in what is now the Mamoré channel and the Mamoré River flowing towards the Iténez. Also the Grande River had a different course, flowing north into the Iténez. (**b**) shows the changes that followed the first uplift, with the Beni River shifting counterclockwise and the formation of a first group of ria lakes (Lake Oceano, Agua Clara and Largo). The crevasse splay of the Beni River in Fig. 3b probably formed during this time too. (**c**) shows how the Beni River shifted further west and how a second uplift formed the lakes Rogaguado and Ginebra. The DFS of the Maniqui River started forming at this time. (**d**) shows the modern setting, as it established during the late Holocene.

mont (Dumont, 1996), the River Beni shifted and occupied the valleys where, later, the lakes Rogaguado and Ginebra formed (Fig. 12b). Interestingly, the western border of the transitional zone coincides with the north–south oriented fault thought to have caused the migration of the deflection point of the Beni River, suggesting that the uplift of the transitional zone controlled the evolution of the Beni River. The same event also reduced the slope of the Maniqui River, inducing the formation of the Maniqui interior delta shown in Fig. 3d. During this time, the Beni River, and probably also the other rivers, was incising the northern LM in relatively deep valleys, also inducing erosion of the lateral valleys. Stratigraphic cores of these structures would allow us to date the underlying palaeosols and the time of the formation of both the Maniqui Delta and the Beni crevasse, allowing us to test whether or not their formation followed neotectonic uplifts. Palaeo-ecological records indicate that during the last glacial period (45.0–12.2 kyr BP) the climate in the LM was much drier than during the Holocene (Burbridge et al., 2004; Whitney et al., 2011). The first uplifting event that can be deduced from the geomorphological evidence happened north of 12.5° S and could be related to the formation of the knickpoint at Guayaramerín (Fig. 5). This uplift caused the backward sedimentation of the Beni River that, damming its lateral valleys, formed the lakes Agua Clara and Largo. This event could also be responsible for the formation of Lake Oceano and the other small ria lakes on the eastern margin of the Mamoré River (Fig. 12b). Further cores from lakes Largo, Agua Clara, Oceano and Ginebra are needed in order to date the beginning of the lacustrine sedimentation in these lakes and establish a more reliable chronology of the uplift events.

### 5.2.3 Mid-Holocene

A second neotectonic event, about 6 kyr BP, formed the knickpoint at Puerto Siles, caused the tilting along the Linea Bala–Rogaguado and formed the lakes Rogaguado and Ginebra (Fig. 12c). The further shift of the Beni River from the Rogaguado and Ginebra valleys towards its current position could have preceded or been synchronous to the Bala–Rogaguado tilting. Lake cores from oxbows located within ancient gallery forests of the Beni River palaeocourses could help determine the timing of the Beni River avulsions and whether the avulsions were synchronous to the formation of the ria lakes. After the first shift of the Beni and before the Mamoré moved to its modern position, the upper course of the modern Mamoré was occupied by an underfit river with a smaller catchment, here called the palaeo-Yacuma in Fig. 12b and c. The palaeo-Yacuma probably transported only fine sediments, because most of its catchment would have been within the LM. The tectonic events that created the lakes Agua Clara and Largo first and the lakes Rogaguado and Ginebra later, probably also dammed this small river, which, during the rainy seasons, became an ephemeral ria lake, resembling the lower sections of the modern Negro, Xingú and Tapajó rivers. During this period, clayish sediments could have deposited within the fluvial valley, forming the thick organic layers that outcrops today along the banks of the Mamoré. These layers have been interpreted as palaeosols formed in the floodplain during periods of reduced river sedimentation, (Plotzki, 2013). Along the bank of the Mamorè, about 50 km north of Santa Ana, the exposed organic layer is about 3 m thick, and radiocarbon data indicates it formed between 12 and 3 kyr BP (Plotzki, 2013). However, the thickness of these palaeosols suggests that they are cumulative palaeosols, where deposition and pedogenesis coexisted (Kraus and Aslan, 1993). Their formation could be better explained by two cyclical processes working at different timescales. On a century to millennium scale, periods of more severe water ponding could have been cyclically alternated with periods of increased drainage. Clays would deposit during periods of water ponding due to downriver uplifts. These uplift events would create knickpoints that the rivers would start to erode immediately; the modern ana-

logues for this situation are the knickpoints at Puerto Siles and Guyaramerín (Fig. 5). The erosion of the knickpoint would increase the drainage, lower the level of the river, reduce the sedimentation and allow for soil formation. On a seasonal scale, clays would have been deposited during the rainy season, when water was ponded within the river valleys; wetlands would have formed during the dry season, following the lowering of the water table. These alternate phases of deposition of clays with soil formation could eventually form very thick layers of organic-rich clays as those found along the Mamoré banks. Evidence that uplift in the northern part of the LM could have induced water ponding, backward sedimentation and elevation of the water table several times during the Holocene, can also be seen in fossil wood (dated 5.575±105, 7.145±150, 9.995±150 and 10.085±150 yr BP) found embedded in clay layers along the rivers Acre, Beni and Madre de Dios (Campbell and Frailey, 1984; Campbell et al., 1985). According to Campbell, the presence of these clayey deposits over a very large area is indicative of a lacustrine depositional environment and is the result of the formation of a large lake, Lake Mojos, which would have had a southward drainage, belonging to the La Plata River basin (Campbell, 1989 cited in Hanagarth, 1993). It has been suggested that during the Holocene, the sedimentation of the Grande and Parapetí rivers closed the southern drainage and that Lake Mojos broke the northern enclosure, leading to the modern situation (Hanagart and Sarmiento, 1990). However, the existence of such a lake is at odds with the general topography of the Cerrado Beniano, which is crossed by many fluvial valleys and not just one, as it would be in the case of Campbell's reconstruction. Moreover, if Lake Mojos drained towards the south, we should find in the Cerrado Beniano a palaeodrainage network which converges toward the south, but, as shown in Fig. 9, this is not the case. As for the palaeosols along the Mamoré, the presence of the layers of clay observed along many rivers in northern LM (Campbell and Frailey, 1984; Campbell et al., 1985), could be the result of water ponding and backwards sedimentations caused by downriver uplifts. If this hypothesis is correct, such deposits could serve as proxies to trace back neotectonic events.

### 5.2.4 Late Holocene

The situation of the Mamoré and Grande rivers did not change until the late Holocene, when the palaeo-Yacuma course was occupied by the Mamoré and the rest of the modern fluvial network was established (Fig. 12d). During the mid- to late Holocene, a large transformation occurred in the central and southern LM. Perhaps due to a climate shift towards wetter conditions (Lombardo et al., 2012; Plotzki et al., 2013), coupled with basinal subsidence (Hanagarth, 1993; Dumont and Fournier, 1994), the Mamoré established its modern position and the Grande River formed a large distributary system (Fig. 12d). It has also been suggested that, when the Grande River became a tributary of the

Mamoré, the increased discharge triggered the Mamoré avulsion (Plotzki et al., 2013). This research shows that during the late Holocene, the western side of the LM was also subject to extensive fluvial depositions and the formation of distributary systems in the case of the rivers Maniqui and Apere, as indicated by the recurrent palaeosols shown in Fig. 4. As for the avulsion of the Grande River and for the fluvial deposition in the western LM, there is insufficient data yet to establish whether it was caused by an increase in precipitation, basinal subsidence or a combination of both.

### 5.3 How did neotectonics affect the LM hydrology during the Holocene?

#### 5.3.1 Hydrology and landscape

The Fitzcarrald Arch's uplift played an important role in the evolution of the Amazon Basin and a major role in increasing its biodiversity (Espurt et al., 2010). However, in the case of the Bolivian Amazon, its effects have not yet been explored. Here, I suggest that the uplift of the Fitzcarrald Arch has determined the evolution of the landscape of the LM at both a regional and a local scale. Tectonics control the differences in soils and vegetation between the hilly landscape of the Cerrado Beniano, the flat and incised transitional zone, and the vast floodplain of the central and southern LM. In the Cerrado Beniano, the uplift causes the outcropping of lateritic crusts and the formation of a hilly topography, while, in the transitional zone, it prevents the deposition of fluvial sediments over the savannahs, with detrimental effects on soil fertility (Lombardo et al., 2013a). The southern LM is a subsiding area, therefore subjected to the deposition of recent fluvial sediments, and hence more fertile. The transition between the Cerrado Beniano and the central/southern LM is also the transition between the uplifting north and the subsiding south. This exerts an important control over the river network of the western LM, determining the underfitting of rivers that flow north at approximately −14° S. When the uplift causes a decrease in river slope, interior deltas and wetlands form, like the large interior delta of the Maniqui River which could have been induced by the uplift of the transitional zone.

The multiplying effect of precipitation on the severity of the floods, where an increase of less than 100 % in mean annual precipitation causes the size of the flooded area to increase by more than 300 %, is caused by the extremely flat landscape in the LM. Precipitation alone would not cause any floods if the LM had a better drainage. The savannah–forest ecotone in the central and southern LM is controlled by the duration and intensity of the seasonal floods (Mayle et al., 2007). Hamilton et al. (2004) have pointed out that the outcrops of the Brazilian Shield in the northern LM constrain the river channels and contribute to a backwater effect at high water. Therefore, as floods determine the savannah–forest ecotone, drainage determines how floods respond to

precipitation and neotectonics determines drainage, it can be concluded that uplifts in northern LM could have been a determining factor in the evolution of the savannah–forest ecotone in the central and southern LM during the Holocene. Uplift events in the northern LM reduced the drainage of the basin, but they were followed by stronger erosion by the rivers going through the uplifted rocks. This study suggests that uplifts were responsible for (1) the creation of interior deltas, (2) the elevation of the river beds with formation of ria lakes in some of the lateral valleys, and (3) the increase in sedimentation of fine silts and clays in the fluvial valleys, with a consequent flattening of the topography and creation of flat valley bottoms. Conversely, periods between different uplifts were characterized by deeper river incision, improved drainage, lowering of the LM water table, reduction of the flooded areas and widespread changes in the process of soil formation and vegetation assemblages. This interplay between uplift and river incision could have been, together with climate, a major player in controlling inundation patterns in the LM and the most important factor determining the evolution of the LM landscape. Moreover, the changes in river slope due to downriver uplifts would have had an impact on the sedimentary load of the Madeira River and, consequently, of the Amazon River, as neotectonics not only affects the shape and behaviour of channels and flood plains but also determines the river's suspended-load transport capacity (Dunne et al., 1998).

#### 5.3.2 Hydrology and people

Likewise, as people lived in the LM since the early Holocene (Lombardo et al., 2013b), it is to be expected that neotectonics had important effects on pre-Columbian peoples too. There is very little data yet about early Holocene human presence in southern Amazonia, the only study published so far (Lombardo et al., 2013b) suggests that hunter-gatherers settled in the southern LM from 10 500 till 4200 cal yr BP, approximately. It would seem that pre-Columbians were able to adapt to changes brought about by mid-Holocene neotectonic events; nevertheless, the formation of the Grande River megafan, approximately 4200 cal yr BP, which buried the hunter-gatherer sites, was followed by a period of about 2000 yr of archaeological hiatus. Although the specific cause behind the Grande River avulsion is unclear, it is evident that it is basinal subsidence that, accommodating the sediment charge, created the preconditions for the formation of megafans. It is likely that the deposition of fertile sediments in southern LM created the necessary conditions for the subsequent development of complex societies in the southeastern LM, the so-called Earthmovers, about 1400 cal yr BP (Lombardo et al., 2012). It has been suggested that, late Holocene pre-Columbian settlement patterns, adaptive strategies and social complexity followed ecological gradients of soil fertility and flood intensity (Lombardo et al., 2011b). The data here presented shows that these ecological gradients are, to a great extent, the expression of the underlying geological control.

## 6  Conclusions

This paper has reviewed the published evidence of neotectonic activity in southern Amazonia and integrated it with new data from the analysis of remote sensing imagery and extensive field work in the LM. Based on the study of modern and palaeorivers, ria lakes, palaeosols and landscape topography in this region, this paper suggests that the uplifting Fitzcarrald Arch is the most likely mechanism behind the observed uplift along the Linea Bala–Rogaguado, the general topography of the LM, the outcropping of the basement in the northern LM, the presence of underfit rivers crossing the savannahs in the transitional zone, the meandering pattern of the Mamoré River and the general shape of the fluvial network. I suggest that the tectonic regime brought about by the Fitzcarrald Arch has determined the hydrology, flooding patterns, the forest-savannah ecotone and overall evolution of the landscape of the LM at both a regional and a local scale. These findings stress that, contrary to what many researchers have assumed, climate and human activities have not been the only factors influencing landscape evolution in the Bolivian Amazon during the Holocene. On the contrary, the interplay between the Fitzcarrald Arch uplift and river incision could have played a fundamental role in the evolution of the LM's landscape since the late Pleistocene. It is unlikely that the different neotectonic elements of the landscape that have been reviewed here were caused by a single uplift. Instead, I propose a reconstruction of the evolution of the landscape of southern Amazonia in which at least two uplift events took place. More data is needed in order to confirm the hypothesis here presented and to better constrain the timing of the different events, such as the formation of ria lakes and interior deltas, river avulsions and deposition of crevasse spays. Nevertheless, the reconstruction here proposed represents a useful working model for future research and highlights the importance of taking into account neotectonics in palaeo-environmental reconstructions of southern Amazonia. In addition, this work highlights the need to disentangle the effects of human populations, climate change and neotectonics on the evolution of the LM, in order to better predict the impact of future climate change in the region and develop better adaptation strategies for local communities.

**Acknowledgements.** The present study has been funded by the Swiss National Science Foundation (SNSF), grants no. 200020 and 141277/1. I would like to thank H. Veit for his support and guidance. Lake Rogaguado has been cored in collaboration with the IRD project HydroGen; M. Grosjean and M-P. Ledru lent me the coring devices and taught me how to use them. L. Rodrigues and A. Giesche helped in several aspects of the fieldwork; J. Bocchetti helped me with several logistical problems in the field, including the fixing of the Waker motor. I am very grateful for the hospitality of the Community of Coquinal and, very specially, to Jonny Ibáñez, who pulled our car out of a crocodile-infested pond in the middle of the night. Further thanks are also due to E. Canal-Beeby, D. Garcia-Castellanos, J. Carson, the editor, S. Mudd, and two anonymous reviewers who helped improve earlier versions of the manuscript.

Edited by: S. Mudd

## References

Aalto, R., Maurice-Bourgoin, L., Dunne, T., Montgomery, D. R., Nittrouer, C. A., and Guyot, J.-L.: Episodic sediment accumulation on Amazonian flood plains influenced by El Niño/Southern Oscillation, Nature, 425, 493–497, doi:10.1038/nature01990, 2003.

Allenby, R. J.: Origin of rectangular and aligned lakes in the Beni Basin of Bolivia, Tectonophysics, 145, 1–20, 1988.

Angermeier, P. L.: The Natural Imperative for Biological Conservation, Conserv. Biol., 14, 373–381, doi:10.1046/j.1523-1739.2000.98362.x, 2000.

Behling, H., Keim, G., Irion, G., Junk, W., and Nunes de Mello, J.: Holocene environmental changes in the Central Amazon Basin inferred from Lago Calado (Brazil), Palaeogeogr. Palaeocl., 173, 87–101, 2001.

Baker, V. R.: Adjustment of fluvial systems to climate and source terrain in tropical and subtropical environments, in: Fluvial Sedimentology, edited by: Miall, A. D., Memoir-Canadian Society of Petroleum Geologists, 211–230, 1977.

Boixadera, J., Poch, R. M., García-González, M. T., and Vizcayno, C.: Hydromorphic and clay-related processes in soils from the Llanos de Moxos (northern Bolivia), Catena, 54, 403–424, doi:10.1016/s0341-8162(03)00134-6, 2003.

Bourrel, L. and Pouilly, M.: Hidrología y dinámica fluvial del Río Mamoré, in: Diversidad biológica en la llanura de inundación del Río Mamoré, edited by: Pouilly, M., Beck, S., Moraes, M., and Ibañez, C., Centro de Ecología Simón I. Patiño, Santa Cruz, Bolivia, 95–116, 2004.

Bourrel, L., Phillips, L., and Moreau, S.: The dynamics of floods in the Bolivian Amazon Basin, Hydrol. Process., 23, 3161–3167, doi:10.1002/hyp.7384, 2009.

Bridge, J. S.: Rivers and floodplains: forms, processes, and sedimentary record, John Wiley & Sons, Oxford, UK, 2009.

Buehler, H. A., Weissmann, G. S., Scuderi, L. A., and Hartley, A. J.: Spatial and Temporal Evolution of an Avulsion on the Taquari River Distributive Fluvial System from Satellite Image Analysis, Journal of Sediment. Res., 81, 630–640, doi:10.2110/jsr.2011.040, 2011.

Burbank, D. W. and Anderson, R. S.: Tectonic geomorphology, John Wiley & Sons, Chichester, UK, 2011.

Burbridge, R. E., Mayle, F. E., and Killeen, T. J.: Fifty-thousand-year vegetation and climate history of Noel Kempff Mercado National Park, Bolivian Amazon, Quaternary Res., 61, 215–230, doi:10.1016/j.yqres.2003.12.004, 2004.

Bush, M. B. and Silman, M. R.: Amazonian exploitation revisited: ecological asymmetry and the policy pendulum, Front. Ecol. Environ., 5, 457–465, doi:10.1890/070018, 2007.

Campbell, K. E. and Frailey, C. D.: Holocene flooding and species diversity in southwestern Amazonia, Quaternary Res., 21, 369–375, doi:10.1016/0033-5894(84)90075-9, 1984.

Campbell, K. E., Frailey, C. D., and Arellano, J. L.: The geology of the Rio Beni: Further evidence for Holocene flooding in Amazonia, Contributions in Science, Natural History Museum of Los Angeles County, 364, 1–18, 1985.

Carson, J. F., Whitney, B. S., Mayle, F. E., Iriarte, J., Prümers, H., Soto, J. D., and Watling, J.: Environmental impact of geometric earthwork construction in pre-Columbian Amazonia, P. Natl. Acad. Sci., 11, 10497–10502, doi:10.1073/pnas.1321770111, 2014.

Costa, J. B. S., Léa Bemerguy, R., Hasui, Y., and da Silva Borges, M. C.: Tectonics and paleogeography along the Amazon river, J. S. Am. Earth Sci., 14, 335–347, doi:10.1016/s0895-9811(01)00025-6, 2001.

Dearing, J. A.: Landscape change and resilience theory: a palaeoenvironmental assessment from Yunnan, SW China, The Holocene, 18, 117–127, doi:10.1177/0959683607085601, 2008.

DeCelles, P. G. and Giles, K. A.: Foreland basin systems, Basin Research, 8, 105–123, doi:10.1046/j.1365-2117.1996.01491.x, 1996.

Dumont, J. F.: Type of lakes as related to neotectonics in western Amazonia, International symposium on the Quaternary of Amazonia, 8–13 November 1992, Manaus, Brazil, 1993.

Dumont, J. F.: Neotectonics of the Subandes-Brazilian craton boundary using geomorphological data: the Marañon and Beni basins, Tectonophysics, 257, 137–151, 1996.

Dumont, J. F. and Fournier, M.: Geodynamic environment of Quaternary morphostructures of the subandean foreland basins of Peru and Bolivia: Characteristics and study methods, Quatern. Int., 21, 129–142, doi:10.1016/1040-6182(94)90027-2, 1994.

Dumont, J. F. and Guyot, J. L.: Ria lac: ou, pourquoi?, Proceedings of the Third International Conference on Geomorphology, 23–29 August, 1993, Hamilton, Canada, 1993,

Dunne, T., Mertes, L. A. K., Meade, R. H., Richey, J. E., and Forsberg, B. R.: Exchanges of sediment between the flood plain and channel of the Amazon River in Brazil, Geol. Soc. Am. Bull., 110, 450–467, doi:10.1130/0016-7606(1998)110< 0450:eosbtf> 2.3.co;2, 1998.

Erickson, C. L.: Amazonia: the historical ecology of a domesticated landscape, in: Handbook of South American archaeology, edited by: Silverman, H. and Isbell, W. H., Springer, Berlin, 157–183, 2008.

Espurt, N., Baby, P., Brusset, S., Roddaz, M., Hermoza, W., Regard, V., Antoine, P. O., Salas-Gismondi, R., and Bolaños, R.: How does the Nazca Ridge subduction influence the modern Amazonian foreland basin?, Geology, 35, 515–518, doi:10.1130/g23237a.1, 2007.

Espurt, N., Funiciello, F., Martinod, J., Guillaume, B., Regard, V., Faccenna, C., and Brusset, S.: Flat subduction dynamics and deformation of the South American plate: Insights from analog modeling, Tectonics, 27, TC3011, doi:10.1029/2007tc002175, 2008.

Espurt, N., Baby, P., Brusset, S., Roddaz, M., Hermoza, W., and Barbarand, J.: The Nazca Ridge and Uplift of the Fitzcarrald Arch: Implications for Regional Geology in Northern South America, in: Amazonia: Landscape and Species Evolution, edited by: Hoorn, C. and Wesselingh, F. P., Wiley-Blackwell Publishing Ltd., Oxford, UK, 89–100, 2010.

Fan, Y., Li, H., and Miguez-Macho, G.: Global Patterns of Groundwater Table Depth, Science, 339, 940–943, doi:10.1126/science.1229881, 2013.

Garcia-Castellanos, D.: Interplay between lithospheric flexure and river transport in foreland basins, Basin Research, 14, 89–104, doi:10.1046/j.1365-2117.2002.00174.x, 2002.

Garcia-Castellanos, D. and Cloetingh, S.: Modeling the Interaction between Lithospheric and Surface Processes in Foreland Basins, in: Tectonics of Sedimentary Basins, John Wiley & Sons Ltd., Chichester, UK, 152–181, 2011.

Gupta, A.: Tropical geomorphology, Cambridge University Press, New York, USA, 2011.

Hajek, E. A. and Edmonds, D. A.: Is river avulsion style controlled by floodplain morphodynamics?, Geology, 42, 199–202, doi:10.1130/g35045.1, 2014.

Hamilton, S. K., Sippel, S. J., and Melack, J. M.: Seasonal inundation patterns in two large savanna floodplains of South America: the Llanos de Moxos(Bolivia) and the Llanos del Orinoco (Venezuela and Colombia), Hydrol. Process., 18, 2103–2116, doi:10.1002/hyp.5559, 2004.

Hanagart, W. and Sarmiento, J.: Reporte preliminar sobre la geoecología de la sabana de Espíritu y sus alrededores (Llanos de Moxos, departamento del Beni, Bolivia), Ecología en Bolivia, 16, 47–75, 1990.

Hanagarth, W.: Acerca de la geoecología de las sabanas del Beni en el noreste de Bolivia, Instituto de ecología, La Paz, Bolivia, 1993.

Hartley, A. J., Weissmann, G. S., Nichols, G. J., and Warwick, G. L.: Large Distributive Fluvial Systems: Characteristics, Distribution, and Controls on Development, J. Sediment. Res., 80, 167–183, doi:10.2110/jsr.2010.016, 2010.

Herzog, S., Maillard Z, O., Embert, D., Caballero, P., and Quiroga, D.: Range size estimates of Bolivian endemic bird species revisited: the importance of environmental data and national expert knowledge, J. Ornithol., 153, 1189–1202, doi:10.1007/s10336-012-0850-2, 2012.

Hogg, A. G., Hua, Q., Blackwell, P. G., Niu, M., Buck, C. E., Guilderson, T. P., Heaton, T. J., Palmer, J. G., Reimer, P. J., Reimer, R. W., Turney, C. S. M., and Zimmerman, S. R. H.: SHCal13 Southern Hemisphere Calibration, 0–50 000 Years cal BP, Radiocarbon, 55, 1889–1903, 2013.

Holbrook, J. and Schumm, S. A.: Geomorphic and sedimentary response of rivers to tectonic deformation: a brief review and critique of a tool for recognizing subtle epeirogenic deformation in modern and ancient settings, Tectonophysics, 305, 287–306, 1999.

Irion, G., Bush, M. B., Nunes de Mello, J. A., Stüben, D., Neumann, T., Müller, G., Morais de, J. O., and Junk, J. W.: A multiproxy palaeoecological record of Holocene lake sediments from the Rio Tapajós, eastern Amazonia, Palaeogeogr. Palaeocl., 240, 523–535, doi:10.1016/j.palaeo.2006.03.005, 2006.

Irion, G., Mello, J. S. N., Morais, J., Piedade, M. F., Junk, W., and Garming, L.: Development of the Amazon Valley During the Middle to Late Quaternary: Sedimentological and Climatological Observations, in: Amazonian Floodplain Forests, edited by: Junk, W. J., Piedade, M. T. F., Wittmann, F., Schöngart, J., and Parolin, P., Ecological Studies, Springer Netherlands, 27–42, 2011.

Jarvis, A., Reuter, H. I., Nelson, A., and Guevara, E.: Hole-filled SRTM for the globe Version 4, available atthe CGIARCSI SRTM 90m Database: http://srtm.csi.cgiar.org (last access: 15 December 2014), 2008.

Junk, W.: Current state of knowledge regarding South America wetlands and their future under global climate change, Aquatic Sciences – Research Across Boundaries, 75, 113–131, doi:10.1007/s00027-012-0253-8, 2013.

Junk, W. J., Bayley, P. B., and Sparks, R. E.: The flood pulse concept in river-floodplain systems, Canadian special publication of fisheries and aquatic sciences, 106, 110–127, 1989.

Kraus, M. J. and Aslan, A.: Eocene hydromorphic Paleosols; significance for interpreting ancient floodplain processes, J. Sediment. Res., 63, 453–463, doi:10.1306/d4267b22-2b26-11d7-8648000102c1865d, 1993.

Langstroth, R. P.: Forest islands in an Amazonian savanna of northeastern Bolivia, University of Wisconsin-Madison, Unpublished PhD dissertation, 1996.

Langstroth, R. P.: Biogeography of the Llanos de Moxos: natural and anthropogenic determinants, Geographica Helvetica, 66, 183–192, 2011.

Latrubesse, E. M. and Franzinelli, E.: The Holocene alluvial plain of the middle Amazon River, Brazil, Geomorphology, 44, 241–257, doi:10.1016/S0169-555X(01)00177-5, 2002.

Latrubesse, E. M., Cozzuol, M., da Silva-Caminha, S. A. F., Rigsby, C. A., Absy, M. L., and Jaramillo, C.: The Late Miocene paleogeography of the Amazon Basin and the evolution of the Amazon River system, Earth-Sci. Rev., 99, 99–124, doi:10.1016/j.earscirev.2010.02.005, 2010.

Lombardo, U. and Prümers, H.: Pre-Columbian human occupation patterns in the eastern plains of the Llanos de Moxos, Bolivian Amazonia, J. Archaeol. Sci., 37, 1875–1885, doi:10.1016/j.jas.2010.02.011, 2010.

Lombardo, U. and Veit, H.: The origin of oriented lakes: Evidence from the Bolivian Amazon, Geomorphology, 204, 502–509, doi:10.1016/j.geomorph.2013.08.029, 2014.

Lombardo, U., Canal-Beeby, E., Fehr, S., and Veit, H.: Raised fields in the Bolivian Amazonia: a prehistoric green revolution or a flood risk mitigation strategy?, J. Archaeol. Sci., 38, 502–512, doi:10.1016/j.jas.2010.09.022, 2011a.

Lombardo, U., Canal-Beeby, E., and Veit, H.: Eco-archaeological regions in the Bolivian Amazon: Linking pre-Columbian earthworks and environmental diversity, Geographica Helvetica, 66, 173–182, 2011b.

Lombardo, U., May, J.-H., and Veit, H.: Mid- to late-Holocene fluvial activity behind pre-Columbian social complexity in the southwestern Amazon basin, The Holocene, 22, 1035–1045, doi:10.1177/0959683612437872, 2012.

Lombardo, U., Denier, S., May, J.-H., Rodrigues, L., and Veit, H.: Human–environment interactions in pre-Columbian Amazonia: The case of the Llanos de Moxos, Bolivia, Quatern. Int., 312, 109–119, doi:10.1016/j.quaint.2013.01.007, 2013a.

Lombardo, U., Szabo, K., Capriles, J. M., May, J.-H., Amelung, W., Hutterer, R., Lehndorff, E., Plotzki, A., and Veit, H.: Early and Middle Holocene Hunter-Gatherer Occupations in Western Amazonia: The Hidden Shell Middens, PLoS ONE, 8, e72746, doi:10.1371/journal.pone.0072746, 2013b.

Lombardo, U., Denier, S., and Veit, H.: Soil properties and pre-Columbian settlement patterns in the Monumental Mounds Region of the Llanos de Moxos, Bolivian Amazon, SOIL Discuss., 1, 81–117, doi:10.5194/soild-1-81-2014, 2014.

Mann, C. C.: Ancient earthmovers of the Amazon, Science, 321, 1148–1152, 2008.

Mayle, F. E., Burbridge, R., and Killeen, T. J.: Millennial-scale dynamics of southern Amazonian rain forests, Science, 290, 2291–2294, doi:10.1126/science.290.5500.2291, 2000.

Mayle, F. E., Langstroth, R. P., Fisher, R. A., and Meir, P.: Long-term forest-savannah dynamics in the Bolivian Amazon: implications for conservation, Philos. T. Roy. Soc. B., 362, 291–307, doi:10.1098/rstb.2006.1987, 2007.

Melack, J. and Hess, L.: Remote Sensing of the Distribution and Extent of Wetlands in the Amazon Basin, in: Amazonian Floodplain Forests, edited by: Junk, W. J., Piedade, M. T. F., Wittmann, F., Schöngart, J., and Parolin, P., Ecological Studies, Springer Netherlands, 43–59, 2011.

Navarro, G.: Clasificación de la vegetación de Bolivia, Centro de ecología difusión fundación Simón I. Patiño, Santa Cruz, Bolivia, 2011.

Pitfield, P. E. J. and Power, G.: Geologic map MagdalenaSD 20-6 con parte de SD 20-2, Ordnance Survey 4/88/880413S, Southampton, UK, 1987.

Plafker, G.: Oriented Lakes and Lineaments of Northeastern Bolivia, Geol. Soc. Am. Bull., 75, 503–522, doi:10.1130/0016-7606(1964)75[503:olalon]2.0.co;2, 1964.

Plafker, G.: Tectonic implications of the oriented lakes and lineaments in northeastern Bolivia, First International Conference on the New Basement Tectonics, Salt Lake City, Utah, USA, 519–527, 1974.

Plotzki, A.: Late Pleistocene to Holocene fluvial dynamics and environmental conditions in the Llanos de Moxos, Bolivian Amazon, PhD, Philosophisch naturwissenschaftlichen Fakultät, Universität Bern, Bern, Switzerland, 2013.

Plotzki, A., May, J. H., and Veit, H.: Past and recent fluvial dynamics in the Beni lowlands, NE Bolivia, Geographica Helvetica, 66, 164–172, 2011.

Plotzki, A., May, J. H., Preusser, F., and Veit, H.: Geomorphological and sedimentary evidence for late Pleistocene to Holocene hydrological change along the Río Mamoré, Bolivian Amazon, J. S. Am. Earth Sci., 47, 230–242, doi:10.1016/j.jsames.2013.08.003, 2013.

Räsänen, M. E., Salo, J. S., and Kalliola, R. J.: Fluvial Perturbance in the Western Amazon Basin: Regulation by Long-Term Sub-Andean Tectonics, Science, 238, 1398–1401, doi:10.1126/science.238.4832.1398, 1987.

Regard, V., Lagnous, R., Espurt, N., Darrozes, J., Baby, P., Roddaz, M., Calderon, Y., and Hermoza, W.: Geomorphic evidence for recent uplift of the Fitzcarrald Arch (Peru): A response to the Nazca Ridge subduction, Geomorphology, 107, 107–117, doi:10.1016/j.geomorph.2008.12.003, 2009.

Roddaz, M., Brusset, S., Baby, P., and Herail, G.: Miocene tidal-influenced sedimentation to continental Pliocene sedimentation in the forebulge–backbulge depozones of the Beni–Mamore foreland Basin (northern Bolivia), J. S. Am. Earth Sci., 20, 351–368, doi:10.1016/j.jsames.2005.11.004, 2006.

Ronchail, J. and Gallaire, R.: ENSO and rainfall along the Zongo valley (Bolivia) from the Altiplano to the Amazon basin, Int. J. Climatol., 26, 1223–1236, doi:10.1002/joc.1296, 2006.

Ronchail, J., Bourrel, L., Cochonneau, G., Vauchel, P., Phillips, L., Castro, A., Guyot, J., and Deoliveira, E.: Inundations in the Mamoré basin (south-western Amazon—Bolivia) and sea-surface temperature in the Pacific and Atlantic Oceans, J. Hydrol., 302, 223–238, doi:10.1016/j.jhydrol.2004.07.005, 2005.

Rossetti, D. F., Zani, H., Cohen, M. C. L., and Cremon, É. H.: A Late Pleistocene–Holocene wetland megafan in the Brazilian Amazonia, Sedimentary Geology, 282, 276–293, doi:10.1016/j.sedgeo.2012.09.015, 2012.

Schumm, S. A., Dumont, J. F., and Holbrook, J. M.: Active tectonics and alluvial rivers, Cambridge University Press, Cambridge, UK, 2002.

Slingerland, R. and Smith, N. D.: River Avulsions and their deposits, Ann. Rev. Earth Planet. Sci., 32, 257–285, doi:10.1146/annurev.earth.32.101802.120201, 2004.

Tinner, W. and Ammann, B.: Long-term Responses of Mountain Ecosystems to Environmental Changes: Resilience, Adjustment, and Vulnerability, in: Global Change and Mountain Regions, edited by: Huber, U., Bugmann, H. M., and Reasoner, M., Advances in Global Change Research, Springer Netherlands, 133–143, 2005.

UNASBVI-JICA: Estudio Hidrogeologico de Beni y Pando, Trinidad –Bolivia, 2009.

Urrego, D. H., Bush, M. B., Silman, M. R., Niccum, B. A., De La Rosa, P., McMichael, C. H., Hagen, S., and Palace, M.: Holocene fires, forest stability and human occupation in south-western Amazonia, J. Biogeogr., 40, 521–533, doi:10.1111/jbi.12016, 2013.

Vegas-Vilarrúbia, T., Rull, V., Montoya, E., and Safont, E.: Quaternary palaeoecology and nature conservation: a general review with examples from the neotropics, Quaternary Sci. Rev., 30, 2361–2388, doi:10.1016/j.quascirev.2011.05.006, 2011.

Walker, J.: Agricultural Change in the Bolivian Amazon, Memoirs in Latin American Archaeology, University of Pittsburgh Latin American Archaeology Publications and Fundación Kenneth Lee, Trinidad, 2004.

Wallace, R., Martinez, J., Lopez-Strauss, H., Barreta, J., Reinaga, A., and Lopez, L.: Conservation Challenges Facing Two Threatened Endemic Titi Monkeys in a Naturally Fragmented Bolivian Forest, in: Primates in Fragments, edited by: Marsh, L. K. and Chapman, C. A., Developments in Primatology: Progress and Prospects, Springer New York, 493–501, 2013.

Weissmann, G. S., Hartley, A. J., Nichols, G. J., Scuderi, L. A., Olson, M., Buehler, H., and Banteah, R.: Fluvial form in modern continental sedimentary basins: Distributive fluvial systems, Geology, 38, 39–42, doi:10.1130/g30242.1, 2010.

Weissmann, G., Hartley, A., Scuderi, L., Nichols, G., Davidson, S., Owen, A., Atchley, S., Bhattacharya, J., Chakraborty, T., and Ghosh, P.: Prograding distributive fluvial systems: geomorphic models and ancient examples, New Frontiers in Paleopedology and Terrestrial Paleoclimatology: SEPM, Special Publication, 104, 131–147, 2013.

Whitney, B. S., Mayle, F. E., Punyasena, S. W., Fitzpatrick, K. A., Burn, M. J., Guillen, R., Chavez, E., Mann, D., Pennington, R. T., and Metcalfe, S. E.: A 45 kyr palaeoclimate record from the lowland interior of tropical South America, Palaeogeogr. Palaeocl., 307, 177–192, doi:10.1016/j.palaeo.2011.05.012, 2011.

Willis, K. J. and Birks, J.-B.: What is natural? The need for a long-term perspective in biodiversity conservation, Science, 314, 1261–1265, 2006.

# 9

# Preservation of terrestrial organic carbon in marine sediments offshore Taiwan: mountain building and atmospheric carbon dioxide sequestration

S.-J. Kao[1,2], R. G. Hilton[3], K. Selvaraj[1,2], M. Dai[2], F. Zehetner[4], J.-C. Huang[5], S.-C. Hsu[1], R. Sparkes[6], J. T. Liu[7], T.-Y. Lee[1], J.-Y. T. Yang[2], A. Galy[6], X. Xu[8], and N. Hovius[9]

[1]Research Center for Environmental Changes, Academia Sinica, Taipei, Taiwan
[2]State Key Laboratory of Marine Environmental Science, Xiamen University, Xiamen, China
[3]Department of Geography, Durham University, Durham, UK
[4]Institute of Soil Research, University of Natural Resources and Life Sciences, Vienna, Austria
[5]Department of Geography, National Taiwan University, Taipei, Taiwan
[6]Department of Earth Sciences, University of Cambridge, Cambridge, UK
[7]Institute of Marine Geology and Chemistry, National Sun Yat-sen University, Kaohsiung, Taiwan
[8]School of Physical Sciences, University of California, Irvine, CA, USA
[9]Geomorphology, GFZ German Research Centre, Telegrafenberg, Potsdam, Germany

*Correspondence to:* S.-J. Kao (sjkao@gate.sinica.edu.tw, sjkao@xmu.edu.cn)

**Abstract.** Geological sequestration of atmospheric carbon dioxide ($CO_2$) can be achieved by the erosion of organic carbon (OC) from the terrestrial biosphere and its burial in long-lived marine sediments. Rivers on mountain islands of Oceania in the western Pacific have very high rates of OC export to the ocean, yet its preservation offshore remains poorly constrained. Here we use the OC content ($C_{org}$, %), radiocarbon ($\Delta^{14}C_{org}$) and stable isotope ($\delta^{13}C_{org}$) composition of sediments offshore Taiwan to assess the fate of terrestrial OC, using surface, sub-surface and Holocene sediments. We account for rock-derived OC to assess the preservation of OC eroded from the terrestrial biosphere and the associated $CO_2$ sink during flood discharges (hyperpycnal river plumes) and when river inputs are dispersed more widely (hypopycnal). The $C_{org}$, $\Delta^{14}C_{org}$ and $\delta^{13}C_{org}$ of marine sediment traps and cores indicate that during flood discharges, terrestrial OC can be transferred efficiently down submarine canyons to the deep ocean and accumulates offshore with little evidence for terrestrial OC loss. In marine sediments fed by dispersive river inputs, the $C_{org}$, $\Delta^{14}C_{org}$ and $\delta^{13}C_{org}$ are consistent with mixing of terrestrial OC with marine OC and suggest that efficient preservation of terrestrial OC ($> 70\%$) is also associated with hypopycnal delivery. Sub-surface and Holocene sediments indicate that this preservation is long-lived on millennial timescales. Re-burial of rock-derived OC is pervasive. Our findings from Taiwan suggest that erosion and offshore burial of OC from the terrestrial biosphere may sequester $> 8\,\mathrm{TgC\,yr^{-1}}$ across Oceania, a significant geological $CO_2$ sink which requires better constraint. We postulate that mountain islands of Oceania provide a strong link between tectonic uplift and the carbon cycle, one moderated by the climatic variability which controls terrestrial OC delivery to the ocean.

## 1 Introduction

Photosynthesis sequesters $CO_2$ within living matter as organic carbon (OC). If a fraction of this productivity escapes respiratory consumption and oxidation, it represents a carbon sink that will reduce greenhouse gas concentrations and influence Earth's radiation energy balance (Sundquist, 1993; Stallard, 1998; Berner, 2006). On geological timescales, the burial of OC in marine sediments is the second largest sink of atmospheric $CO_2$ after carbonate deposition formed from the products of continental silicate weathering (Gaillardet et al., 1999; Hayes et al., 1999, Burdige 2005). The erosion of terrestrial OC and its delivery by rivers to the ocean along with clastic sediments is thought to contribute approximately half of this oceanic OC burial flux (Schlunz and Schneider, 2000; Burdige, 2005; Blair and Aller, 2012), in part because the efficiency of OC burial is closely related to the accumulation rate of the accompanying sediment (Canfield, 1994; Burdige, 2005; Galy et al., 2007a). Therefore $CO_2$-sequestration by OC burial may be sensitive to changes in tectonic and climatic conditions which regulate the erosion and transfer of clastic sediment and terrestrial OC by rivers (Dadson et al., 2003; Hilton et al., 2008, 2012; Milliman and Farnsworth, 2011), giving rise to feedbacks in the global carbon cycle (West et al., 2005) which are not represented in current models of the carbon cycle (Berner, 2006).

The Himalayan orogeny is thought to exert tectonic forcing on the carbon cycle (Gaillardet and Galy, 2008), sequestering $3.7 \pm 0.4\,\mathrm{TgC\,yr^{-1}}$ through erosion of recently photosynthesized OC sourced from vegetation and soil in the terrestrial biosphere ($OC_{biosphere}$) and its preservation and burial in the distant Bengal Fan (France-Lanord and Derry, 1997; Galy et al., 2007a). On mountain islands of Oceania (Taiwan, Philippines, Indonesia, Papua New Guinea and New Zealand), where land–ocean linkages are strong, small mountain rivers drain a larger combined source area than the Himalaya ($\sim 2.7 \times 10^6\,\mathrm{km}^2$ vs. $1.6 \times 10^6\,\mathrm{km}^2$). These rivers transport $\sim 7000\,\mathrm{Tg\,yr^{-1}}$ of clastic sediment (Milliman and Farnsworth, 2011) and an estimated 20–40 % of the global particulate OC flux to the oceans (Lyons et al., 2002). There, convergent plate margins have steep, high standing topography where erosion of $OC_{biosphere}$ occurs at very high rates (up to $\sim 70\,\mathrm{MgC\,km^{-2}\,yr^{-1}}$) and rivers can deliver particulate materials rapidly to the ocean across short floodplains (Dadson et al., 2003; Dadson et al., 2005; Scott et al., 2006; Hilton et al., 2008; Bass et al., 2011; Hilton et al., 2012). These conditions should be conducive to high rates of OC burial and higher OC preservation efficiencies than rivers draining passive margins (Galy et al., 2007a; Bianchi, 2011; Blair and Aller, 2012). However, unlike the Himalayan system, our understanding of the fate of $OC_{biosphere}$ offshore and the resultant $CO_2$ sequestration around these ocean islands remains incomplete (Eglinton, 2008).

Firstly, the incomplete understanding of $OC_{biosphere}$ burial reflects the challenge of accounting for "petrogenic" OC derived from sedimentary rocks ($OC_{petro}$) in river sediments and marine sediments (Blair et al., 2003) which can contribute significantly to the particulate load of mountain rivers (Kao and Liu, 1996; Komada et al., 2004; Leithold et al., 2006; Hilton et al., 2010; Clark et al., 2013). $OC_{petro}$ transfer and re-burial lengthens the residence time of OC in the lithosphere (Galy et al., 2008; Hilton et al., 2011), however the re-burial of $OC_{petro}$ does not represent recent atmospheric $CO_2$

and so must be quantified separately. Secondly, it reflects the difficulty of assessing the range of delivery mechanisms to the ocean by mountain rivers. During floods, high suspended sediment concentrations ($> 40\,\mathrm{g\,L^{-1}}$) can cause the density of the river outflow to surpass that of ambient seawater (hyperpycnal) and result in density currents transporting sediment down submarine canyons into the deep ocean (Mulder and Syvitski, 1995). Previous work has postulated that hyperpycnal discharges are essential for the efficient transfer of terrestrial OC into marine deposits offshore mountain islands (Kao et al., 2006; Hilton et al., 2008). However, large amounts of terrestrial OC and sediment are also delivered to the surface ocean by rivers in hypopycnal plumes with a density lower than seawater. Such plumes disperse fluvial materials over a larger region, which may result in re-suspension and reworking of terrestrial OC (Mulder and Syvitski, 1995; Dadson et al., 2005; Kao and Milliman, 2008). In analogy with passive margin settings, this may lower terrestrial OC burial efficiency (Aller et al., 1996; Aller, 1998; Aller and Blair, 2006; Sampere et al., 2008; Blair and Aller, 2012).

In order to shed some light on the fate of $OC_{biosphere}$ eroded from high standing ocean islands, we consider the mountain island of Taiwan (Fig. 1). In Taiwan, the rapid convergence of the Philippine Sea Plate with the Eurasian continental margin combines with a climate characterized by frequent tropical cyclones, driving high rates of fluvial sediment export to the ocean (Dadson et al., 2003; Kao and Milliman, 2008). Findings from Taiwan are of wider relevance because the steep mountain rivers draining this island are common throughout Oceania (Milliman and Farnsworth, 2011). Steep mountain rivers have short transit times (e.g. Hilton et al., 2008) and deliver most of their sediment loads ($\sim 60$–70 %) under hypopycnal conditions (Dadson et al., 2005; Kao and Milliman, 2008). However, Taiwan's rivers can also produce hyperpycnal plumes (Dadson et al., 2005) allowing us to study terrestrial OC transfer and preservation associated with both modes of fluvial delivery. To assess the offshore transfer of terrestrial OC and its preservation in marine sediments upon deposition, we have collected seafloor sediments and material from sediment traps from (i) the submarine Gaoping Canyon off Southwest Taiwan, which is prone to hyperpycnal inputs; and (ii) the Okinawa Trough, Taiwan Strait and the Gaoping Shelf where hypopycnal inputs are thought to be more important (Fig. 1). To assess longer-term terrestrial OC preservation and burial, we examine sub-surface sediments from these locations and Holocene sediments from the Okinawa Trough. Employing an established approach, we have measured the OC content, stable OC isotopes and radiocarbon content of OC to determine sources of OC in the sediments and to assess the preservation of terrestrial $OC_{biosphere}$ and $OC_{petro}$ offshore (e.g. Komada et al., 2004; Leithold et al., 2006; Galy et al., 2007a, 2008; Hilton et al., 2008, 2010; Blair et al., 2010). The findings from Taiwan are placed in a regional context, and their implications for the global carbon cycle are discussed.

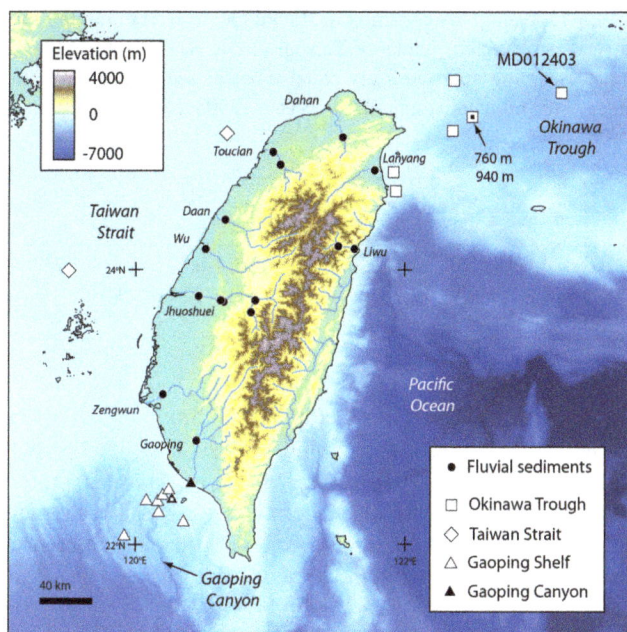

**Figure 1.** Location of terrestrial and marine samples from Taiwan and the surrounding ocean used in this study. River sediments were collected during typhoon floods across the island (black circles) with the sampled river names indicated (Table S1). Marine samples fed by dispersive fluvial inputs (white symbols) were obtained from box core surface sediments, with the location of the longer piston core MD012403 indicated. Sediments were also acquired from within the Gaoping Canyon (black triangles), which is fed by hyperpycnal river plumes. Sediment traps (indicated by squares within symbols) were deployed in the Okinawa Trough (depths provided) and at 608 m in the channel of the Gaoping Canyon.

## 2 Materials and methods

### 2.1 River suspended sediment samples

To characterize the composition of OC input into the ocean by Taiwanese rivers, suspended sediment samples were collected for this study from the primary rivers (Fig. 1) under common flow conditions as well as during tropical cyclone-induced floods, covering water discharges ranging from < 1 to ~ 40 times the long-term average (Table S1), and then complemented with published data (Kao and Liu, 1996; Hilton et al., 2008). For each sample, a known volume (between 250 mL and 1 L) of river water was collected from the surface of the main river channel in a wide-mouthed plastic bottle thoroughly rinsed with river water. The sample was then filtered through 0.7 μm GF/F membrane filters and the contents were dried at 60 °C, weighed to determine total suspended sediment concentration (SSC, g L$^{-1}$) and stored in sealed glass dishes. Water discharge ($Q_w$, m$^3$ s$^{-1}$) was measured by the Water Resources Agency, Taiwan, and reported here where available (Table S1).

## 2.2 Marine sediment samples

To assess the fate of terrestrial OC delivered to the ocean by hyperpycnal discharges, a sediment trap mooring was deployed at 608 m water depth, 42 m above the seafloor, in the submarine Gaoping Canyon, fed by the Gaoping River (Fig. 1). Full details of the collection methods can be found elsewhere (Huh et al., 2009; Liu et al., 2012, 2013). Briefly, the sediment trap mooring and an upward-facing long-range acoustic Doppler profiler were moored in the canyon during the 2008 typhoon season. The mooring was configured with a non-sequential sediment trap, consisting of a conical funnel and core liner, and an intervalometer timer capable of inserting Teflon discs into the collected sediment as embedded time markers (Xu et al., 2010). During deployment, Typhoon Kalmaegi impacted Taiwan (17 July 2008) and the Teflon discs inserted before its landfall and after the flood waters had ceased allowed us to constrain sediment associated with the typhoon event (Liu et al., 2012). Conical sediment traps may result in conservative estimates of accumulation rate due to potential re-suspension of sediment in the funnel (Buesseler et al., 2007). Here we do not rely on accumulation rate data, but note that hydrodynamic sorting may result in a lower percentage of smaller, more buoyant particles present in the trap than the sediment plume during high current velocities. To assess longer-term preservation, a box core was collected by R/V *Ocean Researcher-1* in September and October 2009 at station K1 (160 m water depth), located at the thalweg of the Gaoping Canyon (Fig. 1). The core was sub-sampled at different depths (Table S2). The sediments are thought to represent deposits associated with hyperpycnal river discharge during Typhoon Morakot in August 2009 (Sparkes, 2012; Liu et al., 2013), whose exceptionally heavy rainfall in Taiwan triggered a very large number of landslides (West et al., 2011) and high rates of sediment delivery offshore (Carter et al., 2012). Together, these marine sediments allow us to assess the transfer and deposition of terrestrial materials by river hyperpycnal flows.

To assess the fate of terrestrial OC delivered by more dispersive events (hypopycnal discharges), sediments were collected from marine trap moorings at 760 and 940 m in the southern Okinawa Trough (Fig. 1), where direct hyperpycnal river discharges are less common (Dadson et al., 2005; Hsu et al., 2006; Kao and Milliman, 2008). In addition, seafloor sediment samples collected between 1994 and 2009 with a box corer on R/V *Ocean Researcher-1* and *-2* from the Gaoping Shelf, southern Okinawa Trough and Taiwan Strait were selected (Table S3). Samples were collected from the top 2 cm of these cores using a stainless steel spatula and freeze-dried. We have also examined Holocene sediments in the long piston core MD012403 collected by R/V *Marion Dufresne* in 2001 from a water depth of 1420 m (Kao et al., 2008). The depositional age of these sediments was determined by analysis of the radiocarbon content of planktonic foraminifera so the $\Delta^{14}C_{org}$ at time of deposition can be estimated (Kao et al.,

2008). The methods of sampling terrestrial and marine sediments mean that the study focuses on the transfer and offshore preservation of sand and finer materials. Large woody debris (e.g. logs and trunks) are not likely to be recovered and so their fate remains a question for future research.

## 2.3 Geochemical methods

All marine samples were rinsed with deionized water ($> 18 \text{M}\Omega$) to remove salts. All dried sediment samples were homogenized in an agate mortar. Prior to measurement of the OC concentration ($C_{org}$, %) and analysis of the stable isotopes of OC ($\delta^{13}C_{org}$, ‰), samples were treated with 1 N HCl at $20\,^{\circ}\text{C}$ for 16 h to remove carbonate; the residue was centrifuged and freeze-dried (Kao et al., 2008). $\delta^{13}C_{org}$ analysis was carried out using Carlo-Erba 2100 elemental analyser connected to a Thermo Finnigan Deltaplus Advantage isotope ratio mass spectrometer and reported in $\delta$ notation with respect to the PDB standard and renormalized based on working standards (USGS 40 and acetanilide), with reproducibility better than 0.2‰. Radiocarbon ($^{14}C$) was measured on OC by accelerator mass spectrometry after carbonate removal and graphitization at Woods Hole Oceanographic Institution, USA, Institute of Geological and Nuclear Sciences, New Zealand, and Keck-Carbon Cycle AMS Facility at University of California at Irvine, USA. $^{14}C$ values are given after correction for $^{13}C$ fractionation (normalization to a $\delta^{13}C$ value of $-25$‰), and expressed as percent modern carbon (pMC) comparative to 95 % of the $^{14}C$ activity of the NBS oxalic acid and $\Delta^{14}C$ based on established protocols (Stuiver and Polach, 1977), with precision typically better than 10‰. Samples from the Liwu River in 2004 were analysed by similar methods described elsewhere (Hilton et al., 2008). Inorganic carbon removal by HCl leaching was preferred over HCl vapour to ensure complete removal of dolomite (Galy et al., 2007b), which may be present in Taiwanese bedrock and river sediments (Hilton et al., 2010). As such, following previous work, all OC isotope measurements refer to the acid-insoluble OC (Galy et al., 2007b; Hilton et al., 2010). The deviation between these two methods for terrestrial materials, $\delta^{13}C_{org} \sim \pm 0.2$‰ and $\Delta^{14}C_{org} \sim \pm 10$‰ (Komada et al., 2008), was similar to the precision of the analyses.

## 2.4 Terminology

Previous work quantifying OC transfers from Taiwan and other orogenic belts has used the term "fossil" OC to define OC derived from sedimentary rocks (Kao and Liu, 2000; Galy et al., 2007a; Hilton et al., 2008, 2011; Clark et al., 2013). This is identical to the term $OC_{petro}$, with "petrogenic" used here because of its unambiguous reference to rock-derived OC (Galy et al., 2008). The term "non-fossil" OC has been used to refer to OC derived from vegetation and soil in the solid load of Taiwanese rivers (Hilton et al., 2008,

**Table 1.** Average composition of fluvial OC in rivers draining the eastern and western flanks of Taiwan (Table S1) and the average input assuming approximately equal input of sediment from both sides of the mountain range (Dadson et al., 2003; Kao and Milliman, 2008).

| | $n$ | $C_{org}$ (%) | SD | $\delta^{13}C_{org}$ (‰) | SD | $\Delta^{14}C_{org}$ (‰) | SD |
|---|---|---|---|---|---|---|---|
| Average west | 20 | 0.43 | 0.16 | −25.5 | 0.7 | −646 | 237 |
| Average east | 28 | 0.45 | 0.27 | −24.4 | 1.1 | −677 | 271 |
| Average | 48 | 0.44 | 0.22 | −24.9 | 0.9 | −661 | 254 |

2012) and elsewhere (Clark et al., 2013; Smith et al., 2013) because the output of the mixing analysis is defined as not fossil (Hilton et al., 2010). Here, $OC_{biosphere}$ is used to refer to the same component of OC termed "non-fossil" in previous work (Hilton et al., 2012) because of its clear reference to the source of OC from the terrestrial biosphere.

## 3 Results

### 3.1 Composition of terrestrial OC exported to the ocean

Particulate OC in suspended sediments from the major rivers in Taiwan (Fig. 1) had an average $C_{org} = 0.4 \pm 0.2$ % ($\pm$ SD Table 1), which is at the lower end of values measured in rivers worldwide (Meybeck, 1982; Stallard, 1998) but is consistent with previous measurements on Taiwanese rivers (Kao and Liu, 1996, 2000; Hilton et al., 2008, 2010) and measurements from mountain rivers elsewhere (Komada et al., 2004; Leithold et al., 2006; Clark et al., 2013; Smith et al., 2013). The particulate OC was radiocarbon depleted and $^{13}C$-depleted, with a mean $\Delta^{14}C_{org} = -661 \pm 254$‰ and $\delta^{13}C_{org} = -24.9 \pm 0.9$‰ for the rivers studied here (Table 1). The range in isotopic composition of terrestrial OC define a triangular domain between $\Delta^{14}C_{org}$ and $\delta^{13}C_{org}$ (Fig. 2a) and the measured values are consistent with previous measurements on suspended sediments from Taiwan (Kao and Liu, 2000; Hilton et al., 2008, 2010).

### 3.2 Composition of marine OC

The sediments collected from the trap in the Gaoping Canyon accumulated during Typhoon Kalmaegi (17 July 2008), as constrained by the timing discs deployed by the sediment trap. At that time, SSC in the Gaoping River reached $> 20 \text{g L}^{-1}$ the day after the flood peak. Based on past records of $Q_w$ and SSC (Dadson et al., 2005; Kao and Milliman, 2008), it is highly likely that the Gaoping River surpassed $SSC = 40 \text{g L}^{-1}$ necessary for hyperpycnal discharge in this region during Typhoon Kalmaegi. This is consistent with the very high throughput of sediment in the canyon during the event (Liu et al., 2012). Particulate OC samples from the trap have an average $C_{org} = 0.5 \pm 0.3$ % ($n = 12$), which is within a standard deviation of the mean of western river samples

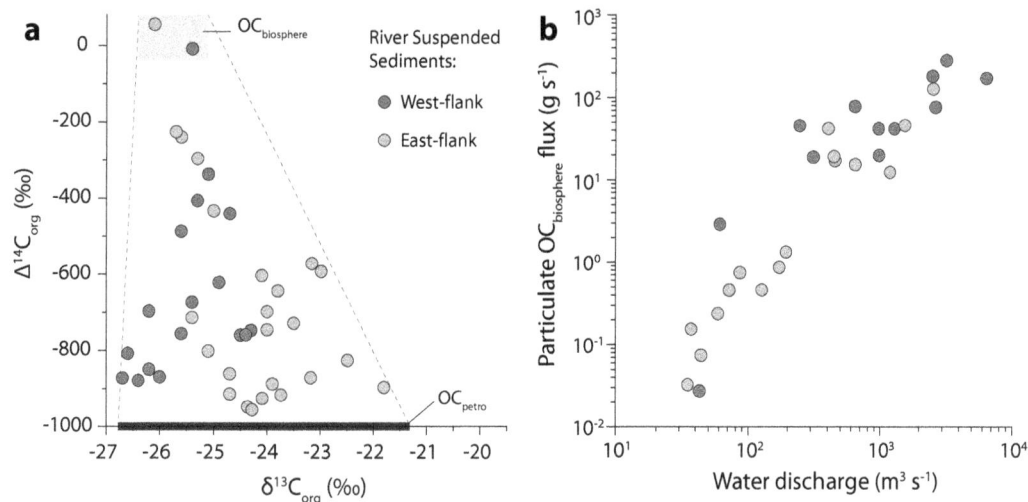

**Figure 2.** Isotopic compositions and flux of particulate OC eroded from the terrestrial biosphere ($OC_{biosphere}$) for Taiwan's rivers. **(a)** Stable and radioactive isotope compositions of organic carbon ($\delta^{13}C_{org}$ and $\Delta^{14}C_{org}$, ‰) of suspended sediments from rivers in Taiwan. Analytical errors are smaller than the point size. Samples from catchments draining the two flanks of the mountain belt define a mixing domain between organic carbon from the terrestrial biosphere ($OC_{biosphere}$) and petrogenic OC ($OC_{petro}$) from bedrocks. **(b)** Instantaneous particulate flux of $OC_{biosphere}$ ($g\,s^{-1}$) as a function of water discharge during floods in the sampled catchments, demonstrating a strong hydrological control on $OC_{biosphere}$ export.

($C_{org} = 0.43 \pm 0.16\,\%$, $n = 20$) and similar to previous measurements in the canyon (Kao et al., 2006). When we include the core samples collected from the canyon following Typhoon Morakot, all of the marine sediments from the Gaoping Canyon ($n = 15$) have a mean $C_{org} = 0.6 \pm 0.4\,\%$, which is only slightly higher than the terrestrial OC (Fig. 3b). Both sets of Gaoping Canyon samples were radiocarbon-depleted (Table S2) and their $\delta^{13}C_{org}$ and $\Delta^{14}C_{org}$ values fall into the triangular domain defined by terrestrial OC carried by Taiwanese rivers (Fig. 3a).

Away from direct hyperpycnal river inputs, marine particulates collected from sediment traps in the Okinawa Trough (Fig. 1) all had higher $\delta^{13}C_{org}$, $\Delta^{14}C_{org}$ and $C_{org}$ values than terrestrial OC and Gaoping Canyon samples (Fig. 4, Table S3). When plotted with seafloor sediments collected from the Okinawa Trough, Taiwan Strait and the Gaoping Shelf (Fig. 1), the samples reveal a significant ($P < 0.0001$, $n = 26$) positive, linear correlation between $\Delta^{14}C_{org}$ and $\delta^{13}C_{org}$ (Fig. 4a). Samples collected from below the sediment water interface on the Gaoping Shelf (Table S3) also plot on this trend, as do Holocene sediments from the Okinawa Trough (Kao et al., 2008). These seafloor sediments and trap samples also define a significant negative correlation ($P = 0.003$, $n = 26$) between $1/C_{org}$ and $\Delta^{14}C_{org}$ (Fig. 5a). A linear trend between isotopes can result from mixing of two dominant sources (a binary mixture), or may reflect a process by which both sets of isotopes are modified (for example by preferential loss of one component of OC). A linear trend between the inverse of concentration and isotope composition may also result from binary mixing, but OC loss can result in a linear

trend which is distinct from mixing. The explanation of these trends will be considered in detail in the Discussion.

## 4 Discussion

### 4.1 Erosion and transfer of terrestrial OC

Physical erosion processes occur at very high rates in mountain landscapes. In Taiwan, suspended sediment yields reach $> 10\,000\,Mg\,km^{-2}\,yr^{-1}$ due to a combination of steep slopes and intense precipitation during tropical cyclones (Dadson et al., 2003; Kao and Milliman, 2008). These factors mean that bedrock landslides are common, delivering clastic sediment to mountain rivers (Hovius et al., 2000). Landslides also erode organic matter from the terrestrial biosphere and supply OC, mixed with clastic sediment, to rivers (Hilton et al., 2008; West et al., 2011). In addition, the high runoff intensity promotes mobilisation of soil organic matter by overland flow processes (Hilton et al., 2012). These previous studies have examined the erosion and transport of $OC_{biosphere}$ in detail and estimated rates of $OC_{biosphere}$ transfer which rank amongst the highest in the world (Kao and Liu, 1996, 2000; Stallard, 1998; Hilton et al., 2008, 2012). Here we summarize the key findings of this previous work in light of the new $^{14}C$ data from Taiwan (Fig. 3) and to inform our assessment of the offshore fate of terrestrial OC (Sects. 4.2 and 4.3).

The geomorphic processes occurring in Taiwan can explain the triangular domain in $\Delta^{14}C_{org}$ and $\delta^{13}C_{org}$ values due to mixing of $OC_{petro}$ and $OC_{biosphere}$ (Hilton et al., 2010). Radiocarbon depletion in the samples can be accounted for by the input of $OC_{petro}$ from bedrock. Due to its geological

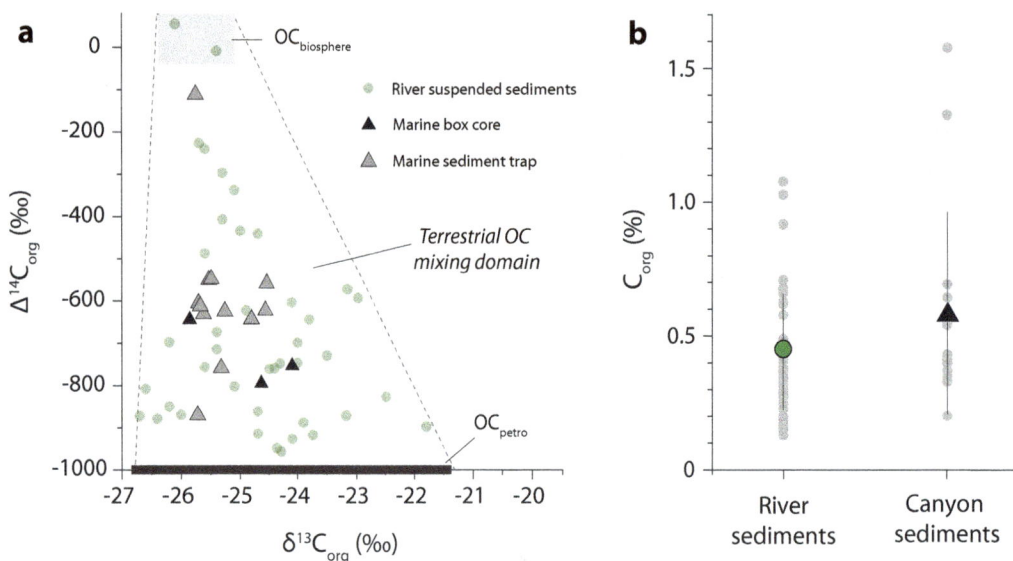

**Figure 3.** Composition of marine sediments in the Gaoping Canyon (Fig. 1) fed by periodic hyperpycnal flows. **(a)** Stable and radioactive isotope compositions of organic carbon ($\delta^{13}C_{org}$ and $\Delta^{14}C_{org}$, ‰) in marine sediments (sediment trap and box core) from the Gaoping Canyon, along with river suspended sediments of Taiwan (as shown in Fig. 2b). **(b)** Organic carbon concentration ($C_{org}$, %), with mean ±standard deviation shown by large symbol and whiskers for river suspended sediments and canyon sediments (trap and core) from the Gaoping Canyon.

age (> 50 ka), $OC_{petro}$ from Mesozoic–Cenozoic sedimentary rocks has no measurable $^{14}C$ and so has $\Delta^{14}C_{org} \sim -1000\,‰$. In addition, $OC_{petro}$ input can account for the range in $\delta^{13}C_{org}$ values at low $\Delta^{14}C_{org}$ in Taiwan (Fig. 2a), where metasedimentary bedrocks have $\delta^{13}C_{org}$ values ranging from $-25\,‰$ up to $-20\,‰$ (Hilton et al., 2010). Hilton et al. (2010) reported that rivers draining the east flank of Taiwan can have higher $\delta^{13}C_{org}$ values ($^{13}C$-enriched) than those draining the west due to different bedrock geology leading to variable $OC_{petro}$ composition. The $^{14}C$-depleted samples are consistent with this observation (Fig. 2a). In contrast, when values of $\Delta^{14}C_{org}$ are higher ($^{14}C$-enriched), the stable isotope composition is much less variable (Fig. 2a). The $\delta^{13}C_{org}$ values are similar to those of C3 biomass and soil in Taiwanese mountain forest (Kao and Liu, 2000; Hilton et al., 2013). Previous work has suggested that surface soil horizons in Taiwan have a $C_{org}$-weighted average $\Delta^{14}C_{org} \sim 0\,‰$ (Hilton et al., 2008), reflecting the high rates of OC export in the source area which can act to limit soil age (Hilton et al., 2012). Addition of this young $OC_{biosphere}$ can explain the range of $\Delta^{14}C_{org}$ values in the suspended sediments (Fig. 2a). However, we note that an expanded soil data set (Hilton et al., 2013) shows that older soil $OC_{biosphere}$ ($\Delta^{14}C_{org} \sim -100\,‰$) may be present in this mountain belt. Its volumetric contribution to the river load is difficult to assess because the $^{14}C$-depletion in the bulk suspended sediment is dominated by $OC_{petro}$ inputs (Fig. 2a). We suggest that organic-compound specific $^{14}C$ measurements may shed light on the erosion and transport of any aged soil OC (e.g. Galy and Eglinton, 2011).

The rapid fluvial transit times in Taiwan, combined with young soils (Hilton et al., 2008), mean that the $^{14}C$ content of particulate OC can provide a proxy for $OC_{petro}$ input to the suspended load, in analogy with work from other small mountain river systems (e.g. Komada et al., 2004; Leithold et al., 2006; Clark et al., 2013). Taking $\Delta^{14}C_{org}$ values for bedrock ($-1000\,‰$) and $OC_{biosphere}$ based on the $C_{org}$-weighted mean of 10 surface soil samples of $0\,‰$ (Hilton et al., 2008) we have used an end member mixing model to quantify the fraction of $OC_{biosphere}$ in river suspended sediments. If older soil organic matter ($\Delta^{14}C_{org} < 0\,‰$) is more important than the existing soil samples suggest (Hilton et al., 2008, 2013), the estimated $OC_{biosphere}$ component will be conservative. Results indicate that the flux of particulate $OC_{biosphere}$ ($g\,s^{-1}$) increases with water discharge (Fig. 2b). This confirms a strong climatic control on the erosion and fluvial transfer of $OC_{biosphere}$ highlighted in several Taiwanese catchments using a slightly different method for quantifying the mixing of $OC_{petro}$ and $OC_{biosphere}$ (Hilton et al., 2010; Hilton et al., 2012). It reflects the activation of erosion processes (overland flow, gully incision, landslides) during heavy precipitation and supply of $OC_{biosphere}$ and sediment to rivers when their transport capacity is high (Hilton et al., 2008, 2012). This mechanism is not unique to Taiwan, and has been observed elsewhere when flux and OC source data are both available in tropical (Bass et al., 2011; Clark et al., 2013; Lloret et al., 2013) and temperate mountain forests (Hatten et al., 2012; Smith et al., 2013). A consequence of this behaviour is that flood events can dominate the particulate flux, with 80–90 % of the decadal transfer of $OC_{biosphere}$

by the Liwu River found to occur during cyclonic storms with return times $> 1$ yr (Hilton et al., 2008). The importance of storm-triggered floods for particulate transfer appears to be a wider feature of small mountain rivers (Townsend-Small et al., 2008; Bass et al., 2011; Lloret et al., 2013).

The suspended sediment samples provide good constraint on the compositional range of terrestrial OC delivered directly to the ocean by hyperpycnal plumes. In addition, because our sample set captures particulate OC from across the mountain belt (Fig. 1), we can assess the likely composition of hypopycnal inputs, which may be expected to be a mixture of sediments sourced from individual river catchments. The average of all river samples (Table 1) is $C_{org} = 0.44 \pm 0.22$ %, $\delta^{13}C_{org} = -24.9 \pm 0.9$ ‰, $\Delta^{14}C_{org} = -661 \pm 254$ ‰. The $\Delta^{14}C_{org}$ value suggests that $OC_{biosphere}$ contributes $\sim 30$ % of the total particulate OC on average with $OC_{petro}$ making up the remaining fraction, in agreement with previous estimates from Taiwan (Kao and Liu, 2000; Hilton et al., 2008; Hilton et al., 2010).

## 4.2 Fate of terrestrial OC offshore Taiwan: hyperpycnal inputs

To constrain the transfer of terrestrial OC to the marine environment, we can use the isotopic and elemental composition of samples from marine sediment traps offshore Taiwan. To assess the longer-term preservation, the composition of seafloor sediments and longer cores can be used. The $\Delta^{14}C_{org}$ and $\delta^{13}C_{org}$ values of terrestrial OC exported by rivers from Taiwan (Fig. 2a) have been characterized more thoroughly than any previous study (Kao and Liu, 1996; Hilton et al., 2008). Recent marine OC can be distinguished from terrestrial OC due to its higher $\delta^{13}C_{org}$ and $\Delta^{14}C_{org}$ values (Hsu et al., 2006). However, the assessment of OC provenance is not the same as quantifying preservation. To do that, we have used $C_{org}$ values of the sediments because $C_{org}$ values are sensitive to changes in the association of OC with clastic particles and can track OC loss (e.g. Galy et al., 2007a; Blair and Aller, 2012; Cathalot et al., 2013). Our investigation of marine samples does not extend to the very deep ocean waters offshore the east coast of Taiwan (Fig. 1). However, $O_2$ concentrations in the unsampled region, which reach water depths $> 4000$ m, are low and comparable to those in the Bay of Bengal at 2000 m water depth (Garcia et al., 2010). OC preservation may be higher in these unsampled areas due to the lower oxidation potential of these deep waters (e.g. Cai and Sayles, 1996; Galy et al., 2007a) than the sites which form the focus of our study (Fig. 1). As a result, our estimates of terrestrial OC preservation efficiency may be conservative.

To assess the fate of terrestrial OC delivered to the ocean by rivers during hyperpycnal discharges, we have examined OC collected by the sediment trap moored in the channel thalweg of the Gaoping Canyon, fed by the Gaoping River (Fig. 1). The sediments trapped during the passage of Typhoon Kalmaegi (17 July 2008) had a range in

$\Delta^{14}C_{org}$ and $\delta^{13}C_{org}$ values, consistent with the mixture of terrestrial $OC_{petro}$ and $OC_{biosphere}$ observed in river sediments (Fig. 3a), with an average $C_{org}$ very close to the river samples (Fig. 3b). These observations suggest that loss of terrestrial OC during transfer to mesopelagic depths ($\sim 600$ m) during this hyperpycnal delivery event has been negligible. The trapped sediment included a "young" organic rich sub-sample ($C_{org} = 1.6$ %, $\Delta^{14}C_{org} = -112$ ‰, Fig. 3b) with some shredded woody debris visible to the naked eye and up to $\sim 1$ cm in size (see also Fig. 4 in Liu et al., 2012). Samples collected from the floor of the Gaoping Canyon after Typhoon Morakot also lie within the terrestrial mixing domain (Fig. 3a). Their $C_{org}$ values also imply little evidence for terrestrial OC loss (Fig. 3b). Thus, we have found that, contrary to a previous study (Kao et al., 2006), loss of $OC_{biosphere}$ (which would systematically lower $\Delta^{14}C_{org}$ and $C_{org}$) is not consistent with the data. Together, the trap and core samples suggest efficient transfer to surface sediments and preservation in the sub-surface ($\sim 100$ %) of terrestrial OC (both $OC_{biosphere}$ and $OC_{petro}$) in a submarine canyon fed by hyperpycnal flows. While the fate of terrestrial OC transported deeper down the canyon (e.g. Carter et al., 2012) remains to be assessed, the low $O_2$ levels (Garcia et al., 2009) and high accumulation rates (Huh et al., 2009) are likely to promote longer-term OC burial.

Moreover, it appears that the natural buoyancy of some macro-particles of $OC_{biosphere}$ can be overcome during hyperpycnal flood discharges, as observed in modern source-to-sink settings elsewhere (Leithold and Hope, 1999) and in the geological record (e.g. Saller et al., 2006). This suggests that the density of the turbid river plume may be high enough to effectively sequester woody debris carried in the sand fraction, while coarser woody material (e.g. logs) float upon discharge to the ocean (West et al., 2011). In addition, water logging of sand-sized woody debris may occur prior to entrainment or during transport, as observed in the sand-sized bedload of larger fluvial systems (Bianchi et al., 2007). In the short mountain rivers of Taiwan, it is unclear whether this mechanism operates; the observation warrants further investigation of the transport of macro-particles of $OC_{biosphere}$ in mountain rivers (e.g. Turowski et al., 2013).

## 4.3 Fate of terrestrial OC offshore Taiwan: hypopycnal inputs

The fate of terrestrial OC away from direct hyperpycnal inputs can be examined using marine samples collected from a wider region around Taiwan (Fig. 1). The core and trap samples from the Okinawa Trough, Taiwan Strait and the Gaoping Shelf displayed significant trends between $\Delta^{14}C_{org}$ and $\delta^{13}C_{org}$ (Fig. 4a), which distinguish them from the Gaoping Canyon samples (Fig. 3a). However, compared to recent marine OC ($OC_{marine}$) from the western Pacific (Hsu et al., 2006; Table 2), they were variably depleted in both $^{13}C$ and $^{14}C$ (Fig. 4a). The values cannot be explained by aging and

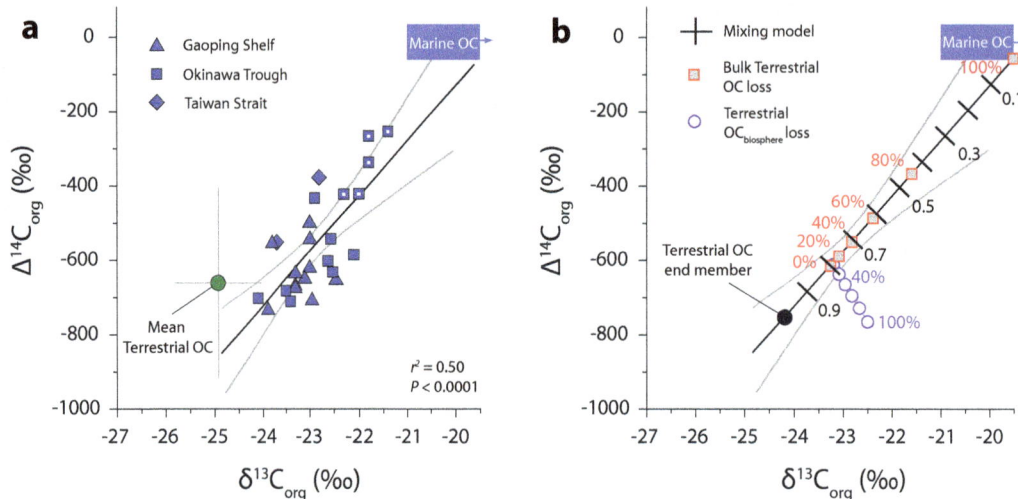

**Figure 4.** Stable and radioactive isotopic compositions of organic carbon ($\delta^{13}C_{org}$ and $\Delta^{14}C_{org}$, ‰) in marine sediments offshore Taiwan fed by dispersive terrestrial inputs (Fig. 1). **(a)** White dots denote samples from the trap moorings. The mean terrestrial OC composition delivered by rivers (green circle, whiskers ± SD) and the expected composition of recent marine OC (blue box) are shown. The samples display a positive linear relationship (black line, with 95 % confidence intervals in grey). **(b)** Linear relationship displayed in the samples along with the isotopic composition of OC predicted by (i) mixing marine OC and terrestrial OC (black line and dashes with fraction of terrestrial OC); (ii) loss of terrestrial $OC_{biosphere}$ starting at fraction terrestrial OC = 0.8 (circles with % loss); and (iii) bulk terrestrial OC loss (squares with % loss).

**Table 2.** Compositions used in the mixing and terrestrial OC loss models.

|  | $\Delta^{14}C_{org}$ (‰) | $\delta^{13}C_{org}$ (‰) | $C_{org}$ (%) |
|---|---|---|---|
| $OC_{marine}$ | −59[a] | −19.5 | 30 |
| River OC | −753[b] | −24.2[b] | 0.48[b] |
| River $OC_{petro}$ | −1000 | −23.5[b] | – |
| River $OC_{biosphere}$ | 0[c] | −26.0[b] | – |

[a] From open marine surface trap samples (Hsu et al., 2006). [b] Indicative values used to examine the nature of trends in the data (Figs. 4b and 5b) informed by the measured compositions (Table 1, Figs. 2a, 4a and 5a). [c] Measured mean of bulk soils from Taiwan weighted by organic carbon content (Hilton et al., 2008).

re-suspension of $OC_{marine}$ because this $^{14}$C-depletion only results in $\Delta^{14}C_{org}$ values of approximately −50‰ to −100‰ in this setting (Hwang et al., 2010). Instead, the linear trend between $\Delta^{14}C_{org}$ and $\delta^{13}C_{org}$ values may be indicative of binary mixing between end members with distinct compositions (Komada et al., 2004; Clark et al., 2013). The best fit to the data intersects the average of measured riverine OC (itself a mixture of $OC_{petro}$ and $OC_{biosphere}$) and the values expected for recent $OC_{marine}$ (Fig. 4a). Thus, the first order pattern in the samples collected away from direct hyperpycnal inputs can be explained by mixing $OC_{marine}$ with terrestrial OC.

However, loss of terrestrial OC during marine transfer and deposition may have caused $\Delta^{14}C_{org}$ and $\delta^{13}C_{org}$ values to evolve towards the composition of $OC_{marine}$. The linear trend

suggests that if this loss has occurred, it has done so in a relatively short period of time, because otherwise $\Delta^{14}C_{org}$ would vary with time and produce a non-linear relationship with $\delta^{13}C_{org}$. To assess the possible loss of terrestrial OC in the marine realm, we model a scenario of instantaneous loss (see Supplement). The results indicate that preferential loss of $OC_{biosphere}$ (e.g. Kao et al., 2006; Cathalot et al., 2013) produces a negative, linear trend between $\Delta^{14}C_{org}$ and $\delta^{13}C_{org}$ (Fig. 4b), which is not consistent with the data. On the other hand, bulk loss of terrestrial OC (both $OC_{biosphere}$ and $OC_{petro}$) can produce the observed positive, linear trend between $\Delta^{14}C_{org}$ and $\delta^{13}C_{org}$ (Fig. 4b).

To constrain whether mixing or loss is the dominant control on the isotopic composition of the marine samples, we have turned to $C_{org}$. The model of bulk terrestrial OC loss, which can explain the $\Delta^{14}C_{org}$ and $\delta^{13}C_{org}$ values (Fig. 4b), cannot reproduce the negative linear relationship between $1/C_{org}$ and $\Delta^{14}C_{org}$. The modelled bulk terrestrial OC loss results in a trend which is perpendicular to that observed in the samples (Fig. 5b). Thus, it appears that the patterns in the data are not consistent with either selective (i.e. $OC_{biosphere}$, Fig. 4b) or pervasive (i.e. $OC_{biosphere}$ and $OC_{petro}$, Fig. 5b) loss of terrestrial OC. Thus, the only way to account for the first order trends in the measured isotopic and elemental composition of OC ($\delta^{13}C_{org}$, $\Delta^{14}C_{org}$ and $C_{org}$) in the offshore sediments is through a mixture of $OC_{marine}$ and riverine OC (itself a mixture of $OC_{petro}$ and $OC_{biosphere}$) (Figs. 4 and 5).

The scatter around the linear trends in the data may reflect second-order temporal (or spatial) variations in the $\Delta^{14}C_{org}$,

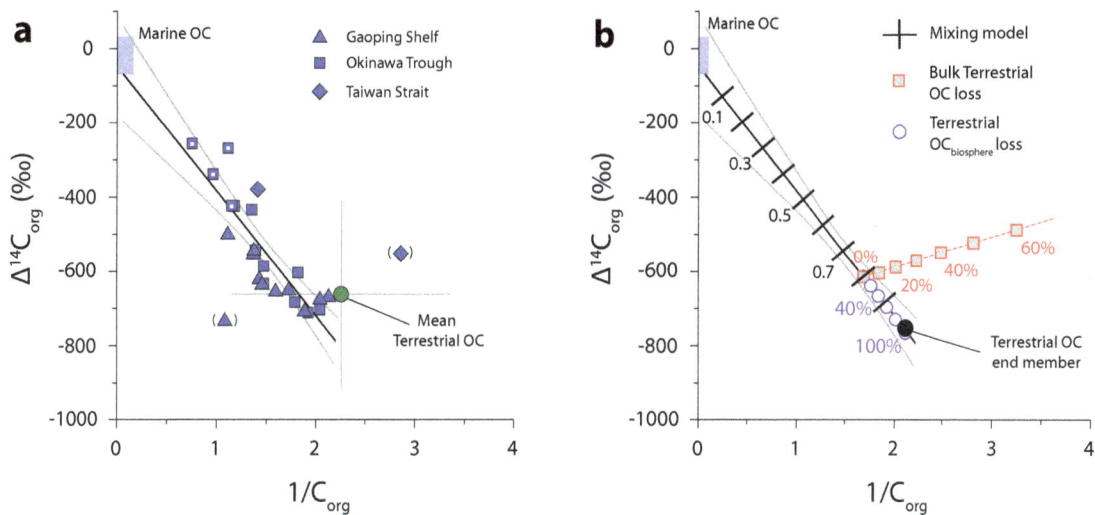

**Figure 5.** Radiocarbon contents of organic carbon (expressed as $\Delta^{14}C_{org}$, ‰) versus the inverse of OC concentration ($1/C_{org}$) in marine sediments offshore Taiwan fed by dispersive terrestrial inputs (Fig. 1) with symbols as in Fig. 4. (**a**) All samples show a negative relationship between the variables ($r = -0.6$; $P = 0.003$) with a linear fit to all samples apart from those in brackets shown by the blank line ($r^2 = 0.7$; $P = 0.0001$, grey line is the 95 % confidence interval). (**b**) Linear relationship displayed in the samples along with the OC content predicted by (i) mixing marine OC and terrestrial OC (black line and dashes with fraction of terrestrial OC); (ii) loss of $OC_{biosphere}$ (starting at fraction terrestrial OC = 0.8, circles with % loss); and (iii) bulk terrestrial OC loss (squares with % loss).

$\delta^{13}C_{org}$ and $C_{org}$ values of the marine and terrestrial OC end members (Table 1), which can explain the sample compositions. However, the scatter may also reflect some terrestrial OC loss. The instantaneous loss model provides constraint on the percentage of terrestrial OC loss, which is compatible with the variability around the linear trends (Fig. 5). For bulk OC loss, the scatter corresponds to ~ 20 % loss (Fig. 5b). To acknowledge the uncertainty on this value, and to provide a conservative estimate of terrestrial OC burial efficiency, we suggest < 30 % loss (i.e. preservation efficiency > 70 %) has occurred in these deposits. Note that this sample set is dominated by surface marine sediments (Table S3) and so the preservation efficiency refers to the land-to-ocean transfer and deposition of terrestrial $OC_{biosphere}$ and $OC_{petro}$. However, sub-surface sediments from the Gaoping Shelf and the Holocene sediments from the Okinawa Trough (Table S3) are consistent with the trends (Figs. 4a and 5a). While the limited number of these subsurface samples ($n = 7$) makes it difficult to draw irrefutable conclusions, it appears that high preservation efficiencies are also a feature of sedimentary burial on longer timescales in this setting (Kao et al., 2008).

Efficient preservation of terrestrial $OC_{biosphere}$ and $OC_{petro}$ in both hyperpycnal and hypopycnal marine sediments (Figs. 3, 4 and 5) is consistent with the high sedimentation rates in the deep ocean basins surrounding Taiwan, which result from the tectonic and climate setting. Sedimentation rates in the southern Okinawa Trough reach > 1 mm yr$^{-1}$, sustained by fluvial sediment delivery (Hsu et al., 2006). Indeed, if the total suspended sediment flux of ~ 380 Tg yr$^{-1}$ (Dadson et al., 2003) is mostly deposited within ~ 100 km

of the coast (over ~ 150 000 km$^2$), then the average sedimentation rate would be ~ 2 mm yr$^{-1}$ (density of 2.2 g cm$^{-3}$). The rapid accumulation of clastic sediment limits the time over which OC is exposed to O$_2$ (Burdige, 2005; Galy et al., 2007a). Since the O$_2$ penetration depth in muddy marine sediments is typically on the order of 1–10 mm (Hedges and Keil, 1995; Cai and Sayles, 1996), $OC_{biosphere}$ deposited offshore Taiwan is probably exposed to O$_2$ for only a matter of years.

Our findings are consistent with marine sediments undergoing rapid accumulation elsewhere, with bulk OC preservation efficiencies of 70–100 % (Galy et al., 2007a; Blair and Aller, 2012). These tend to exceed terrestrial organic carbon preservation rates in other depositional settings (Burdige, 2005; Blair and Aller, 2012). Unlike steep mountain rivers, large river deltas on passive margins can experience successive resuspension and deposition episodes within mobile muds (e.g. the Amazon Delta). In these systems, marine organic material can be entrained into mobile deposits, driving early diagenetic reactions and facilitating loss of refractory terrestrial OC during sedimentary refluxing and suboxic diagenesis (Aller, 1998; Aller and Blair, 2006). While conditions in large rivers on passive margins can promote rapid export and deposition of marine OC (Bianchi et al., 2006), resuspension and re-working of sediments on shallow-sloping deltas can also promote incineration of terrestrial OC (Aller et al., 1996, 2008; Aller and Blair, 2006), even following a rapid sediment accumulation event (Sampere et al., 2008). In contrast, apart from the ~ 100 m deep Taiwan Strait (Fig. 1), which may be analogous to some of these passive margin settings, rivers export materials to deep basins around the island. The rapid accumulation rates offshore Taiwan and the delivery of terrestrial OC to water depths > 500 m in $O_2$ poor waters (Garcia et al., 2009) are conditions more analogous to the Bengal Fan system, where terrestrial OC burial efficiencies are also very high (Galy et al., 2007a).

## 5  Implications for the global carbon cycle

After accounting for addition of terrestrial $OC_{petro}$ and $OC_{marine}$ to the offshore sediments (e.g. Fig. 4), we can now assess how the erosion of terrestrial biomass ($OC_{biosphere}$) drives sequestration of atmospheric $CO_2$. Our findings suggest that in the Taiwan sediment routing system, rivers deliver sediments which contain on average 0.15 % of terrestrial $OC_{biosphere}$ (average $C_{org} \times$ average fraction of $OC_{biosphere}$) in hypo- and hyperpycnal river plumes. Given the suspended sediment flux from Taiwan to the ocean of $384 \, Tg \, yr^{-1}$ (Dadson et al., 2003; Dadson et al., 2005), this abundance of $OC_{biosphere}$ and the estimated preservation efficiencies of terrestrial OC of > 70 % (see Sect. 4.2), we calculate an $OC_{biosphere}$ burial flux of $0.5–0.6 \, TgC \, yr^{-1}$ in basin fills derived from Taiwan. This may be a lower bound if material coarser than ~ 500 µm (i.e. large woody debris, whose fate remains poorly constrained) contributes importantly to $OC_{biosphere}$ transfer and burial (West et al., 2011). Normalized over Taiwan's mountain island surface area ($35\,980 \, km^2$), this estimated flux represents an $OC_{biosphere}$ burial yield of $13–16 \, MgC \, km^{-2} \, yr^{-1}$. $CO_2$ sequestration associated with physical erosion of Taiwan appears to be seven times more efficient per $km^2$ than the Himalayan erosion system, which has a burial yield of $\sim 2 \, MgC \, km^{-2} \, yr^{-1}$ (Galy et al., 2007a). Our analysis suggests that the rock-derived $OC_{petro}$ is reburied offshore Taiwan at a rate of $0.9–1.1 \, TgC \, yr^{-1}$, similar to the total $OC_{petro}$ buried annually in the Bay of Bengal (Galy et al., 2008). $OC_{biosphere}$ burial from this single mountain island represents ~ 1 % of the estimated total annual OC burial in the oceans (Schlunz and Schneider, 2000; Burdige 2005) from only 0.02 % of Earth's continental surface.

The islands of Oceania are sediment production hotspots, with suspended sediment yields typically > $1000 \, Mg \, km^{-2} \, yr^{-1}$ (Milliman and Farnsworth, 2011). The transfer of $OC_{biosphere}$ together with clastic sediment may enhance the terrestrial OC burial efficiency (Canfield, 1994; Burdige, 2005; Blair and Aller, 2012), even when materials are delivered by hypopycnal river plumes (see Sect. 4.3). As in Taiwan, terrestrial productivity is high across tropical Oceania and mountain forests contain large stores of $OC_{biosphere}$ in standing biomass and soil (Dixon et al., 1994). This permits us to extrapolate our observations to provide a tentative estimate of the $CO_2$ sink associated with the oceanic burial of terrestrial $OC_{biosphere}$. Assuming a linear relationship between sediment yield and $OC_{biosphere}$ yield (e.g. Hilton et al., 2012), the terrestrial $OC_{biosphere}$ content

from Taiwan can be combined with the Oceania sediment export of $\sim 7000 \, Tg \, yr^{-1}$ (Dadson et al., 2003; Milliman and Farnsworth, 2011) and the range of preservation efficiencies obtained here (70–100 %) to estimate a terrestrial $OC_{biosphere}$ burial flux of $8–11 \, TgC \, yr^{-1}$ from the mountain islands of Oceania. This estimate is likely to be conservative because the high sediment yields in Taiwan of $\sim 9000 \, Mg \, km^{-2} \, yr^{-1}$ result in a lower percent of $OC_{biosphere}$ in sediments when compared to other mountain rivers (Leithold et al., 2006; Hilton et al., 2012; Clark et al., 2013; Smith et al., 2013). Alternatively, it could be assumed that the $OC_{biosphere}$ burial yield from Taiwan ($13–16 \, MgC \, km^{-2} \, yr^{-1}$) holds over the Oceania area ($2.7 \times 10^6 \, km^2$), giving a terrestrial $OC_{biosphere}$ burial flux of $35–40 \, TgC \, yr^{-1}$. This value could be viewed as an upper bound, since the erosion rate of $OC_{biosphere}$ from Taiwan may be toward the high end of global values (Hilton et al., 2012). Both estimates do not consider $OC_{marine}$ burial associated with these sediments, which could increase $CO_2$ sequestration (Fig. 4a). The lower conservative estimate of $OC_{biosphere}$ burial by the erosion of Oceania is globally significant. It represents ~ 10 % of estimates of the total OC burial in clastic sediments in the oceans (Schlunz and Schneider, 2000; Burdige 2005; Blair and Aller, 2012). Adjustment of these global estimates is warranted and requires further observational constraint on the processes and magnitude of this significant flux in the global carbon cycle.

Our findings suggest that mountain building in Oceania can result in a globally important geological $CO_2$ sink through erosion of the terrestrial biosphere, $OC_{biosphere}$ transport by mountain rivers and preservation in marine sediments from hyperpycnal but, importantly, also hypopycnal delivery events (Figs. 4 and 5). This region provides a strong link between tectonic uplift and the carbon cycle, which should influence atmospheric $CO_2$ concentrations on geological timescales. Importantly, $CO_2$ sequestration by erosion of $OC_{biosphere}$ should be sensitive to the coverage of terrestrial biomass in the tropics, which is in part moderated by the available supply of $CO_2$ for productivity (Norby et al., 2005). In addition, the amount and variability of runoff control the erosion and export of terrestrial $OC_{biosphere}$ (Fig. 2b) and clastic sediment by small mountain rivers (Dadson et al., 2003; Hilton et al., 2008, 2012). Therefore, islands of Oceania have the potential to introduce stabilizing feedbacks in the carbon cycle on geological timescales, which are presently not considered in Earth System models (Berner, 2006; Archer et al., 2010) and not invoked to explain the evolution of atmospheric $CO_2$ levels in the Cenozoic (e.g. Pagani et al., 2009). One aspect of this may be the link between warming ocean temperatures and the occurrence of extreme tropical cyclones in the western Pacific (Elsner et al., 2008) which deliver $OC_{biosphere}$ and sediment efficiently to the ocean, as previously hypothesized (Hilton et al., 2008). However, a wider response across Oceania may be felt due to $CO_2$ fertilization of tropical forests (Norby et al., 2005), while ocean warming also increases runoff and runoff variability in the tropics away from tropical cyclone influence (Milly et al., 2005). The corresponding increase in terrestrial $OC_{biosphere}$ export to the oceans from tropical islands may help to mitigate an increase in greenhouse gas concentration, with enhanced $CO_2$ sequestration by terrestrial $OC_{biosphere}$ burial in the ocean. These feedbacks may play a significant role in linking tectonics and climate and their impact on the long-term carbon cycle, all of which deserve further attention.

**Acknowledgements.** We thank J. Chien (RCEC, AS) for isotope analysis and K. T. Jiann (NSYSU) for offering Taiwan Strait samples, and also V. Galy, D. Burdige and three anonymous referees for their comments, which improved the manuscript. This work was funded by Taiwan (NSC-97-2611-M-001-001-MY2, NSC-97-2628-M-001-025, NSC-99-2116-M-001-011), China (NSFC 41176059, 973 Project 2009CB421200 and 2009CB421206, 111 Program B07034) and the Natural Environment Research Council (NERC), UK (Radiocarbon Allocation 1203.1006, 1228.0407).

Edited by: V. Galy

# References

Aller, R. C.: Mobile deltaic and continental shelf muds as suboxic, fluidized bed reactors, Mar. Chem., 61, 143–155, 1998.

Aller, R. C. and Blair, N. E.: Carbon remineralization in the Amazon-Guianas tropical mobile mudbelt: A sedimentary incinerator, Continental Shelf Res., 26, 2241–2259, 2006.

Aller, R. C., Blair, N. E., Xia, Q., and Rude, P. D.: Remineralization rates, recycling and storage of carbon in Amazon shelf sediments, Continental Shelf Res., 16, 753–786, 1996.

Aller, R. C., Blair, N. E., and Brunskill, G. J.: Early diagenetic cycling, incineration, and burial of sedimentary organic carbon in the central Gulf of Papua (Papua New Guinea), J. Geophys. Res., 113, F01S09, doi:10.1029/2006JF000689, 2008.

Archer, D., Eby, M., Brovkin, V., Ridgwell, A., Cao, L., Mikolajewicz, U., Caldeira, K., Matsumoto, K., Munhoven, G., Montenegro, A., and Tokos, K.: Atmospheric lifetime of fossil fuel carbon dioxide, Annu. Rev. Earth Planet. Sci., 37, 117–134, 2009.

Bass, A. M., Bird, M. I., Liddell, M. J., and Nelson, P. N.: Fluvial dynamics of dissolved and particulate organic carbon during periodic discharge events in a steep tropical rainforest catchment, Limnol. Oceanogr., 56, 2282–2292, 2011.

Berner, R. A.: GEOCARBSULF: A combined model for Phanerozoic atmospheric $O_2$ and $CO_2$, Geochim. Cosmochim. Acta, 70, 5653–5664, 2006.

Bianchi, T. S.: The role of terrestrially derived organic carbon in the coastal ocean: A changing paradigm and the priming effect, Proc. Nat. Acad. Sci., 108, 19473–19481, 2011.

Bianchi, T. S., Allison, M. A., Canuel, E. A., Corbett, D. R., McKee, B. A., Sampere, T. P.,Wakeham, S. G., and Waterson, E.: Rapid export of organic matter to the Mississippi Canyon, EOS, 87, 572–573, 2006.

Bianchi, T. S., Galler, J. J., and Allison, M. A.: Hydrodynamic sorting and transport of terrestrially-derived organic carbon in sediments of the Mississippi and Atchafalaya Rivers, Estuarine Coastal Shelf Sci., 73, 211–222, 2007.

Blair, N. E. and Aller, R. C.: The fate of terrestrial organic carbon in the marine environment, Annu. Rev. Mar. Sci., 4, 17.1–17.23, 2012.

Blair, N. E. Leithold, E. L., Ford, S. T., Peeler, K. A., Holmes, J. C., and Perkey D. W.: The persistence of memory: The fate of ancient sedimentary organic carbon in a modern sedimentary system, Geochim. Cosmochim. Acta, 67, 63–73, 2003.

Blair N. E., Leithold, E. L., and Aller, R. C.: From bedrock to burial: the evolution of particulate organic carbon across coupled watershed-continental margin systems, Mar. Chem., 92, 141–156, 2004.

Blair, N. E., Leithold, E. L., Brackley, H., Trustrum, N., Page, M., and Childress, L.: Terrestrial sources and export of particulate organic carbon in the Waipaoa sedimentary system: Problems, progress and processes, Mar. Geol., 270, 108–118, 2010.

Buesseler, K. O., Antia, A. N., Chen, M., Fowler, S. W., Gardner, W. D., Gustafsson, O., Harada, K., Michaels, A. F., Rutgers van der Loeff, M., Sarin, M., Steinberg, D. K., and Trull, T.: An assessment of the use of sediment traps for estimating upper ocean particle fluxes, J. Marine Res., 65, 345–416, 2007.

Burdige, D. J.: Burial of terrestrial organic matter in marine sediments: A re-assessment, Global Biogeochem. Cycles, 19, GB4011, doi:10.1029/2004GB002368, 2005.

Cai, W. J. and Sayles, F. L.: Oxygen penetration depths and fluxes in marine sediments, Mar. Chem., 52, 123–131, 1996.

Canfield, D. E.: Factors influencing organic-carbon preservation in marine-sediments, Chem. Geol., 114, 315–329, 1994.

Carter, L., Milliman, J. D., Talling, P. J., Gavey, R. and Wynn R. B.: Near synchronous and delayed initiation of long run-out submarine sediment flows from a record-breaking river flood, offshore Taiwan, Geophys. Res. Lett., 39, L12603, doi:10.1029/2012GL051172, 2012.

Cathalot, C., Rabouille, C., Tisnérat-Laborde, N., Toussaint, F., Kerhervé, P., Buscail, R., Loftis, K., Sun, M-Y., Tronczynski, J., Azoury, S., Lansard, B., Treignier, C., Pastor, L., and Tesi, T.: The fate of river organic carbon in coastal areas: a study in the Rhône River delta using multiple isotopic ($\delta^{13}C$, $\Delta^{14}C$) and organic tracers, Geochim. Cosmochim. Acta, 118, 33–55, doi:10.1016/j.gca.2013.05.001, 2013.

Clark, K. C., Hilton, R. G., West, A. J., Malhi, Y., Grocke, D. R., Bryant, C. L., Ascough, P. L., Robles, A., and New, M.: New views on old carbon in the Amazon river: Insight from the source of organic carbon eroded from the Peruvian Andes, Geochemistry, Geophysics, Geosystems, 14, 1644–1659, doi:10.1002/ggge.20122, 2013.

Dadson, S. J., Hovius, N., Chen, H., Dade, W. B., Hsieh, M. L., Willett, S. D., Hu, J. C., Horng, M. J., Chen, M. C., Stark, C. P., Lague, D., and Lin, J. C.: Links between erosion, runoff variability and seismicity in the Taiwan orogen, Nature, 426, 648–651, 2003.

Dadson, S. J., Hovius, N., Pegg, S., Dade, W. B., Horng, M. J., and Chen, H.: Hyperpycnal river flows from an active mountain belt, J. Geophys. Res. Earth Surf., 110, F04016, doi:10.1029/2004JF000244, 2005.

Dixon, R. K., Brown, S., Houghton, R. A., Solomon, A. M., Trexler, M. C., and Wisniewski, J.: Carbon pools and flux of global forest ecosystems, Science, 263, 185–190, 1994.

Eglinton, T. I.: Tempestuous transport, Nature Geosci., 1, 727–728, 2008.

Elsner, J. B., Kossin, J. P., and Jagger, T. H.: The increasing intensity of the strongest tropical cyclones, Nature, 455, 92–95, 2008.

France-Lanord, C. and Derry, L. A.: Organic carbon burial forcing of the carbon cycle from Himalayan erosion, Nature, 390, 65–67, 1997.

Gaillardet, J. and Galy, A.: Himalaya–Carbon Sink or Source? Science, 320, 1727–1728, 2008.

Gaillardet, J., Dupré, B., Louvat, P., and Allegre, C. J.: Global silicate weathering and $CO_2$ consumption rates deduced from the chemistry of large rivers, Chem. Geol., 159, 3–30, 1999.

Galy, V., France-Lanord, C., Beyssac, O., Faure, P., Kudrass, H., and Palhol, F.: Efficient organic carbon burial in the Bengal fan sustained by the Himalayan erosional system, Nature, 450, 407–410. 2007a.

Galy, V., Bouchez, J., and France-Lanord, C.: Determination of total organic carbon content and $\delta^{13}C$ in carbonate rich detrital sediments, Geostand. Geoanal. Res., 31, 199–207, 2007b.

Galy, V., Beyssac, O., France-Lanord, C., and Eglinton, T. I.: Recycling of graphite during Himalayan erosion: A geological stabilization of carbon in the crust, Science, 322, 943–945, 2008.

Garcia, H. E., Locarnini, R. A., Boyer, T. P., Antonov, J. I., Baranova, O. K., Zweng, M. M., and Johnson, D. R.: World Ocean Atlas 2009, Volume 3: Dissolved oxygen, apparent oxygen utilization, and oxygen saturation. S. Levitus, Ed. NOAA Atlas NESDIS 70, US Government Printing Office, Washington, DC, 344 pp, 2010.

Hatten, J. A., Goñi, M. A., and Wheatcroft, R. A.: Chemical characteristics of particulate organic matter from a small mountainous river in the Oregon Coast Range, USA, Biogeochem., 107, 43–66, 2012.

Hayes, J. M., Strauss, K., and Kaufman, A. J.: The abundance of $^{13}C$ in marine organic matter and isotopic fractionation in the global biogeochemical cycle of carbon during the past 800 Ma, Chem. Geol., 161, 103–125, 1999.

Hedges, J. I. and Keil, R. G.: Sedimentary organic matter preservation: An assessment and speculative synthesis, Mar. Chem., 49, 81–115, 1995.

Hilton, R. G., Galy, A., Hovius, N., Chen, M. C., Horng, M. J., and Chen, H.: Tropical-cyclone-driven erosion of the terrestrial biosphere from mountains, Nature Geosci., 1, 759–762, 2008.

Hilton, R. G., Galy, A., Hovius, N., Horng, M. J., and Chen, H.: The isotopic composition of particulate organic carbon in mountain rivers of Taiwan, Geochim. Cosmochim. Acta, 74, 3164–3181, 2010.

Hilton, R. G., Galy, A., Hovius, N., Horng, M. J., and Chen, H.: Efficient transport of fossil organic carbon to the ocean by steep mountain rivers: An orogenic carbon sequestration mechanism, Geology, 39, 71–74, 2011.

Hilton, R. G., Galy, A., Hovius, N., Kao, S. J., Horng, M. J., and Chen, H.: Climatic and geomorphic controls on the erosion of terrestrial biomass from subtropical mountain forest, Global Biogeochem. Cycles, 26, GB3014, doi:10.1029/2012GB004314, 2012.

Hilton, R. G., Galy, A., West, A. J., Hovius, N., and Roberts, G. G.: Geomorphic control on the $\delta^{15}N$ of mountain forests, Biogeosciences, 10, 1693–1705, 2013, http://www.biogeosciences.net/10/1693/2013/.

Hovius, N., Stark, C. P., Chu, H. T., and Lin, J. C.: Supply and removal of sediment in a landslide-dominated mountain belt: Central Range, Taiwan, J. Geol., 108, 73–89, 2000.

Hsu, S. C., Kao, S. J., and Jeng, W. L.: Quantitative links among fluvial sediment discharge, hemipelagic trap flux and sediment accumulations and implications for climate change, Deep Sea Res. I, 53, 241–252, 2006.

Huh, C. A., Liu, J. T., Lin, H.-L., and Xu, J. P.: Tidal and flood signatures of settling particles in the Gaoping submarine canyon (SW Taiwan) revealed from radionuclide and flow measurements, Mar. Geol., 267, 8–17, 2009.

Hwang, J., Druffel, E. R. M., and Eglinton, T. I.: Widespread influence of resuspended sediments on oceanic particulate organic carbon: Insights from radiocarbon and aluminum contents in sinking particles, Global Biogeochem. Cy., 24, GB4016, doi:10.1029/2010GB003802, 2010.

Kao, S. J. and Liu, K. K.: Particulate organic carbon export from a subtropical mountainous river (Lanyang Hsi) in Taiwan, Limnol. Oceanogr., 41, 1749–1757, 1996.

Kao, S. J. and Liu, K. K.: Stable carbon and nitrogen isotope systematics in a human-disturbed watershed (Lanyang-Hsi) in Taiwan and the estimation of biogenic particulate organic carbon and nitrogen fluxes, Global Biogeochem. Cy., 14, 189–198, 2000.

Kao, S. J. and Milliman, J. D.: Water and sediment discharge from small mountainous rivers, Taiwan: The roles of lithology, episodic events, and human activities, J. Geol., 116, 431–448, 2008.

Kao, S. J., Shiah, F. K, Wang, C. H., and Liu, K. K.: Efficient trapping of organic carbon in sediments on the continental margin with high fluvial sediment input off southwestern Taiwan, Continental Shelf Res., 26, 2520–2537, 2006.

Kao, S. J., Dai, M. H., Wei, K. Y., Blair, N. E., and Lyons, W. B.: Enhanced supply of fossil organic carbon to the Okinawa Trough since the last deglaciation, Paleoceanography, 23, PA2207, doi:10.1029/2007PA001440, 2008.

Komada, T., Druffel, E. R. M., and Trumbore, S. E.: Oceanic export of relict carbon by small mountainous rivers, Geophys. Res. Lett., 31, L07504, doi:10.1029/2004GL019512, 2004.

Komada, T., Anderson, M. R., and Dorfmeier, C. L.: Carbonate removal from coastal sediments for the determination of organic carbon and its isotopic signatures, $\delta^{13}C$ and $\Delta^{14}C$: Comparison of fumigation and direct acidification by hydrochloric acid, Limnol. Oceanogr., 6, 254–262, 2008.

Leithold, E. L. and Hope, R. S.: Deposition and modification of a flood layer on the Northern California shelf: Lessons from and about the fate of terrestrial particulate organic carbon, Mar. Geol., 154, 183–195, 1999.

Leithold, E. L., Blair, N. E., and Perkey, D. W.: Geomorphologic controls on the age of particulate organic carbon from small mountainous and upland rivers. Global Biogeochem. Cycles, 20, GB3022, doi:10.1029/2005GB002677, 2006.

Liu, J. T., Wang, Y.-H., Yang, R. J., Hsu, R. T., Kao, S.-J., Lin, H.-L., and Kuo, F. H.: Cyclone-induced hyperpycnal turbidity currents in a submarine canyon, J. Geophys. Res., 117, C04033, doi:10.1029/2011JC007630, 2012.

Liu, J. T., Kao, S.-J., Huh, C.-A., and Hung, C.-C.: Gravity flows associated with flood events and carbon burial: Taiwan as instructional source area, Annu. Rev. Mar. Sci., 5, 47–68, 2013.

Lloret, E., Dessert, C., Pastor, L., Lajeunesse, E., Crispi, O., Gaillardet, J., and Benedetti, M. F.: Dynamic of particulate and dissolved organic carbon in small volcanic mountainous tropical watersheds, Chem. Geol., 351, 229–244, doi.org/10.1016/j.chemgeo.2013.05.023, 2013.

Lyons, W. B., Nezat, C. A., Carey, A. E., and Hicks, D. M. Organic carbon fluxes to the ocean from high-standing islands, Geology, 30, 439–442, 2002.

Meybeck, M.: Carbon, nitrogen and phosphorus transport by world rivers, Am. J. Sci., 282, 401–450, 1982.

Milliman, J. D. and Farnsworth, K. L.: River Discharge to the Coastal Ocean: A Global Synthesis, Cambridge University Press, Cambridge, UK, 2011.

Milly, P. C. D., Dunne, K. A., and Vecchia, A. V.: Global pattern of trends in streamflow and water availability in a changing climate, Nature, 438, 347–350, 2005.

Mulder, T. and Syvitski, J. P. M.: Turbidity currents generated at river mouths during exceptional discharges to the world oceans, J. Geol., 103, 285–299, 1995.

Norby, R. J., DeLucia, E. H., Gielen, B., Calfapietra, C., Giardina, C. P., King, J. S., Ledford, J., McCarthy, H. R., Moore, D. J. P., Ceulemans, R., De Angelis, P., Finzi, A. C., Karnosky, D. F., Kubiske, M. E., Lukac, M., Pregitzer, K. S., Scarascia-Mugnozza, G. E., Schlesinger, W. H., and Oren R.: Forest response to elevated $CO_2$ is conserved across a broad range of productivity, PNAS, 102, 18052–18056, 2005.

Pagani, M., Caldeira, K., Berner, R. A., and Beerling, D. J.: The role of terrestrial plants in limiting atmospheric $CO_2$ decline over the past 24 million years, Nature, 460, 85–88, 2009.

Saller, A., Lin, R., and Dunham, J.: Leaves in turbidite sands: The main source of oil and gas in the deep-water Kutei Basin, Indonesia, AAPG Bull., 90, 1585–1608, 2006.

Sampere, T. P., Bianchi, T. S., Wakeham, S. G., and Allison, M. A.: Sources of organic matter in surface sediments of the Louisiana Continental Margin: Effects of primary depositional/transport pathways and Hurricane Ivan, Cont. Shelf Res., 28, 2472–2487, 2008.

Schlünz, B. and Schneider, R. R.: Transport of terrestrial organic carbon to the oceans by rivers: Re-estimating flux and burial rates, Int. J. Earth Sci., 88, 599–606, 2000.

Scott, D. T., Baisden, W. T., Davies-Colley, R., Gomez, B., Hicks, D. M., Page, M. J., Preston, N. J., Trustrum, N. A., Tate, K. R., and Woods, R. A.: Localized erosion affects national carbon budget, Geophys. Res. Lett., 33, L01402, doi:10.1029/2005GL024644, 2006.

Smith, J. C., Galy, A., Hovius, N., Tye, A., Turowski, J. M., and Schleppi, P.: Runoff-driven export of particulate organic carbon from soil in temperate forested uplands, Earth Planet. Sci. Lett., 365, 198–208, 2013.

Sparkes, R. B.: Marine sequestration of particulate organic carbon from mountain belts, Ph.D. thesis, University of Cambridge, Cambridge, UK, 2012.

Stallard, R. F.: Terrestrial sedimentation and the carbon cycle: Coupling weathering and erosion to carbon burial, Global Biogeochem. Cy., 12, 231–257, 1998.

Stuiver, M. and Polach, H. A.: Discussion: Reporting of $^{14}$C data, Radiocarbon, 19, 355–363, 1977.

Sundquist, E. T.: The global carbon dioxide budget, Science, 259, 934–941, 1993.

Townsend-Small, A., McClain, M. E., Hall, B., Noguera, J. L., Llerena, C. A., and Brandes, J. A.: Suspended sediments and organic matter in mountain headwaters of the Amazon River: Results from a 1-year time series study in the central Peruvian Andes, Geochim. Cosmochim. Acta, 72, 732–740, doi:10.1016/j.gca.2007.11.020, 2008.

Turowski, J.M., Badoux, A., Bunte, K., Rickli, C., Federspiel, N., and Jochner, M.: The mass distribution of coarse particulate organic matter exported from an Alpine headwater stream, Earth Surf. Dynam., 1, 1–11, 2013.

West, A. J., Galy, A., and Bickle, M.: Tectonic and climatic controls on silicate weathering, Earth Planet. Sci. Lett., 235, 211–228, 2005.

West, A. J., Lin, C. W., Lin, T. C., Hilton, R. G., Liu, S. H., Chang, C. T., Lin, K. C., Galy, A., Sparkes, R. B., and Hovius, N.: Mobilization and transport of coarse woody debris to the oceans triggered by an extreme tropical storm, Limnol. Oceanogr., 56, 77–85, 2011.

Xu, J. P., Swarzenski, P. W., Noble, M. A., and Li, A.-C.: Event-driven sediment flux in Hueneme and Mugu submarine canyons, southern California, Mar. Geol., 269, 74–88, 2010.

# 10

# An overview of underwater sound generated by interparticle collisions and its application to the measurements of coarse sediment bedload transport

**P. D. Thorne**

National Oceanography Centre, Joseph Proudman building, 6 Brownlow Street, Liverpool L3 5DA, UK

*Correspondence to:* P. D. Thorne (pdt@noc.ac.uk)

**Abstract.** Over the past 2 to 3 decades the concept of using sound generated by the interparticle collisions of mobile bed material has been investigated to assess if underwater sound can be utilised as a proxy for the estimation of bedload transport. In principle the acoustic approach is deemed to have the potential to provide non-intrusive, continuous, high-temporal-resolution measurements of bedload transport. It has been considered that the intensity of the sound radiated should be related to the amount of mobile material and the frequency spectrum to the size of the material. To be able to fully realise this use of acoustics requires an understanding of the parameters which control the generation of sound as particles impact. In the present work the aim is to provide scientists developing acoustics to measure bedload transport with a description of how sound is generated when particles undergo collision underwater. To investigate the properties of the sound generated, examples are provided under different conditions of impact. It is considered that providing an overview of the origins of the sound generation will provide a basis for the interpretation of acoustic data, collected in the marine environment for the study of bedload sediment transport processes.

## 1 Introduction

Quantifying bedload transport in rivers, estuarine and coastal environments is generally challenging due to the difficulties of obtaining accurate measurements. For measuring the bedload transport of coarse sediments, such as gravels and cobbles, a number of methodologies have been developed. Many measurements have utilised box or tray samplers, and this approach is still common today. Hubbell (1964), Engel and Lam Lau (1981), Ergenzinger and de Jong (2003), Bunte et al. (2008) and Holmes (2010), along with many others, have considered this technique. Some of the major shortcomings of this direct sampling method are the need for operators in the field, the limited range of conditions when measurements are possible, dangerous conditions during large floods, the impact on the flow introduced by the presence of the sampler, the lack of spatial/temporal resolution, the variable efficiency of samplers, the problems in obtaining continuous records and other difficulties specific to the particular samplers.

To circumvent some of these difficulties, alternative measurement technologies have been investigated. Dorey et al. (1975) investigated, with limited success, the feasibility of using sidescan sonar to track acoustically transponding pebbles. The tagging of gravel particles radioactively was developed by Crickmore et al. (1972) and applied with success to monitoring the movement of gravel. Reid et al. (1984) utilised artificial pebbles constructed with a ferrite rod at its centre and deployed them in a brook. Particle mobility was detected using an electromagnetic sensing system installed in the bed and transport compared with the bed shear. Geophones and particle impacts on pipes, columns and plates are now commonly used to estimate coarse gravel transport (Turowski and Rickenmann, 2009; Gray et al., 2010; Rickenmann et al., 2012). In recent years acoustic Doppler velocity profiles, ADCPs, have been used to measure apparent bedload velocity using a combination of ADCP bottom tracking velocity and boat velocity derived from differential global positioning systems, DGPSs (Rennie and Church, 2010). Recently the measurement of seismic noise near rivers has been investigated to assess bedload transport (Burtin et al., 2011; Tsai et al., 2012). A workshop on contemporary methodologies for the monitoring of bedload transport has recently been published online by Rickenmann et al. (2013).

Another approach adopted, and the one focussed upon here, has been to monitor the movement of gravel by recording the acoustic sediment-generated noise, SGN, arising from particle–particle collisions as bedload transport occurs. Observations of this type have continued to be reported over the past 5 decades (Bedeus and Ivicsics, 1963; Johnson and Muir, 1969; Tywoniuk and Warnock, 1973; Jonys, 1976; Richards and Milne, 1979; Thorne et al., 1984; William et al., 1989; Mason et al., 2007; Barton et al., 2010; Camenen et al., 2012; Basset et al., 2013). The appeal of the acoustic approach is that it offers the potential to obtain, with very little interference with the state of the bed and the flow, the initiation of particle movement, continuous temporal records, sub-second assessment of mass transport rates and estimates of mobile particle size.

To be able to utilise and interpret SGN with any degree of confidence requires an understanding of the sound source generation. The source arises from the impact of two or more particles as interparticle collisions occur as the bed becomes mobile and bedload transport occurs. The generation of sound by impacting bodies has primarily been examined in air, usually by parties interested in machine noise emissions (Banerji, 1916, 1918; Koss and Alfredson, 1973; Koss, 1974a, b; Akay and Hodgson, 1978a, b; Akay, 1978). It is this work which was adapted for the study of acoustic radiation by colliding bodies underwater (Thorne and Foden, 1988; Thorne, 1990). The source of radiation has been labelled rigid body radiation due to the origin of the pressure disturbance being generated by the acceleration of the body, rather than due to the natural modes of vibration of the body (Koss and Alfredson, 1973). To solve the problem, each sphere is treated as an independent source which generates a transient that can be described by an impulse solution convolved with the acceleration time history during the collision. The impact process is assumed to be elastic so that a Hertzian acceleration description can be employed (Goldsmith, 1960). The sound field is then obtained by summing the transients radiated from each sphere with due allowance for the time difference for the sound to propagate from each sphere to the field measurement point.

The use of SGN to acoustically measure bedload transport and the underlying theory of rigid body radiation is distributed among acoustic, geological, hydraulic, geophysical and sedimentological journals. The aim of the present work is to bring together an overview of SGN and its underlying theoretical basis, in a form which scientists interested in using acoustics for bedload transport would find useful. Here the main solutions from rigid body radiation analysis are simplified to make the topic more accessible and the outputs focussed more closely than previously on bedload transport.

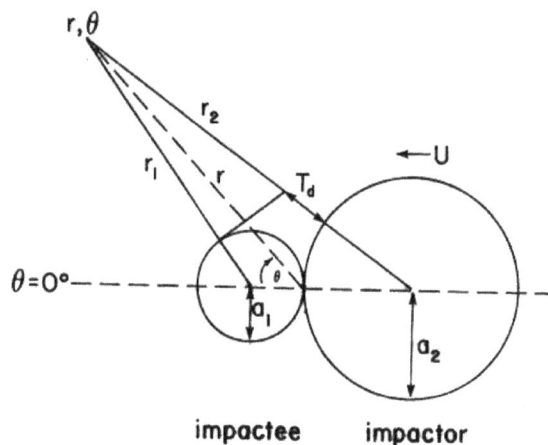

**Figure 1.** Geometry for the theory. The impactor of radius $a_2$ collides with velocity $U$ and with the impactee of radius $a_1$. $\theta$ is the angle between the line of sphere movement and the direction to the field point, and $r$ is the range to the field point. $T_d$ is the difference in the arrival time of the sound radiated by the impactor relative to the impactee.

Therefore, using the rigid body radiation theory, initial calculations have been carried out to elucidate the origins of the structure of the radiated signal in the time and frequency domain. To investigate how changes in impact parameters affected the time domain signal and the frequency spectrum, a series of calculations were conducted. The intention of this work was to provide broad illustrations of the radiated signal response due to variation in sphere collision impact parameters. Some general features are identified and considered in the light of using SGN for the measurement of bedload transport. Some modelling of multiple impacts is presented as an analogy to the type of data that may be collected in a coastal or riverine environment. It is hoped that the present overview will stimulate further interest in SGN by making the acoustic analysis more straightforward and by illustrating its potential capability.

## 2 Theory for the underwater sound generated by impacting spheres

The background theory for impacting spheres in water was developed in Thorne and Foden (1988), and only the results from the theoretical analysis are presented here. The geometry for the theory is given in Fig. 1. When solid elastic spheres collide, the main source of sound generation is due to the rigid body radiation associated with the acceleration of the impactee of radius $a_1$ and the deceleration of the impactor of radius $a_2$ following impact. These produce sound waves which radiate from the spheres into the water. To simplify the analysis presented here, the first conditions assumed is $\rho_s/2\rho_o \ll 1$, where $\rho_s$ is the density of spheres and $\rho_o$ the density of water. This condition is not strongly adhered to for marine gravels with densities of the order of $2500 \, \mathrm{kgm^{-3}}$; however it considerably reduces the complexity of the time domain solution and simply leads to an overestimate of the signal levels by 10–20 %. The second condition is $r/a \gg 1$, where $r$ is the distance from the impacting spheres to the location at which the radiated sound is observed and $a$ is the radius of the spheres; this condition can generally be readily adhered to. The pressure, $P_s(t)$, in the time domain for a single particle undergoing Hertzian impact (Goldsmith, 1960), with a half sine wave acceleration profile (Koss and Alfredson, 1973), can then be expressed as the convolution integral of the radiated pressure due to a unit impulse acceleration and the acceleration time history (Thorne and Foden, 1988); the solution is given by

$0 \leq t \leq t_o$

$$P_s(t) = P_{to} \left\{ \left(2\xi^2 - 1\right) \cos \pi \tau + 2\xi \, \sin \, \pi \tau - \left[\left(2\xi^2 + 1\right)\right. \right.$$
$$\left. \sin \pi \xi \tau + \left(2\xi^2 - 1\right) \cos \pi \xi \tau\right] e^{-\pi \xi \tau} \right\}, \qquad (1a)$$

$t > t_o$

$$P_s(t) = P_{to} \left\{ [\left(1 - 2\xi^2\right) \cos \pi \xi (\tau - 1) - \left(2\xi^2 + 1\right) \right.$$
$$\sin \pi \xi (\tau - 1)] e^{-\pi \xi (\tau - 1)} - [\left(2\xi^2 + 1\right) \sin \pi \xi \tau$$
$$\left. + \left(2\xi^2 - 1\right) \cos \pi \xi \tau] e^{-\pi \xi \tau} \right\}, \qquad (1b)$$

where $t_o$ is the duration time of the collision, $\xi$ is a non-dimensional parameter relating the collision time to the sphere size, $\tau$ is a normalised time parameter and $P_{to}$ is a pressure amplitude scaling factor; these four parameters are defined below. The pressure wave, $P_p(t)$, from the impactor and the impactee pair, is the combined pressure wave radiated from each sphere with due allowance for the time delay of the impactor signal relative to the impactee signal as given below:

$$P_p(t) = P_s(t, a_1) + P_s[(t - T_d), a_2]. \qquad (1c)$$

$T_d$ is the difference in the arrival time of the sound radiated by the impactor relative to the impactee. Equation (1) provides the time domain waveform for the sound radiated when one sphere impacts with another. A number of parameters are defined below:

$$P_{to} = \frac{a_j A_j \rho_o c U \cos \theta}{2r(4\xi^4 + 1)},$$

$$t_o = 4.53 \left[ \left( \frac{1 - \sigma_1^2}{E_1 \pi} + \frac{1 - \sigma_2^2}{E_2 \pi} \right) \left( \frac{m_1 m_2}{m_1 + m_2} \right) \right]^{0.4}$$
$$\left( \frac{a_1 + a_2}{U a_1 a_2} \right)^{0.2},$$

$$A_1 = 2m_2/(m_1 + m_2), A_2 = -(m_1/m_2)A_1,$$
$$\xi = ct_o/\pi a_j, \tau = t/t_o.$$

Here $m$ is the sphere mass, $E$ Young's modulus, $\sigma$ Poisson's ratio for the spheres and subscripts (1) and (2) refer to the

impactee and the impactor; $j = 1$ or 2. $T_d \approx a_1 (1 + \pi/2)/c$ for $\theta = 0$ and $T_d \approx 2a \cos\theta/c$ when $a_1 = a_2$ and $\theta > 0$, where $\theta$ is the angle between the line of sphere movement and the direction to the field point, $c$ is the velocity of sound in water, $U$ is the impact velocity, $r$ is the range to the field point and $t$ is time. $A_j U \pi/(2t_o)$ is the maximum acceleration of each sphere. Polar co-ordinates were selected for the analysis because of the axial symmetry of the radiated sound about the direction of polar angle, $\theta$, in the azimuth direction.

A similar expression can be obtained for the frequency spectrum when two spheres collide by using the time convolution theorem. This provides the spectrum for the collision from the product of the Fourier transform of the radiated pressure due to a unit impulse acceleration, with the Fourier transform of the acceleration time history; this results in the expression given below (Thorne and Foden, 1988):

$$P_s(f) = P_{fo} \frac{1 + e^{-i\pi\varepsilon}}{1 - \varepsilon^2} \frac{\xi a_j + ir\varepsilon}{2\xi^2 - \varepsilon^2 + 2i\xi\varepsilon}, \qquad (2a)$$

$$P_p(f) = P_s(f, a_1) + P_s(f, a_2) e^{-i\omega T_d}, \qquad (2b)$$

where

$$P_{fo} = \frac{a_j A_j \rho_o c U \cos \theta}{2r^2 \omega_o},$$

$$\varepsilon = f/f_o, \quad f_o = 1/2t_o, \quad \omega_o = 2\pi f_o, \quad i = \sqrt{-1}.$$

To illustrate the sound generated when one particle impacts with another, Fig. 2 shows how the structure of the pressure wave is formed. The predicted underwater rigid body radiation for two spheres impacting was computed using Eqs. (1) and (2). Glass sphere properties were chosen for the calculations to represent marine shingle as both materials are often composed of silicates. For the calculation presented in Fig. 2, the velocity of sound and the density of water were respectively taken to be $1480 \, \mathrm{ms^{-1}}$ and $1000 \, \mathrm{kgm^{-3}}$, for the colliding spheres $a_1 = a_2 = 0.02 \, \mathrm{m}$, $\rho_s = 2500 \, \mathrm{kgm^{-3}}$, $U = 0.3 \, \mathrm{ms^{-1}}$, $r = 0.3 \, \mathrm{m}$, $\theta = 0°$, $E = 7.10^{10} \, \mathrm{Nm^{-2}}$ and $\sigma = 0.2$ (Kaye and Laby, 1986). In Fig. 2a, the time series solution is presented. The first signal to arrive is from the accelerating impactee; this consists initially of a compressive peak from the surface of the sphere closer to the receiver, followed by a rarefaction from the rear surface. After a delay, $T_d$, the signal from the impactor arrives. This is phase-inverted due to the impactor decelerating in the direction towards the receiver, thereby giving initially a rarefaction followed by a compression peak. The total signal from the two spheres can be seen to have an "M" waveform. The spectrum, obtained using Eq. (2), can be seen in Fig. 2b. The spectrum is oscillatory in form peaking at $f \approx 1.7 f_o$, with nulls at $f = (2n + 1)f_o$, where n is an integer and with the spectrum reducing to negligible values above $7f_o$. The oscillations in the spectrum are associated with the term arising from the Fourier transform of the half sinusoidal acceleration profile. Software to calculate the

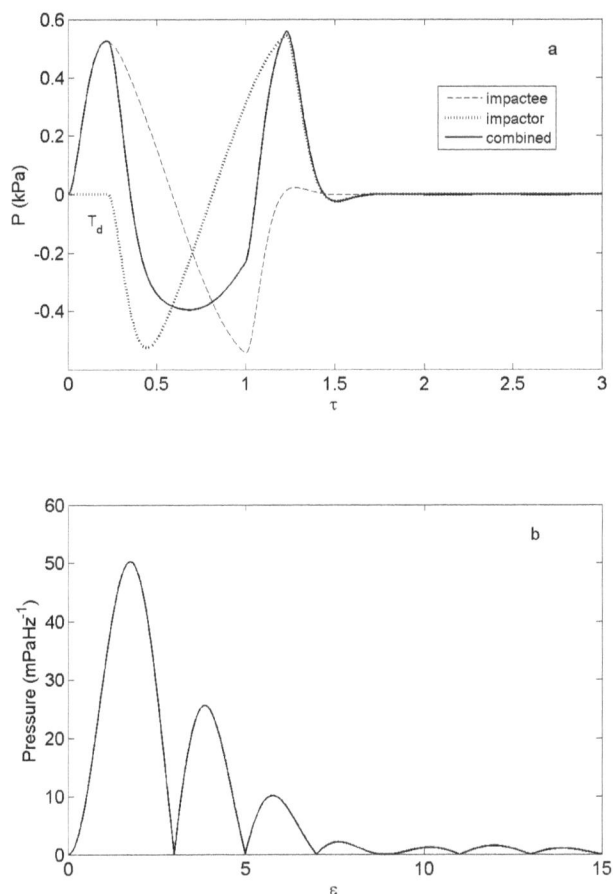

**Figure 2.** Calculations for two glass spheres of the same size impacting with radius $a = 0.02$ m and impact velocity $U = 0.3$ ms$^{-1}$. **(a)** The time domain solution using Eq. (1) for the pressure, $P$, with normalised time $\tau = t/t_o$ and **(b)** the frequency domain solution using Eq. (2) for the spectral pressure with normalised frequency $\varepsilon = f/f_o$.

impact radiation is given at http://noc.ac.uk/using-science/products/software/csr-acoustic-inversions Program (4) glass_sphere_impact_paper.m. This software defines all the terms in Eqs. (1) and (2) and allows users to investigate the properties of the sound radiated as parameters are changed for the impacting spheres.

## 3 Prediction for the radiated sound and comparisons with laboratory observations

### 3.1 (i) Single particle pair impacts

The aim here is to provide insight into the structure of the rigid body radiation field as impact parameters were varied. Therefore a number of calculations for the time domain waveform and the frequency spectrum were carried out. Initially spheres of equal size were considered with radii of $0.005, 0.015$ and $0.05$ m. These spheres were chosen to represent a broad range of particle sizes from fine pebbles to cob-

bles. This was considered to be representative of the common sizes of materials on which the acoustic technique would be used for bedload measurements. The results are considered as indicative of the signals that would arise from individual particle pair impacts in coarse sediment bedload transport. For the calculations Eqs. (1) and (2) were evaluated with all the others parameter having the same values as used to obtain Fig. 2. Figure 3 shows the results of the computations. Considering the time domain waveforms shown in plots 3a–c, it can be observed that both the duration and amplitude, $P_p$ (t), of the waveform increase with sphere size. For spheres of equal size $A_2 = -A_1 = 1$, $\xi$ is constant, and therefore it is the linear dependence of $P_{to}$ on $a$ which produces the increase in amplitude with sphere size. The duration of the waveform is controlled by $t_o$ and $T_d$, and for equally sized spheres both are a linear function of $a$. Therefore the waveform duration linearly increases with sphere size. For the spectra shown in plots 3d–f the spectrum shifts to lower frequencies and increases in amplitude as sphere size increases. The shift to lower frequencies is due to the inverse relationship between $f_o$ and $t_o$ and hence $f_o \alpha 1/a$. For the spectral amplitude, $P_p$ (f), it is a function of $a/\omega_o$, with $\omega_o$ being dependent on $1/a$ through $t_o$; hence $P_p$ (f) is proportional to $a^2$ as seen in the plots. In Thorne and Foden (1988) a number of comparisons were made between predictions and measurements of impacting steel spheres underwater; the results supported the rigid body impact model. Two cases of glass spheres impacting underwater were also made, and the results for $a_1 = a_2 = 0.015$ m case are compared with predictions in Fig. 3b and e, where it can be seen that the predictions from rigid body radiation compare very favourably with the observations.

In the marine environment particles of different size will generally be impacting, and it is therefore interesting to examine this case. To assess the radiated sound field from spheres of different size impacting, calculations were carried out using the same parameters as used for Fig. 3, but with the individual impacting spheres having a different radius. The results are presented in Fig. 4 and can be directly compared with the plots in Fig. 3. In the frequency spectrum plots in Fig. 4 the vertical solid and dashed lines respectively show the location of the spectral peak for the same size spheres impacting with radii $a_1$ and $a_2$. For the case of $a_1 = 0.005$ m and $a_2 = 0.015$ m the time domain waveform and frequency spectrum are given in plots 4a and 4d. Relative to the case of $a_1 = a_2 = 0.005$ m shown in Fig. 3a and d, the time domain signal is 50 % longer in duration and 25 % higher in amplitude, while the location of the peak frequency in the spectrum has reduced by 20 %, lying between the two vertical lines, and with a spectral amplitude which has doubled. Comparing the results for $a_1 = 0.005$ m and $a_2 = 0.015$ m with $a_1 = a_2 = 0.015$ m in Fig. 3b and e, the converse is the case, with the duration and amplitude of the time domain waveform being reduced and the location of the peak frequency in the spectrum doubled and the spectral amplitude

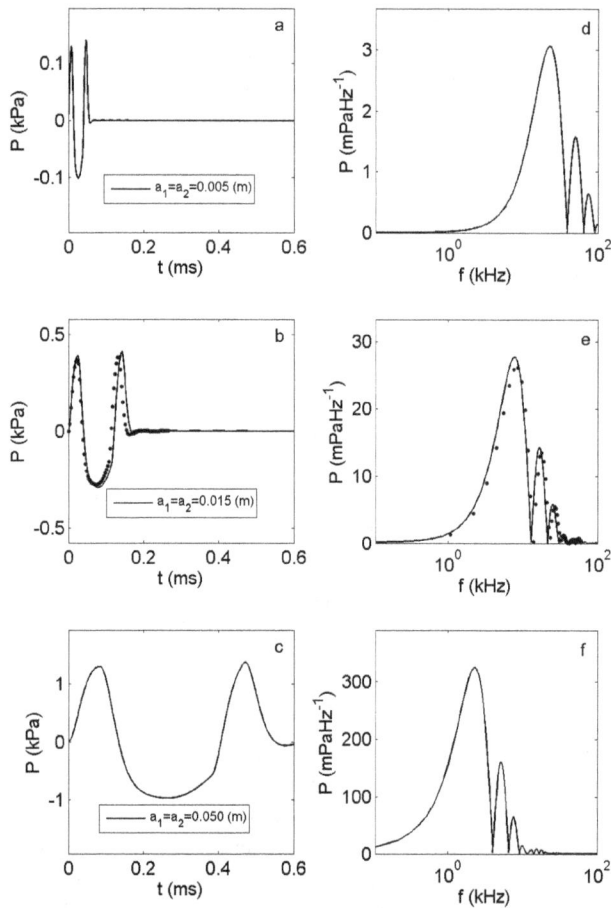

**Figure 3.** Calculations for glass spheres of the same size impacting as the sphere size was increased. **(a)–(c)** Time domain waveform calculated using Eq. (1) and **(d)–(f)** frequency spectrum calculated using Eq. (2). Measurements (●) are from Thorne and Foden (1988).

**Figure 4.** Calculations for glass spheres of different sizes impacting. **(a)–(c)** Time domain waveform calculated using Eq. (1) and **(d)–(f)** frequency spectrum calculated using Eq. (2). The vertical lines are the location of the peak frequencies for spheres of equal size impacting have respectively radii of the impactee, $a_1$ (-), and the impactor, $a_2$ (- -).

significantly lower. Colliding spheres of different size therefore have an admixture of the radiated sound properties of the accelerating impactee, $a_1$, and the decelerating impactor, $a_2$. Plots 4b, e, c and f further illustrate this admixture when compared with results presented in Fig. 3. Such admixtures will almost certainly be present in riverine bedload transport as the bed will invariably by composed of a broad range of particle sizes.

To assess how changes in the impact velocity, $U$, and the field point angle, $\theta$, affected the time domain waveform and the frequency spectrum, a number of calculations were carried out, and the results are presented in Fig. 5. For these calculations $a_1 = a_2 = 0.015$ m, and apart from the changes in $U$ and $\theta$ all other parameters were the same as used for the calculations shown in Fig. 2. As shown in plots 5a and b, an increase in $U$ has a marginal influence on the duration of the time domain waveform and the frequency spectrum covered, with the main response being an increase in signal amplitude. The former is due to the weak dependence of $t_o$ on $U^{-0.2}$, while inspection of Eqs. (1) and (2), for $\xi > 1$,

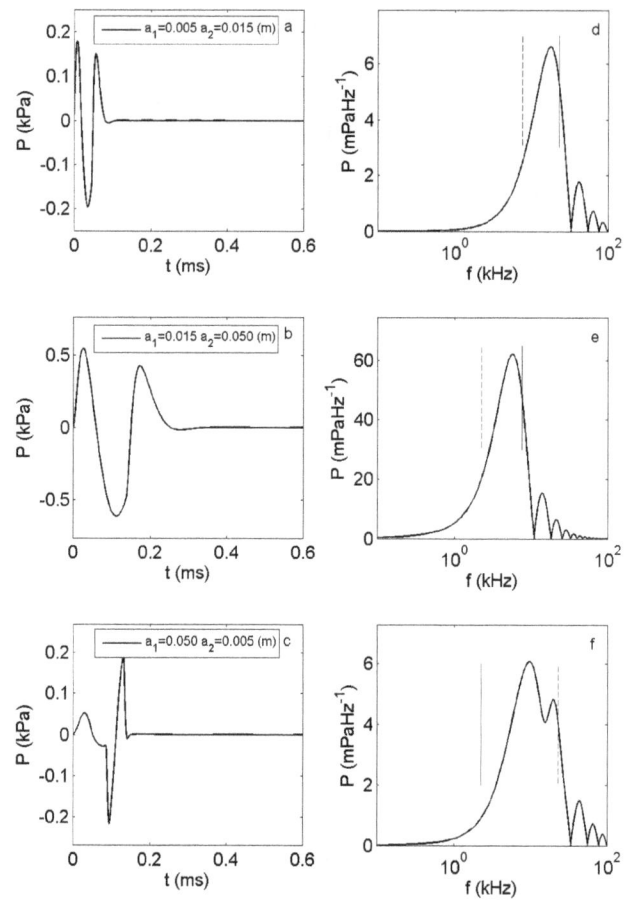

shows the signal amplitude can be approximated as being proportional to $U/\xi^2$, which due to $\xi$ dependence on $t_o$ results in a signal amplitude related to $U^{1.4}$. For the changes in $\theta$, the response is observed in plots 5c and d, where an increase in $\theta$ has the primary response of decreasing the signal amplitude and a second-order effect of reducing the duration of the time domain waveform and broadening the frequency spectrum, although the location of the peak frequency in the spectrum remains essentially unchanged. The significant reduction in signal amplitude is owing to the $\cos(\theta)$ term in $P_{to}$, $P_{fo}$ and $T_d$, which leads to the signal amplitude being dipole in form, reducing approximately as $\cos^2(\theta)$, while the decrease in pulse length and broadening of the spectrum is due to the reduction in $T_d$.

The implications from the recently computed results presented in Figs. 3–5 are that, to first order, the duration of the time domain waveform, the width of the spectrum and the frequency at which the spectrum peaks are principally con-

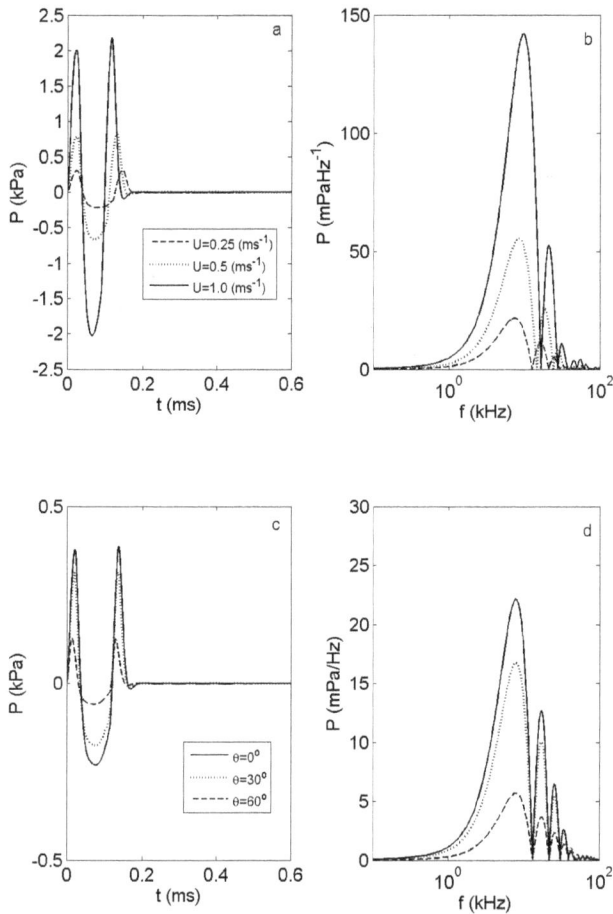

**Figure 5.** Calculations for the time domain waveform using Eq. (1) and the frequency spectrum using Eq. (2) for $a = 0.0015$ m as the velocity, $U$, increased (**a** and **b**) and as the angle between the direction of impact and the measurement position, $\theta$, increased (**c** and **d**).

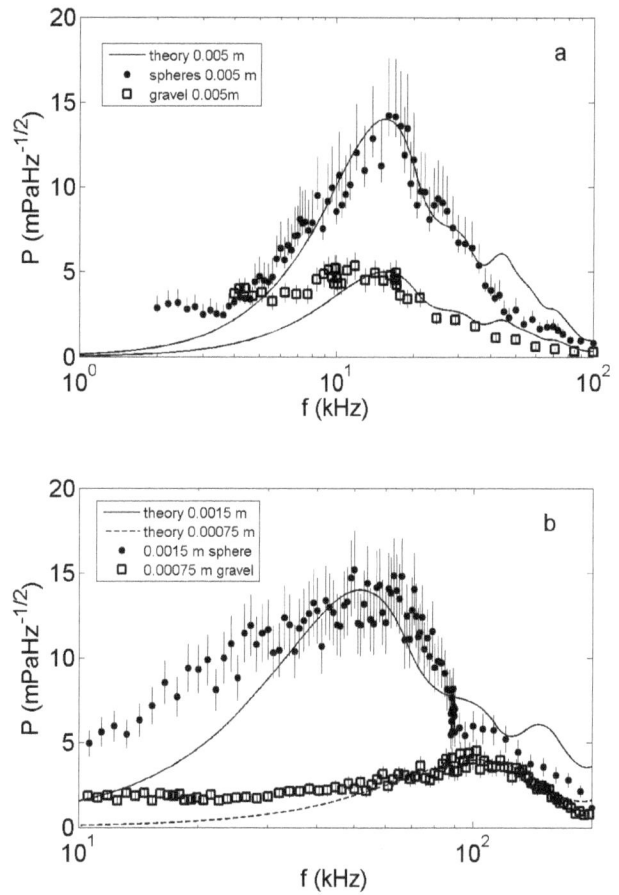

**Figure 6.** Comparison of the measured and calculated spectra using Eq. (2) with smoothing, for quasi-bedload conditions in a rotating drum for; (**a**) 0.005 m radius glass spheres and gravel and (**b**) 0.0015 m radius glass spheres and 0.00075 mm radius gravel. Measurements (•) are from Thorne (1985), and (□) are from Thorne (1986a).

trolled by the size of the spheres impacting, while the amplitude of the time domain waveform and frequency spectrum depend upon sphere size and impact velocity, with the additional factor of the dipole structure for the amplitude of the radiated field.

## 3.2 (ii) Multiple particle pair impacts

Laboratory measurements on multiple particles impacting have been carried out by Jonys (1976), Millard (1976) and Thorne (1985, 1986a). The works of Thorne incorporated the results of the earlier works and are still the most comprehensive study of the underwater sound radiated from multiple collisions of quasi-bedload transport, and therefore the measurements from these two papers are used here to illustrate the salient acoustic features.

The instrumentation used has been previously described (Thorne, 1985); therefore only a brief description is given here. Sediments were agitated in a vertical wooden drum 1 m in diameter and 0.5 m deep; this rotated about a horizontal

axis and was totally submerged underwater in a concrete tank ($3 \times 2 \times 2$ m). The drum was lined with a 0.002 m thick expanded polystyrene sheet to ensure that the only significant collision noise generated was due to interparticle collisions. The front of the drum was open, apart from a small lip around the circumference to retain the sediments, thereby allowing hydrophones (underwater sound receivers) to be placed at any position inside the drum while the material was being agitated. The output from the hydrophones was fed into a low-noise amplifier, then through a filter with a passband between 1 kHz and 600 kHz, and in parallel into an signal envelope detector and a spectrum analyser. The results were corrected for background noise and the instrumentation frequency response. To account for the fact that measurements were conducted in reverberant tank, rather than the free-field conditions in a riverine environment, measurements were made using a known broadband noise source in the tank (Thorne, 1985; see Appendix). This analysis showed that measure-

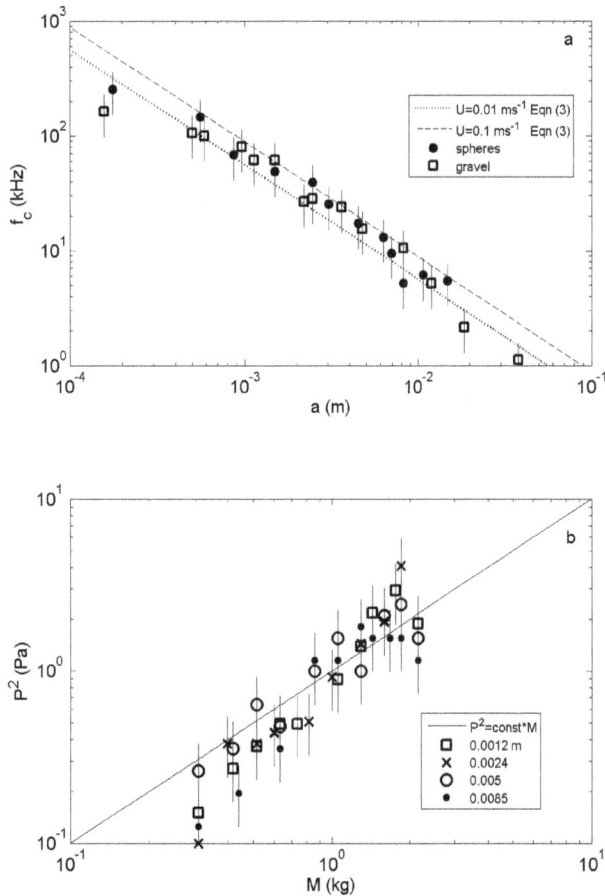

**Figure 7. (a)** Calculated characteristic central frequency, $f_c$, obtained using Eq. (3) and measurements with particle radius, $a$. **(b)** The mean square pressure, $P^2$, with mass, $M$, compared with Eq. (4). Measurements were collected under quasi-bedload conditions. Measurements are from Jonys (1976), Millard (1976), Thorne (1985) and Thorne (1986a).

ments in a confined reverberant environment can yield effectively free-field amplitude and spectral levels. The measurements in the rotating drum were intended to simulate quasi-bedload transport and impacts. Measurements were carried out on glass spheres ranging in radius between $a = 0.0018$ and 0.015 m, and gravels between $a = 0.0015$ and 0.0125 m. In Figs. 6 and 7 some of the measurements are compared with predictions from the rigid body model of acoustic radiation due to impact.

Presented in Fig. 6 are the spectrum for particles centred on nominal radii of 0.00075, 0.0015 and 0.005 m. In Fig. 6a data for spheres and gravel of radius of 0.005 m are shown; both data sets are broadband in nature, with spectral peaks frequencies being around 15 kHz. The fact that spectral levels are lower for the gravel is not necessarily significant because measurements were made under different conditions of mass and rotation speed. Owing to the impact velocities, $U$, the number of particles impacting, $N$, and the value

of the angle between the direction of impact and the measurement position, $\theta$, not being well specified in the rotating drum experiments, the output from the rigid body radiation calculations were scaled to the observations. Further using Hertzian impact theory as opposed to non-Hertzian (Riner and Petculescu, 2010) and assuming the particles are spherical as opposed to, say, more cylindrical shape (Wang and Tong, 1992) may be somewhat restrictive for comparisons with the gravel data, however, it provides internal consistency between the sphere and gravel analysis. The comparison is therefore between the modelled and measured spectral form. Further, when multiple particle size pairs of slightly different effective sizes and velocities are impacting simultaneously, marginally different overlapping frequency spectra are summed at the receiver. Therefore for the rigid body calculations the spectra have been smoothed to reduce the oscillations in the spectrum observed in the previous figures associated with only a single size particle pair impacting. This application of rigid body radiation theory to quasi-bedload transport is an important step in acoustic modelling in the riverine environment and the inversion of the acoustic measurements to obtain particle size and bedload transport. As can be seen in Fig. 6a the smoothed rigid body radiation spectra compare reasonably well in form with the sphere and gravel spectra, although, particularly for the gravel, the reduction in the spectral amplitude at lower frequencies seen in the data is smaller than predicted by the model. This was a common and unresolved observation for gravel across most of the measurements (Thorne 1986a, 1990). In Fig. 6b, results for 0.0015 m glass spheres and 0.00075 m gravel are shown, and the measurements clearly show a separation in the spectra, with the smaller gravel material having a maximum in the spectrum at twice the frequency of that of the glass spheres. Comparison of the predicted spectrum with the observations again captures the broadband nature of the sound, the location of the maximum in the spectrum, the roll-off at the higher frequencies and the smaller material having a maximum amplitude at a higher frequency; however, as with the cases in Fig. 6a, the modelled lower frequency components are underestimated.

To represent the variation of the spectra with particle size, the frequency at which the spectrum nominally peaks, or the centroid of the spectrum, has been used to define a characteristic central frequency. This is illustrated in Fig. 7a, where results from the studies of Jonys (1976), Millard (1976) and Thorne (1985, 1986a) are presented. What can be clearly seen is that the characteristic central frequency has to first order an inverse dependency on particle size. From rigid body radiation theory the frequency at which the spectrum peaks, $f_{pk}$, can be seen in Fig. 2b to be approximately given by $1.7 f_o$, where $f_o = 1/2t_o$. For simplicity, if the assumption is made that two identical particles are impacting, the expression for $t_o$ can be simplified (see the definition below Eq. 1), and this allows $f_{pk}$ to be expressed as

$$f_{pk} = 0.15 \left\{ \frac{E}{\rho_s(1-\sigma^2)} \right\}^{0.4} \frac{U^{0.2}}{a} \qquad (3)$$

All the parameters in Eq. (3) were known apart from the particle impact velocities in the rotating drum. Typical circumferential drum speeds of $0.3\,\mathrm{ms^{-1}}$ were used so impact velocities in the range of $0.01$–$0.1\,\mathrm{ms^{-1}}$ would not seem unreasonable and were therefore used in the evaluation of Eq. (3) to obtain the lines in Fig. 7a. The data generally lie between the two lines, thereby indicating that Eq. (3) is probably a reasonable description for a characteristic central frequency for the broadband impact spectrum, with, guided by Fig. 2b, the significant section of the spectrum lying in the frequency band between approximately $f_{pk}/4$ and $4f_{pk}$.

In the marine environment the amount of material transported as bedload will vary over time depending on the size of the sediments on the bed and the hydrodynamic conditions. To simulate this variability, a series of measurement were carried out on gravels of different radii, with $a = 0.0012, 0.0024, 0.005$ and $0.0085$ m where the mass, $M$, of sediments in the drum was increased at constant rotation speed. Treating the interparticle impacts in the drum as similar random, independent noise sources, the total signal can be expressed as (Beranek, 1971)

$$P^2 = \sum_{i=1}^{N} P_i^2 \approx N\overline{P_i^2}, \qquad (4)$$

where $N$ is the number of sources, $P_i^2$ are the individual source pressure levels squared, the over-bar represents a mean value and $P^2$ is the total mean squared pressure. As $N$ is proportional to $M$, $P^2$ should be approximately linearly dependent on $M$. Figure 7b shows the results of the measurements and the line represents Eq. (4). Although there is some variability in the four data sets, the general slope of the data is consistent with a linear relationship between $P^2$ and $M$. The scatter in the data is associated with variable impact velocities about a mean, the different size particles within a samples and non-sphericity of the particles impacting.

In general therefore, it can be seen that the relatively simple rigid body radiation model captures in broad terms the form of the spectrum for large numbers of particles impacting in a quasi-bedload manner, although for reasons not resolved in the present analysis the lower frequency components are somewhat underpredicted. The spectrum can be broadly specified as having a characteristic central frequency which is inversely related to particle size and with a bandwidth nominally between $f_{pk}/4$ and $4f_{pk}$. Finally to represent increases in bedload, the amount of material in the drum was gradually increased at a constant rotation speed, which resulted in the mean squared pressure increasing linearly with mass. These results do indicate that there is the possibility that inversion algorithms could be developed in

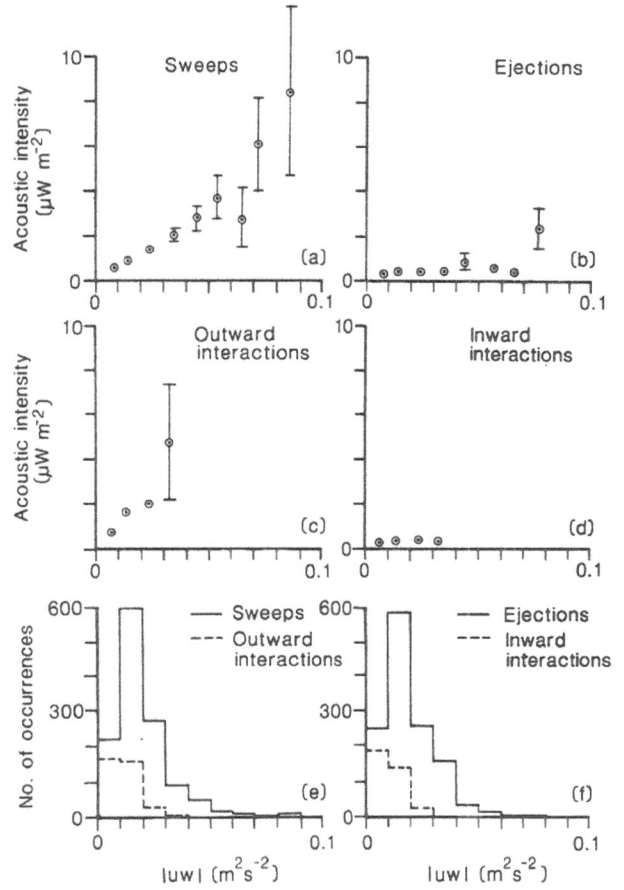

**Figure 8.** Comparison of the proxy for sediment transport, acoustic intensity, with the magnitude of the kinematic stress, |uw|, due to (a) sweeps, (b) ejections, (c) outward interactions, (d) inward interactions and (e, f) the number of events for the four kinematic stress quadrature components. Measurements are from Heathershaw and Thorne (1985).

the future for estimating bedload parameters from the signature of the SGN signal.

## 4 Example from a field study

There have been a number of field trials of the SGN technique, and the results have been variable (Bedeus and Ivicsics, 1963; Tywoniuk and Warnock, 1973; Jonys 1976; Richards and Milne, 1979; Thorne et al., 1989; Williams et al., 1989; Voulgaris et al., 1995; Mason et al., 2007; Barton et al., 2010; Belleudy, 2010; Bassett et al., 2013). One of the more successful and interesting studies was to utilise the non-intrusive high-temporal-resolution measuring capability of SGN to examine the relationship between turbulent bursting in tidal flows and the bedload transport of coarse gravels (Heathershaw and Thorne, 1985; Thorne et al., 1989). Using an instrumented frame, concurrent measurements of the three orthogonal components of the turbulent flow and instantaneous bedload transport were collected above a gravel bed in a tidally dynamic environment with currents peaking at around $1.0\,\mathrm{ms}^{-1}$ at 1 m above the bed. High-resolution bedload transport measurements were derived from two hydrophones measuring the SGN and calibrated in situ for bedload transport using visual measurements of sediment transport using an underwater video camera (Thorne, 1986b). The flow and acoustic measurements were respectively collected at 0.33 and 0.24 m above the bed, and the synchronised data were digitised at 5.0 Hz and recorded. From the turbulent flow measurements the kinematic Reynolds stress, $\tau_{uw}$, was calculate using $\tau_{uw} = -uw\,\mathrm{m}^2\mathrm{s}^{-2}$, where u and w were respectively the horizontal and vertical fluctuating turbulent components of the flow. From the acoustic calibration it was shown that the SGN acoustic intensity, $I\,\mu\mathrm{Wm}^{-2}$, which is proportional to the squared pressure, was an acceptable surrogate for bedload transport (Thorne, 1986b). Applying a quadrature analysis to the kinematic Reynolds stress, events in the flow were identified as sweeps $(u > 0, w < 0)$, ejections $(u < 0, w > 0)$, outward $(u > 0, w > 0)$ and inward $(u < 0, w < 0)$ interactions and compared with the corresponding values for $I$. The results of the analysis are shown in Fig. 8.

The measurements demonstrated quite clearly in Fig. 8a and b that the acoustic intensity, and hence gravel transport, associated with sweep events were substantially higher than the intensity levels during ejection events at high stress values. This difference increased as the magnitude of the kinematic stress increased. From this it was concluded that of the two types of motion that contributed to the bulk of the kinematic Reynolds stress, ejections and sweeps, only sweeps were capable of supporting appreciable coarse sediment movement. It was also noted that unexpectedly outward interaction events, although weaker and less frequent than sweeps, as shown in Fig. 8c and e, were capable of supporting greater sediment movement than sweeps for the same stress levels. This is despite the fact that they make a negative contribution to the Reynolds stress. Correspondingly, there was little sediment movement associated with inward interactions. The results showed that horizontal turbulent velocity fluctuations, u, may have greater dynamical significance in terms of coarse sediment movement than the instantaneous contributions, $-uw$, to the Reynolds stress. Also vertical fluctuations were considered as important provided they were associated with increases in u. This was the case for outward interactions, Fig. 3c, where $w > 0$ indicated additional lift on exposed gravel particles by fluid moving away from the bed. This, in turn was considered to account for sediment transport rates being higher than those of similarly sized sweeps. The observed close dependence on u was considered explainable if the gravel were moved principally by form drag acting on the flow-normal projected area of exposed particles. To examine this, a correlation analysis between the total horizontal flow, $U$, and $I$, and Reynolds stress, $-\rho_0\overline{uw}$, and $I$ were carried out. The results of the analysis showed that in

all cases sediment movement was better correlated with form drag than with the instantaneous stress.

This study illustrated that SGN can provide detailed high-temporal-resolution measurements of sediment response to turbulent flow conditions and showed for the first time that the bedload movement of seabed gravels is caused principally by sweep-type motions in the bottom boundary layer and to a lesser extent by outward interactions. This observation could be explained if form drag rather than shear stress were assumed to be the principle cause of gravel movement. It was speculated that such relationships between sediment transport and turbulent motions could lead to a new generation of sediment transport equations which accounted for the turbulent bursting process (Clifford et al., 1995; Williams, 1996; Sumer et al., 2003).

## 5   Discussion and conclusion

The aim of the present paper has been to provide scientist and engineers who are interested in the measurement of coarse sediment bedload transport with an overview of the background to sediment-generated noise. The new calculations presented here on sphere impacts and comparisons with quasi-bedload transport have illustrated that the rigid body radiation approach should provide a first-order framework for understanding and interpreting SGN collected in riverine and coastal environments. When the bed becomes mobile, interparticle collisions occur which radiate sound into the water, and this SGN can been used as a proxy for bedload transport rates. To understand and predict the sound field generated by the collision of particles, a theoretical framework based on rigid body radiation has been presented. Initially predictions were made with colliding pairs of glass spheres, and the impact of sphere size, impact velocity and field point angle was examined to assess the effect these had on the measured time domain signal and the frequency spectrum. Limited comparison with available data was carried out to assess the veracity of the theory. To move beyond simple two-particle impacts, larger numbers of particles were impacted using a rotating drum arrangement; this experimental configuration was employed to simulate quasi-bedload conditions. In these studies data were collected on glass spheres and natural gravels. Spectral analysis of the measurements showed comparable spectra to the two sphere impact results, and rigid body radiation gave reasonable first-order agreement with the rotating drum data. Assessment of a characteristic central frequency for the spectra showed a clear inverse relationship with the size of the impacting particles, and an expression derived from the impact duration of the collision time, $t_0$, given by Eq. (3), provided a reasonable description for the observations. To establish a relationship between the amount of material impacting in the rotating drum and the mean square signal level recorded, measurement were carried out for a number of different particle sizes at constant drum rotation speed. The results showed that the mean square signal level was proportional to the amount of material in the drum and hence ostensibly the number of collisions. To explain the observations, the interparticle impacts in the drum were considered to be similar random independent noise sources which summed linearly with the mean square pressure.

The outcome from the two sphere impact studies and the measurements in the rotating drum indicated that the relatively simple Eqs. (1)–(4), derived from rigid body radiation, and the linear summation of mean square pressures, provide a framework for a first-order understanding of SGN. The results showed that, for pairs of spheres impacting, the amplitude of the signal is a function of the sphere size, impact velocity and the location of the position of observation. For measurements of bedload in the field, the location of the receiver will normally remain fixed relative to the bed and the size of the bedload material will be nominally constant; therefore the signal amplitude will essentially depend on the impact velocity and the number of particles impacting. If it is assumed that impact velocity is proportional to the velocity on the mobile material, then the mean square signal amplitude should be acting ostensibly as a nominal proxy for the bedload transport, and this has been reported in a number of studies (Johnson and Muir, 1969; Thorne, 1986b; Barton et al., 2010). The form of the spectrum has been shown to be primarily dependent on the size of the impacting particles, with the impact velocity and measurement location having only second-order effects. The form of the spectrum is therefore a reasonably robust indicator of the size of the mobile material and as such has been used to estimate the size of the bedload material (Thorne, 1986a; Mason et al., 2007; Belleudy et al., 2010; Basset et al., 2013).

One of the more common difficulties in the application of SGN to the measurement of coarse sediment bedload transport is the level of the background aquatic soundscape (Wenz, 1972; Thorne, 1986b; Vracar and Mijic, 2011). Contributions from biophony (sounds from aquatic animals), geophony (sounds from natural abiotic phenomenon) and anthrophony (sounds from manmade activities) can make interpretation and assessment of the SGN problematic. To-date most SGN measurements have been collected using nominally omnidirectional hydrophones. Looking to the future, the mounting of such hydrophones in acoustically reflective housings to increase directionality (as with an omnidirectional bulb in a car headlight), and thereby rejecting erroneous background noise, could be an interesting step forward. Also given that predictions can be made for the spectrum of the sound from a knowledge of particle size, this may be used with bandpass filtering to enhance the SGN signal relative to the general soundscape. One area which is still deficient is rigorous assessments of the SGN technique using independent measurements of coarse sediment bedload transport. Further studies in flumes and in the field would establish, with greater veracity than available at present, the capabilities and uncertainties in the application of SGN to the robust measurement of bedload transport and particle size.

## Appendix A

**Table A1.** List of variables.

| | |
|---|---|
| $a$ | Sphere radius (m) |
| $a_1$ | Impactee radius (m) |
| $a_2$ | Impactor radius (m) |
| $a_j$ | j=1 represents the impactee and $j = 2$ represents the impactor (m) |
| $c$ | Velocity of sound in water $(\text{ms}^{-1})$ |
| $A$ | Mass acceleration factor (-) |
| $E$ | Young's modulus for the spheres (Pa) |
| $f$ | Frequency of sound (Hz) |
| $f_o$ | $1/(2t_o)$ (Hz) |
| $i$ | $\sqrt{-1}$ |
| $f_{pk}$ | Peak frequency in the theoretical impact spectrum (Hz) |
| $m$ | Sphere mass (kg) |
| $m_1$ | Mass of the impactee (kg) |
| $m_2$ | Mass of the impactor (kg) |
| $P_i$ | Pressure from individual impacts of bedload material (Pa) |
| $P^2$ | Sum of $P_i^2$ over $N$ individual simultaneous impacts $(\text{Pa}^2)$ |
| $P_s(t)$ | Pressure time series for a single sphere accelerating (Pa) |
| $P_p(t)$ | Pressure time series for a pair of spheres impacting (Pa) |
| $P_{to}$ | Pressure amplitude term (Pa) |
| $P_s(f)$ | Spectral pressure distribution for a single sphere accelerating $(\text{Pa Hz}^{-1})$ |
| $P_p(f)$ | Spectral pressure distribution for a pair of spheres impacting $(\text{Pa Hz}^{-1})$ |
| $P_{fo}$ | Spectral pressure amplitude term $(\text{Pa Hz}^{-1})$ |
| $t$ | Time (s) |
| $t_o$ | Impact duration time (s) |
| $T_d$ | Delay time between the impactee and impactor to the field point (s) |
| u | Horizontal fluctuating turbulent components of the flow $(\text{ms}^{-1})$ |
| $U$ | Impact velocity of the spheres $(\text{ms}^{-1})$ |
| w | vertical fluctuating turbulent components of the flow $(\text{ms}^{-1})$ |
| $\varepsilon$ | Normalised frequency $f/f_o$ (-) |
| $\theta$ | Angle between the line of sphere movement and the direction to a radiated field point (-) |
| $\xi$ | $ct_o/\pi a$ (-) |
| $\pi$ | 3.141592 (-) |
| $\rho_o$ | Density of water $(\text{kgm}^{-3})$ |
| $\rho_s$ | Density of spheres $(\text{kgm}^{-3})$ |
| $\sigma$ | Poisson's ratio for the spheres (-) |
| $\tau$ | Normalised time $t/t_o$ (-) |
| $\tau_{uw}$ | Kinematic Reynolds stress $-uw$ $(\text{m}^2\,\text{s}^{-2})$ |
| $\omega_o$ | Angular frequency $2\pi f_o$ (Hz) |

**Acknowledgements.** This overview on sediment-generated noise and coarse bedload transport originated from an invitation to the author to provide a keynote presentation at the International Workshop of Acoustic and Seismic Monitoring of Bedload and Mass Movements held in Zurich, Switzerland, 4–7 September 2013. The author thanks the organisers of the workshop and in particular Jonathan Laronne for the invitation. The preparation of this manuscript was carried out following an invite to make a submission to a special issue section on "Acoustic and seismic monitoring of bedload and mass movements" as part of the journal Earth Surface Dynamics. The study was supported by funding from the Natural Environmental Research Council, UK, National Capability.

Edited by: J. Turowski

# References

Akay, A.: A review of impact noise, J. Acoust. Soc. Am., 64, 977–987, 1978.

Akay, A. and Hodgson, T. H.: Acoustic radiation from the elastic impact of a sphere with a slab, Appl. Acoust., 11, 185–304, 1978a.

Akay, A. and Hodgson, T. H.: Sound radiation from an accelerated or decelerated sphere, J. Acoust. Soc. Am. 63, 313–318, 1978b.

Banerji, S.: On aerial waves generated by impact, Philos. Mag. J. Sci., 6.32, 96–111, 1916.

Banerji, S.: On aerial waves generated by impact, Part II, Philos. Mag. J. Sci., 6.35, 97–111, 1918.

Barton, J. S., Slingerland, R. L., Pittman, S., and Gabrielson, T. B.: Monitoring coarse bedload transport with passive acoustic instrumentation: A field study. Published online in 2010 by USGS as part of US Geological Survey Scientific Investigations Report 2010-5091, available at: http://pubs.usgs.gov/sir/2010/5091/papers/listofpapers.html, 38–51, 2010.

Bassett, C., Thomson, J., and Polagye, B.: Sediment-generated noise and bed stress in a tidal channel, J. Geophys. Res.-Oceans, 118, 2249–2265, doi:10.1002/jgrc.20169, 2013.

Bedeus, K. and Iviscics, L.: Observations of the noise of bedload, Proc. Intl. Assoc. Sci. Hydrol., 65, 384–390, 1963.

Belleudy, P., Valette, A., and Graff, B.: Passive hydrophone monitoring of bedload in river beds: First trials of signal spectral analysis. Published online in 2010 by USGS as part of US Geological Survey Scientific Investigations Report 2010-5091, available at: http://pubs.usgs.gov/sir/2010/5091/papers/listofpapers.html, 67–84, 2010.

Beranek, L. L.: Noise and Vibration Control, Published by Mcgraw-hill, New York, Chap. 2, 1971.

Bunte, K., Abt, S. R., Potyondy, J. P., and Swingle, K. W.: A Comparison of Coarse Bedload Transport Measured with Bedload Traps and Helley-Smith Samplers, Geodinamica Acta, 21/1-2, 53–66, 2008.

Burtin, A., Cattin, R., Bollinger, L., Vergne, J., Steer, P., Robert, A., Findling, N., and Tiberi, C.: Towards the hydrologic and bed load monitoring from high-frequency seismic noise in a braided river: The 'torrent de St Pierre', French Alps, J. Hydrol., 408, 43–53, 2011.

Camenen, B., Jaballah, M., Geay, T., Belleudy, P., Laronne ,J. B., and Laskowski, J. P.: Tentative measurements of bedload transport in an energetic alpine gravel bed river, in: River flow 2012 edited by: Murillo, R., Taylor and Francis Group London, 379–386, ISBN 978 0-415-62129-8, 2012.

Clifford, N. J., Richards, R. A., Brown, R. A., and Lane, S. N.: Scales of variation of suspended sediment concentration and turbidity in a glacial meltwater stream, Geograf. Annal., 77A, 45–65, 1995.

Crickmore, M. J., Waters, C. B., and Price, W. A.: The measurement of offshore shingle movement. Proceedings of the Thirteenth Coastal Engineering Conference held at Vancouver, Canada, 2, 1005–1025, 1972.

Dorey, A. P., Finch, A. R., and Dyer, K. R.: A miniature transponding pebble for studying gravel movement. In Conference proceedings on Instrumentation in Oceanography held at University College of North Wales, Bangor, 327–332, 1975.

Ergenzinger, P. and de Jong, C.: Perspectives on bed load measurement, in: Erosion and sediment transport measurement in rivers: Technological and methodological advances (Proceedings of the Oslo Workshop, June 2002), IAHS Publ. 283, 113-125, 2003.

Engel, P. and Lam Lau, Y.: The efficiency of basket type bedload samplers. Erosion and sediment transport measurements. Proc of the Florence Symposium, IAHR Publ., 133, 27–34, 1981

Goldsmith, W.: Impact, Impact Published by Edward Arnold, London, Chap. 4, 1960.

Gray, J. R., Laronne, J. B., and Marr, J. D. G.: Surrogate Bedload Monitoring Techniques, US Geological Survey Scientific Investigations Report 2010–5091, 37 p., available at: http://pubs.usgs.gov/sir/2010/5091, The 26 included papers are available at: http://pubs.usgs.gov/sir/2010/5091/papers/listofpapers.html, 2010.

Heathershaw, A. D. and Thorne, P. D.: Sea-bed noises reveal role of turbulent bursting phenomenon in sediment transport by tidal currents, Nature, 316, 339–342, 1985.

Holmes, R. R.: Measurement of Bedload Transport in Sand-Bed Rivers: A Look at Two Indirect Sampling Methods. Published online in 2010 by USGS as part of US Geological Survey Scientific Investigations Report 2010-5091, available at: http://pubs.usgs.gov/sir/2010/5091/papers/listofpapers.html, 236–252, 2010.

Hubbell, D. W.: Apparatus and techniques for measuring bedload, paper N 182/B/1748, US Geological Survey Water-Supply, US Geological Survey, Washington, 1–74, 1964.

Johnson, P. and Muir, T. C.: Acoustic detection of sediment movement, J. Hydraul. Res., 7, 519–540, 1969.

Jonys, C. K.: Acoustic measurement of sediment transport, Scientific Series no. 66, Department of Fisheries and the Environment, Inland Waters Directorate CCIW Branch, Burlington, Ontario, Canada, 1–114, 1976.

Kaye, G. W. C. and Laby, T. H.: Tables of Physical and Chemical Constants, Published by Longman, New York, 15th Edn., 1986.

Koss, L. L.: Transient sound from colliding spheres – inelastic collisions, J. Sound Vib., 36, 555–562, 1974a.

Koss, L. L.: Transient sound from colliding spheres – Normalized results, J. Sound Vib., 36, 541–553, 1974b.

Koss, L. L. and Alfredson, R. J.: Transient sound radiated by spheres undergoing an elastic collision, J. Sound Vib., 27, 59–75, 1973.

Mason, T., Priestley, D., and Reeve, D. E.: Monitoring near shore shingle transport under waves using a passive acoustic technique, J. Acoust. Soc. Am., 122, 737–746, doi:10.1121/1.2747196, 2007.

Millard, N. W.: Noise generated by moving sediments. Proceedings of Conference on Recent Developments in Underwater Acoustics, Institute of Acoustics, Held at AUWE, Portland, 31 March–1 April, paper 3.5, 1976.

Reid, I., Brayshaw, A. C., and Frostick, L. E.: An electromagnetic device for automatic detection of bedload motion and its field application, Sedimentology, 31, 269–276, 1984.

Rennie, C. D. and Church, M.: Mapping spatial distributions and uncertainty of water and sediment flux in a large gravel bed river reach using an acoustic Doppler current profiler, J. Geophys. Res., 115, F03035, doi:10.1029/2009JF001556, 2010.

Richards, K. S. and Milne, L. M.: Problems in the calibration of an acoustic device for the observation of bedload transport, Earth Surf. Process., 4, 335–346, 1979.

Rickenmann, D., Turowski, J., Fritschi, B., Klaiber, A., and Bedload, L. A.: transport measurements at the Erlenbach stream with geophone sand automated basket samplers, Earth Surf. Process. Landf., 37, 1000–1011, 2012.

Rickenmann, D., Laronne, J. B., Turowski, J. M., and Vericat, D.: International Workshop of Acoustic and Seismic Monitoring of Bedload and Mass Movements, 4–7 September, Birmensdorf, Zürich, Switzerland, Abstracts, available at: http://www.wsl.ch/dienstleistungen/publikationen/pdf/12904.pdf, 2013.

Riner, J. and Petculescu, A.: Non-Hertzian behavior in binary collisions of plastic balls derived from impact acoustics, J. Acoust. Soc. Am., 128, 132–136, 2010

Sumer, B. M., Chua, L. H. C., Cheng, N. S., and Fredsoe, J.: Influence of turbulaence on bed load sediment transport, J. Hydraul. Eng., 12, 585–596, 2003.

Thorne, P. D.: The measurement of acoustic noise generated by moving artificial sediments, J. Acoust. Soc. Am., 78, 1013–1023, 1985.

Thorne, P. D.: Laboratory and marine measurements on the acoustic detection of sediment transport, J. Acoust. Soc. Ame., 80, 899–910, 1986a.

Thorne, P. D.: An intercomparison between visual and acoustic detection of seabed gravel movements, Mar. Geol., 72, 11–31, 1986b.

Thorne, P. D.: Seabed generation of ambient noise, J. Acoust. Soc. Am., 87, 149–153, 1990.

Thorne, P. D. and Foden, D. J.: Generation of underwater sound by colliding spheres, J. Acoust. Soc. Am., 84, 2144–2152, 1988.

Thorne, P. D., Heathershaw, A. D., and Troiano, L.: Acoustic detection of seabed gravel movement in turbulent tidal currents, Mar. Geol., 54, M43–M48, 1984.

Thorne, P. D., Williams, J. J., and Heathershaw, A. D.: In situ acoustic measurements of marine gravel thereshold and transport, Sedimentology, 36, 61–74, 1989.

Tsai, V. C., Minchew, B., Lamb, M. P., and Ampuero, J. P.: A physical model for seismic noise generation from sediment transport in rivers, Geophys. Res. Lett., 39, L02404, doi:10.1029/2011GL050255, 2012.

Turowski, J. M. and Rickenmann, D.: Tools and cover effects in bedload transport observations in the Pitzbach, Austria, Earth Surf. Process. Landf., 34, 26–37, 2009.

Tywoniuk, N. and Warnock, R. G.: Acoustic detection of bedload transport. Proceedings of the 9th Canadian Hydrology Symposium, Ottawa, Ontario, Canada, 728–749, 1973.

Voulgaris, G., Wilkin, M. P., and Collins, M. B.: The in situ passive acoustic measurement of shingle movement under waves and currents: instrument (TOSCA) development and preliminary results, Continental Shelf Res., 15, 1195–1211, 1995.

Vracar, M. S. and Mijic, M.: Ambient noise in large rivers, J. Acoust. Soc. Am., 130, 1787–1791, 2011.

Wang, Y. F. and Tong, Z. F.: Sound radiated from the impact of two cylinders, J. Sound Vibration, 159, 295–303, 1992.

Wenz, G. M.: Review of underwater acoustic research:Noise, J. Acoust. Soc. Am., 51, 3, 1010–1024, 1972.

William, J. J.: Turbulent flow in rivers, in: Advances in Fluvial Dynamics and Stratigraphy. edited by: Carling, P. A. and Dawson, M. R., Published by John Wiley and Sons Ltd., 1–32, 1996.

Williams, J. J., Thorne, P. D., and Heathershaw, A. D.: Comparison between acoustic measurements and predictions of the bedload transport of marine gravels, Sedimentology, 36, 973–979, 1989.

# 11

# Coastal vulnerability of a pinned, soft-cliff coastline – I: Assessing the natural sensitivity to wave climate

**A. Barkwith**[1], **C. W. Thomas**[1], **P. W. Limber**[2,*], **M. A. Ellis**[1], **and A. B. Murray**[2]

[1]British Geological Survey, Keyworth, Nottingham, NG12 5GG, UK
[2]Nicholas School of the Environment, Duke University, Durham, NC, USA
[*]now at: Dept. of Geological Sciences, University of Florida, Gainesville, FL, USA

*Correspondence to:* A. Barkwith (andr3@bgs.ac.uk)

**Abstract.** The impact of future sea-level rise on coastal erosion as a result of a changing climate has been studied in detail over the past decade. The potential impact of a changing wave climate on erosion rates, however, is not typically considered. We explore the effect of changing wave climates on a pinned, soft-cliff, sandy coastline, using as an example the Holderness coast of East Yorkshire, UK.

The initial phase of the study concentrates on calibrating a numerical model to recently measured erosion rates for the Holderness coast using an ensemble of geomorphological and shoreface parameters under an observed offshore wave climate. In the main phase of the study, wave climate data are perturbed gradually to assess their impact on coastal morphology. Forward-modelled simulations constrain the nature of the morphological response of the coast to changes in wave climate over the next century. Results indicate that changes to erosion rates over the next century will be spatially and temporally heterogeneous, with a variability of up to ±25 % in the erosion rate relative to projections under constant wave climate. The heterogeneity results from the current coastal morphology and the sediment transport dynamics consequent on differing wave climate regimes.

## 1 Introduction

The coastal zone and immediate hinterland is a highly important socio-political domain (Pendleton, 2010). It is also amongst the most vulnerable, particularly when climate change alters sea level, weather systems and wave climates. Understanding the geomorphological response and sensitivity of coastal regions to these changes are key society-relevant scientific inquiries. Many studies have focussed on observation and monitoring, in order to understand the key processes and the rates at which they happen, particularly with regard to erosion or accretion along low-lying "soft" coasts dominated by weakly or unconsolidated sediments. Numerical modelling, parameterised in part by observational data, is increasingly used to study both coastal processes and the response of coastal evolution to climatic changes, under current conditions and those which might pertain in the future.

In this paper, we report a numerical modelling study of a soft-cliff, sandy coastline, which is pinned in place at the up-drift end by a rocky headland that resists erosion. We use the Holderness Coast, eastern England, as an example (Fig. 1). Whilst well studied and monitored (Scott Wilson, 2009; Quinn et al., 2009; Montreuil and Bullard, 2012), the possible future states of this coastline have received only minimal investigation using numerical modelling (Castedo et al., 2012). Efforts to understand this coastline are vital as it is among the most rapidly eroding of coastlines in Europe, with concomitant and serious threats to people, property, the local economy and infrastructure along its length. Valentin (1971) and de Boer (1964) showed that shoreline retreat at Holderness has been of the order of kilometres since the sea reclaimed the North Sea Basin at the end of the Quaternary. Many ancient settlements recorded in texts, such as the 12th century *Domesday Book* and old maps, have been lost to the sea. Current settlements and infrastructure contin-

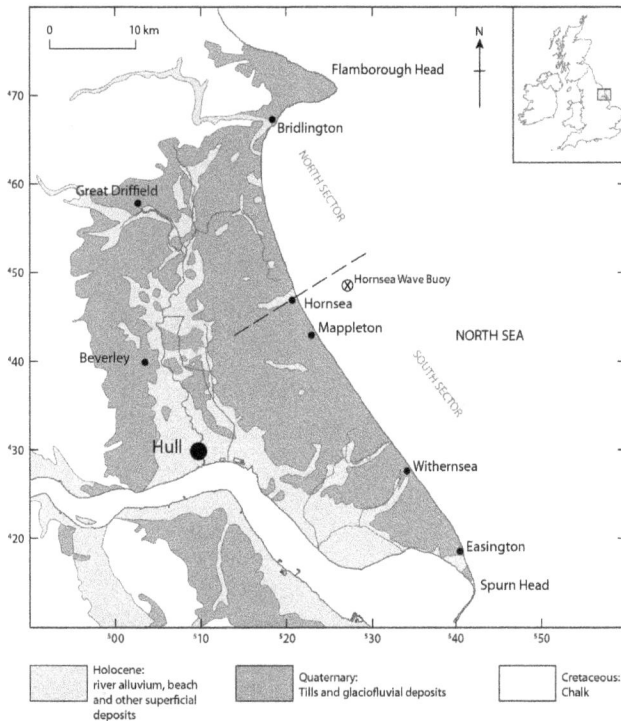

**Figure 1.** Geological composition of the Holderness coast (main) and the location of the region within the UK (insert). Also indicated are the positions of the Hornsea wave buoy, from which wave climate was recorded, and the division into northern and southern coastline regions, as referenced by the sea wall at Hornsea (dashed line), to aid analysis.

tinue to be lost, damaged or under imminent threat as the coastline retreats westwards. On the human level, loss of land and property can often be catastrophically rapid because of the episodic nature of cliff collapse over short timescales.

In order to understand the mechanisms and retreat rates within the Holderness littoral cell, and to develop a practical coastal management strategy (see Scott Wilson, 2009), most recent studies have focussed on monitoring and measurement of coastal position and beach profiles over several years (Quinn et al., 2009; Brown et al., 2012; Montreuil and Bullard, 2012). Modern observational techniques often use lidar scanning systems to provide accurate measurements of cliff retreat and volume loss. Many data sets of coastal change now exist to support coastal management decisions. However, such studies do not necessarily reflect what will happen in the future. The relatively short timescales over which these studies have been made inevitably represent only recent "snapshots" of geomorphological processes. Such geomorphological processes are stochastic, and therefore short records may not represent long-term-averaged conditions. These studies have limited predictive value, especially if the factors that control coastal evolution change significantly in the coming decades. Mesoscale modelling can allow long-term behavioural trends to be identified and explore the role

of changes in driving forces. In addition, numerical modelling can be used to understand the morphological sensitivity of the Holderness coast under different climate change scenarios. Through modelling, the impact of such changes on settlements, land and infrastructure in the longer term can be assessed, even if only to show that current coastal recession, both in rate and form, is likely to continue in the forthcoming decades.

In order to investigate how a pinned, soft-cliff, sandy coastline might respond to future changes in hydrodynamic driving processes, and the rates at which such changes may occur, we have applied the numerical coastal evolution model, CEM (Ashton et al., 2001; Ashton and Murray, 2006a, b), to the Holderness coast. The model allows us to postulate how spatially and temporally sensitive the Holderness coastline is to differing wave climate scenarios. Changes to offshore wave height or approach angle modify gradients in alongshore transport, determining beach volume flux rates and subsequently cliff erosion rates. Previous work has shown that changing the distribution of wave-approach angles can change the shape of a sandy coastline (Slott et al., 2006; Moore et al., 2013); here we investigate how wave climate change scenarios affect the evolution of a soft-cliff coastline.

## 2   Geomorphology and wave climate of the Holderness coastline

### 2.1   Geomorphology

The Holderness coast stretches ∼ 60 km from the chalk cliffs at Flamborough Head in the north to Spurn Head in the south (Scott Wilson, 2009; Quinn et al., 2009). The coastline is cut mainly in glacial till deposited during Devensian glaciations (c. 35 to 11.5 ka BP). Cliffs range from 2 to 35 m in height. The glacial till is composed of heterolithic clay, sand and gravel resting on a chalk platform sloping gently to the east (Catt, 2007). Coastal defences, including groynes, rock revetments and concrete sea walls, protect the larger towns and villages along the coastline. Recent recession rates, ranging between c. 1 m yr$^{-1}$ and c. 5 m yr$^{-1}$ depending on time and location, have been documented by Quinn et al. (2009) and Montreuil and Bullard (2012). South of Flamborough Head, wave-driven cliff erosion has created one of the fastest-retreating coastlines in Europe (IECS, 1994). With little external sediment transported into the Holderness coastline from the north (May, 1980), material derived from eroded cliffs supplies the bulk of the sediment flux southwards along the coastline. The Humber Estuary forms the southern boundary of the sediment cell, acting as a sink for sediment transported along the coast. The narrow sand and gravel spit at Spurn Head extends south-westwards across the mouth of the Humber for 3.5 km, and is known to have a complex, dynamically evolving morphology influenced by the inter-

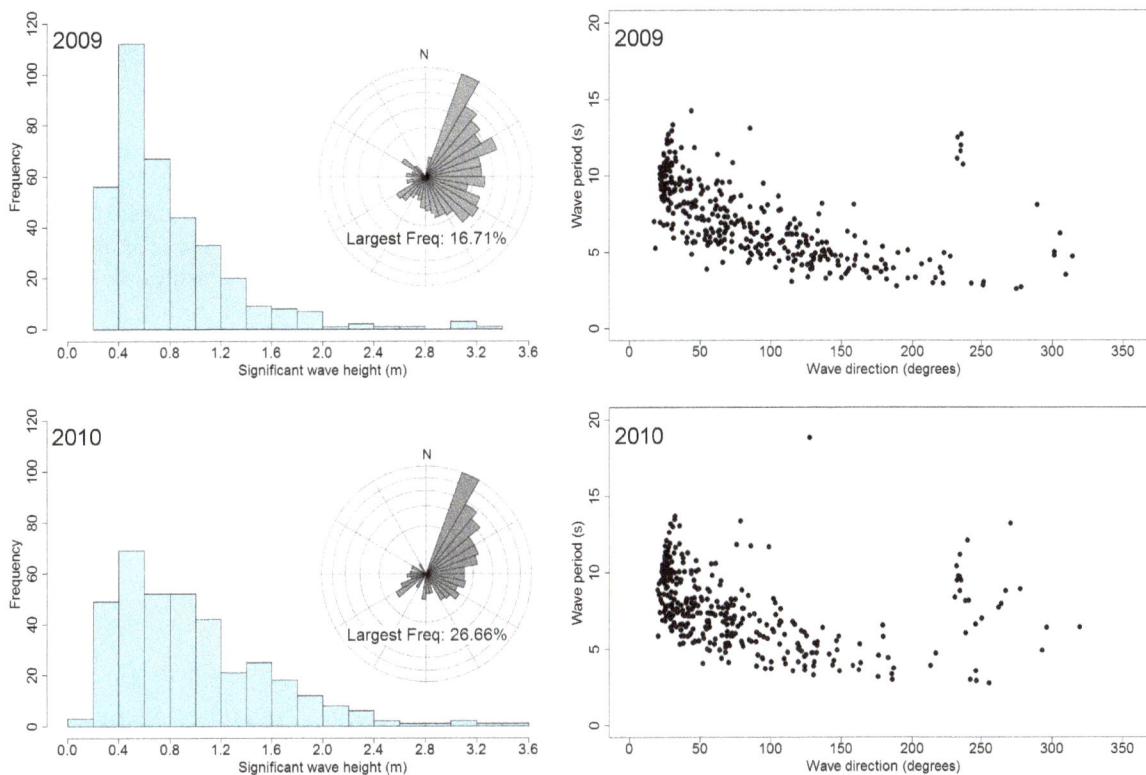

**Figure 2.** Wave climates used in the modelling. The data are daily averages, calculated from the data recorded every 30 min by the Hornsea WaveRider III buoy (CCO, 2013). The rose diagrams show the direction from which the wave is travelling. The "petals" are in 10° intervals and area-scaled by frequency percentage. The dominance of waves travelling from northeasterly directions is clear, particularly during 2010. Note the low frequencies of waves travelling in offshore directions. The histograms show that significant wave height data are positively skewed, but with marked variations in height frequencies over each year. Neither are well fitted to standard distributions, either in raw or transformed form. Overall, there were greater frequencies of higher waves in 2010, suggesting more unsettled weather than in 2009. $N = 365$ in both roses and histograms.

action of estuarine, tidal and longshore systems (Ciavola, 1997).

## 2.2 Recent historical wave climate

The current wave climate off the Holderness coast is recorded by the Hornsea Directional Waverider III buoy (CCO, 2013). This buoy has recorded mean half-hourly significant wave height, period and direction since June 2008. Wave climate is characterised by a north-easterly wave approach (Fig. 2), with a mean period of 7.3 s (2.6 to 18.8 s) and a seasonally variable significant wave height (0.2 to 3.5 m) with an annual mean of 0.9 m.

Daily averaged data for the years 2009 and 2010 are shown in Fig. 2; the term "historic" denotes this data set for the remainder of this manuscript. The dominant NNE mode exhibits frequency between 16 and 25 %. This mode was particularly strong in 2010. In 2009, modes from the ENE and ESE were more prominent. The differences in wave directions between the two years will be reflected in the modelled sedi-

ment fluxes and resultant coastal evolution. Significant wave heights were higher more frequently in 2010 than in 2009 (Fig. 2). The data show that seas off Holderness were rougher and more focussed in direction during 2010, and thus likely to cause more erosion and sediment transport than in 2009. A generally monotonic decline in wave period as the wave direction rotates clockwise from north (0°) can be observed in the buoy data (Fig. 2). The distribution of wave period with wave direction is similar for both 2009 and 2010, although the greater spread in wave direction is evident in the 2009 data. Wave periods lie within largely the same range in both years: between 4 and 14 s. On a few days, offshore waves, derived from the SW, have periods in the 8 to 14 s range. The longest period waves are mainly derived from the northeasterly direction, the dominant mode in wave direction. We infer these waves to be the long-period swell waves derived from North Atlantic low-pressure systems tracking across the north-eastern Atlantic, the waves refracting around northern Britain and down the North Sea. For just a few days each year, there are long-period waves travelling from a south-westerly direction.

## 2.3   Representing the future wave climate

Possible future wave climates for the North Sea have been studied in detail over the last decade, the motivation being to examine the effects of climate change on coastal flooding (Woth et al., 2006; Grabemann and Weisse, 2008) and ocean infrastructure (Wang et al., 2004).

Oceanic modes can increase or decrease the strength of incoming winds over Europe and have a dominant effect on the wave climate. For Europe, the North Atlantic Oscillation (NAO) is the dominant mode, modifying the path of the prevailing westerly winds and the position of storm tracks with a quasi-decadal frequency (Hurrell, 1995; Hurrell and van Loon, 1997). The NAO, and subsequently the intensity, frequency and tracks of storms, is likely to be affected by changes in climate over the coming century (Woollings et al., 2010). This is currently being studied using a multi-system modelling approach; greenhouse gas emission scenarios are used to force global circulation and regional climate models (IPCC, 2007), producing the atmospheric variables (wind speed and direction) needed in the derivation of wave climate using a wave model. The impact of future climate states on the NAO have not been quantified with any degree of certainty (Woollings, 2010); the triggering mechanisms for phase-switching, oceanic–atmospheric interactions that drive the system and teleconnections with other systems are poorly understood (Bladé et al., 2012).

Insufficient knowledge of the NAO combined with the range of greenhouse gas predictions and the variety of models used for research has produced a wide range of possible perturbations to future wind patterns and hence to future wave climates. Sutherland and Wolf (2002) linked a wave climate model to a coupled global climate model and found that predicted wave height changes are within ±5 % of current significant wave height. Due to the uncertainties inherent in the modelling, Hulme et al. (2002) provided a qualitative assessment of future wave climates in the Atlantic, suggesting modifications to the NAO would produce more westerly wave directions, therefore reducing wave height in the North Sea Basin. Further study by DEFRA (2010) on the impacts of climate change on wave climate highlights an increase in North Sea wave heights between 1973 and the mid-1990s, though more recent trends are unclear. Future UK wave climate projections were found to be very sensitive to the climate model scenarios for storms, which themselves are uncertain (DEFRA, 2010).

There are several ways to represent the future wave climate for the Holderness coast in the simulation phase of this study. Numerically simulated wave climate data could provide the required driving data; however, as discussed, there is a great deal of uncertainty currently associated with representing the large-scale systems which influence these data

sets. With minimal re-engineering, observational wave climate data may be used to drive the future coastal modelling. Long-term observational data, which can be perturbed, exist for the North Sea Basin at several locations. However, the North Sea wave climate is non-linear and highly spatially heterogeneous, and it is therefore difficult to extrapolate non-local data to a specified location. Although only recorded over a 2-year period (at the time of the study), the historic recorded wave climate provides the best data set to drive model simulation, due to the locality of the observation point. For the calibration and baseline simulations (described in Sect. 4), these data are not perturbed. In the future scenarios, the offshore wave direction data are rotated by angles of between 0 and 20° in clockwise and anticlockwise directions, selected at random for each run. Similarly, significant wave heights are perturbed by up to ±0.4 m. To account for decadal to centennial-scale changes in wave climate, these adjustments to wave height and angle are applied linearly.

## 3   Modelling

Previous simulations of Holderness coastal morphology have focussed on cliff stability modelling at sub-centennial timescales. These studies employ two-dimensional cross-section models to consider rotational (e.g. Gibbons, 2004; Quinn et al., 2010) and translational (Robertson, 1990) cliff failure, with cliff topple as the prime coastal recession mechanism for the latter. Castedo et al. (2012) combine these failure mechanisms with the hydrodynamics and geotechnical characteristics of the coast determined at several locations, using observed recession rates to calibrate parameters. Future erosion rates at each location were calculated for the remainder of this century and found to have a quasi-linear response to sea-level rise. Potential changes in wave climate and their possible impact on the evolution of coastal morphology and retreat into the future have not been investigated. Two-dimensional planform models of coastal morphology allow the influence of wave climate variability on erosion and accretion rates along the coast to be explored. This section describes the model used in this study and its underlying conceptual framework.

We have adapted the coastline evolution model originally developed by Ashton et al. (2001), Ashton and Murray (2006a, b), and Valvo et al. (2006) to allow sediment inputs derived from cliff retreat (Fig. 3). Wave-generated erosion of a sea cliff may be spatially and temporally variable on short timescales (i.e. focused at the cliff toe, causing undercutting and subsequent overhang collapse; Young and Ashford, 2008); however at the decadal scale cliff retreat can be treated as a process considered to occur evenly over the entire cliff profile (Walkden and Hall, 2005; Limber and Murray, 2011, 2014; Limber et al., 2014). The rate of cliff retreat is thus time-averaged, and implicitly includes shorter-term changes such as storm-induced erosion (Sallenger et al.,

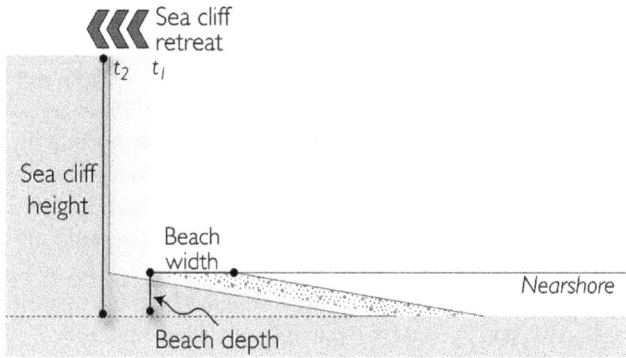

**Figure 3.** Cross-sectional view of the shoreface and cliff retreat variables, as defined for the modified coastal evolution model used in this study.

2002). For simplicity, model cliff topography is uniform, reflecting the mean cliff height of the Holderness coast.

Beach geometry and rates of sandy shoreline change are also averaged over short-term events (List et al., 2006). As the shoreline position changes, beach geometry remains constant, sediment is spread over the entire beach profile, and bathymetry contours are shore-parallel (Ashton and Murray, 2006a). The change in sandy coastline position ($\eta_b$) through time is governed by

$$\frac{\partial \eta_b}{\partial t} = -(1 - \gamma H C)\frac{\mathrm{d}\eta_c}{\mathrm{d}t} + S - \left(\frac{1}{D}\frac{\partial Q_s}{\partial x}\right), \qquad (1)$$

where $\gamma$ is a beach geometry constant that converts the volume of material eroded from the sea cliff into a beach width, $H$ is sea-cliff height divided by the depth to which the beach extends, $C$ is the proportion of sea-cliff material that is coarse enough to contribute to beach width (Limber et al., 2008), $S$ is a beach sediment loss rate, $D$ is the water depth (closure depth) to which shore-parallel bathymetry contours extend, $Q_s$ is alongshore sediment transport (Ashton and Murray, 2006a); and $x$ is alongshore position.

Equation (1) is discretised into uniform cells. The first term on the right-hand side represents sediment input into the coastal system as cliffs erode and rock is weathered into mobile sediment. There is an additional cliff retreat rate term ($\eta_c$) because the beach is pinned to the cliff as it retreats landward. The beach acts as a protective cover, reducing wave impact at the cliff toe. Accordingly cliff retreat rate is highest when local beach width ($w$) at a particular location is zero, decreasing exponentially as beach width increases (Sallenger et al., 2002; Valvo et al., 2006; Lee, 2008). To represent wave energy attenuation as waves refract towards the coastline (Adams et al., 2002), cliff retreat rate also depends on the mean daily breaking wave angle. The flux of coastal wave energy is maximised when waves approach a model cell orthogonally, and decreases as the incident wave angle increases. Cliff retreat through time is thus a function of wave

angle and beach width calculated by

$$\frac{\mathrm{d}\eta_c}{\mathrm{d}t} = \cos(\varphi - \theta)\mathrm{Er}_0 e^{-\frac{w(t)}{w_{\text{scale}}}}, \qquad (2)$$

where $\varphi$ is the incident angle of the deep-water wave; $\theta$ is the orientation of the coastline for a particular model cell; $E_0$ is the time-averaged, bare-rock cliff retreat rate; and $w_{\text{scale}}$ is a length-scale constant dependent on beach width, which provides near-complete cover from wave attack, so that cliff retreat becomes negligible (i.e. $\sim 1$ % of the maximum value; Sallenger et al., 2002; Limber and Murray, 2011). Different lithologies can be represented in the model by varying $\mathrm{Er}_0$ and $C$: $\mathrm{Er}_0$ represents erosional resistance, and $C$ reflects the fraction of fine-grade sediment in the fallen material. More resistant lithologies (the chalk at Flamborough Head) have a lower $\mathrm{Er}_0$ than rocks more susceptible to erosion (the glacial till along the Holderness embayment; Limber et al., 2014). Through a calibration process (described in Sect. 4), site-specific, uniform values for $\mathrm{Er}_0$ and $C$ can be set using long-term field observations of cliff retreat (e.g. Hapke and Reid, 2007). Although the model does not explicitly model the smaller-scale structural variations that affect the retreat rate of the rock, such as joints and fractures (Clark and Johnson, 1995; Trenhaile et al., 1998; Dickson et al., 2004), the long-term cliff retreat rate allows for implicit representation (e.g. structurally weaker rocks will have higher rates of retreat). This assumes a relatively even distribution of these features within a given rock type over a given spatial scale.

The second term in Eq. (1) represents constant beach sediment losses ($S$) through time due to, for example, the attrition and subsequent offshore transport of beach sediment in the surf zone, or as a human impact, such as sand mining (Thornton et al., 2006; Limber and Murray, 2011; Limber et al., 2014).

The final term represents the gradient in wave-driven alongshore sediment flux that causes large-scale, long-term shoreline change (erosion, accretion, the formation of capes and spits). Sediment flux is calculated via the common Coastal Engineering Research Center (CERC) sediment transport equation, as discussed at length elsewhere (Komar, 1971; Ashton et al., 2001; Ashton and Murray, 2006a, b; Valvo et al., 2006; List and Ashton, 2007; van den Berg et al., 2012). The magnitude of sediment flux is a function of incoming wave angle relative to the orientation of the coastline, and sediment transport occurs at a greater rate when wave height and period (through effects on shoaling and refraction) increase. Therefore, wave climate characteristics will have a marked effect on how sediment is distributed along a coastline and ultimately how large-scale coastal morphology will evolve.

A factor not expressed within Eq. (1) is the wave shadowing influence from protruding sections of coastline. The area covered by the shadowed zone is dependent on the incoming angle of the offshore wave, with respect to the shoreline, and the size of the headland. In the shadow, it is assumed that

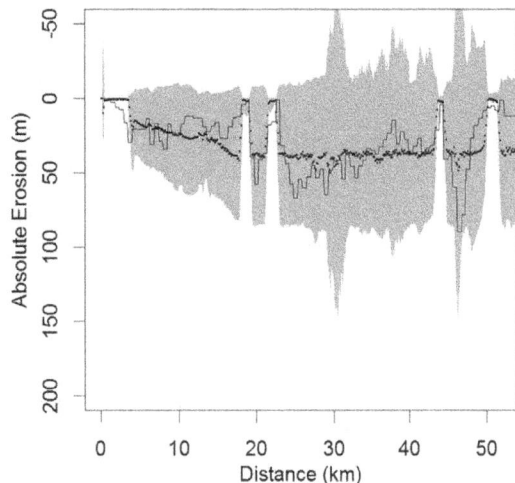

**Figure 4.** Range of simulated coastline retreat over a 15-year period, as captured during the ensemble calibration process (grey shadowing). The observed rates of change between 1995 and 2010 (modified from Montreuil and Bullard, 2012) are given as the solid black line and the ensemble member with the lowest RMSE plotted as a series of points for comparison.

there is no sediment transport when waves impinge beyond the critical angle at which shadowing occurs; however, sediment can be transported into this zone when waves approach at angles for which there is no shadowing (Ashton and Murray, 2006a).

In the original model, offshore wave climate is represented by a fixed offshore wave height and period, and by a four-bin probability density function (PDF), which defines the degree of asymmetry in wave direction and the fraction of high-angle waves (see Ashton et al., 2001). Recent wave climate records are available for the North Sea, off the coast of the Holderness cell, allowing a realistic representation of current wave climate (see Sect. 2). The model was therefore adapted to use observed wave records to drive the simulation. The major caveat in this conversion is that waves observed as propagating in an offshore direction, as determined by the average orientation of the coast, are converted to a null wave angle and height so that no sediment is transported during that time step.

## 4  Calibration and setup

The model is discretised into 100 m square cells, representing the region from Flamborough Head in the north to the Humber Estuary in the south, and a daily time step used to drive each simulation. The eastern boundary is approximately 40 km east of Hull and the western boundary 20 km to the west. The northern and southern boundaries of the model contain a mixture of nodes representing land, beach and sea, and use a specified boundary condition that allows a sand flux out of, but not into, the system. The western (land) and

eastern (sea) boundaries are set with a no-flow condition such that sediment cannot be created, removed or passed through these interfaces. The Humber Estuary is represented as a sediment sink in the model. In reality, the Humber River transports sediment into the North Sea Basin, an area outside the model domain. For the purpose of this study we ignore the spit at Spurn Head, and this region is therefore excluded from both the calibration process and data analysis. The daily offshore wave height and angle, which provide energy for sediment transport in the model, are abstracted from the historic wave data, and cycled for the length of each simulation.

### 4.1  Calibration

Lithological and shoreface properties have been measured at several locations along the Holderness coastline (Newsham et al., 2002); however, these observations are spatially limited and are not likely to represent the coastline as a whole. To derive these parameters we ran the model 2000 times with different initial shoreface and lithological properties, and compare the simulated recession to observed rates using the root-mean-square error (RMSE). Observed recession rates used for this comparison are those compiled by Montreuil and Bullard (2012), spanning the 15-year period 1995–2010. To improve the RMSE between simulated and observed rates, larger coastal defences were represented in the simulation as slow-eroding surfaces. Each model run in the calibration is allocated 10 years of spin-up time, enough for each simulation to reach a steady-state condition, before the 15 years of simulation. Attenuation of steady state ensures that beach sediment, which is initialised as uniform along the coast, is distributed and that any small-scale roughness in the coastal profile used to initialise the coastline shape can be removed. The Monte Carlo (Metropolis and Ulam, 1949; Robert and Casella, 2004) approach adopted for the calibration process ensures that, for the main phase of simulation, the model is initialised with recession rates that are closest to those observed.

The closest simulated match to the observed recession data, and that selected to represent the lithological and shoreface parameters in the simulations used to project future coastal retreat, has an RMSE of 13.20 m and a mean error of 0.81 m (Fig. 4). The agreement between simulated and observed erosion rates is spatially variable along the coast. In the model we assume a homogeneous geology and topography and that the simulated coastline has a relatively uniform retreat rate around the designated coastal defences. Observational data show variable erosion along the coast and accelerated erosion down-drift of coastal defences. The difference in down-drift erosion around defences is due to the representation of wave refraction within the model (see Ashton and Murray, 2006a). The non-refraction differences are attributable to factors that may be considered either temporal or spatial. Temporally, the measured retreat rate reflects short-lived events such as storms and sand-bar placement, whereas the simulated retreat is averaged. Spatially, the susceptibility of the coastline to erosion is treated homogeneously in the model (apart from the division into hard and soft rock), whereas in reality, the geology is heterogeneous and many small-scale, engineered coastal defences are interspersed along the length of the coast. This heterogeneity is composed of regions more (due to joints and stream heads) and less (due to higher cohesion or hard coastal defences) susceptible to erosion. Erosional homogeneity is also affected by cliff failure and the subsequent protection this affords the new cliff face. These processes can be considered as implicit in the model if the process occurs evenly along the coast.

## 4.2 Future simulation

To assess the sensitivity of the coastline to possible future wave climates up to 2100, an ensemble of 1350 model runs is undertaken, with the historic offshore wave climate perturbed by up to $\pm 20°$ in rotation and up to $\pm 0.4$ m significant wave height. These modifications to the historic wave climate, which is cycled at the end of each 2-year period, are applied linearly over the 90-year period from 2010 to 2100. Changes in offshore wave direction and significant wave height are selected at random between the bounding levels for each future simulation, with the initial state represented by the historic wave climate data. In addition to these perturbed future simulations, a baseline simulation is undertaken to gauge recession rates. This baseline simulation represents a future which continues to receive the same, unperturbed, historic wave climate and uses the same initialisation parameters as the future ensemble.

As we are assessing the response of a natural coastline to wave climate, the baseline and future ensemble simulations are undertaken without coastal defences. The rest of the model domain, grid-spacing and time-stepping attributes for these future simulations are identical to the calibration setup. To ensure the model is initialised from a steady state, a 10-year spin-up phase (as assessed during the calibration), starting from the current coastline position under current wave conditions, is performed before each simulation. The output from the spin-up period is omitted from the results and analysis. Erosion rates and sensitivity to wave climates are presented and discussed with reference to the start date 1 January 2010.

## 5   Results and analysis

Our analysis initially examines the spatial distribution of absolute erosion (recession) along the Holderness coast, and compares this recession relative to the baseline. By spatially averaging the relative erosion for each ensemble member, and plotting this value against the wave perturbation factors, the influences of rotating the wave climate and changing the wave height are examined. Finally, temporal analysis highlights the increasing diversity of the ensemble-relative erosion through the simulation period.

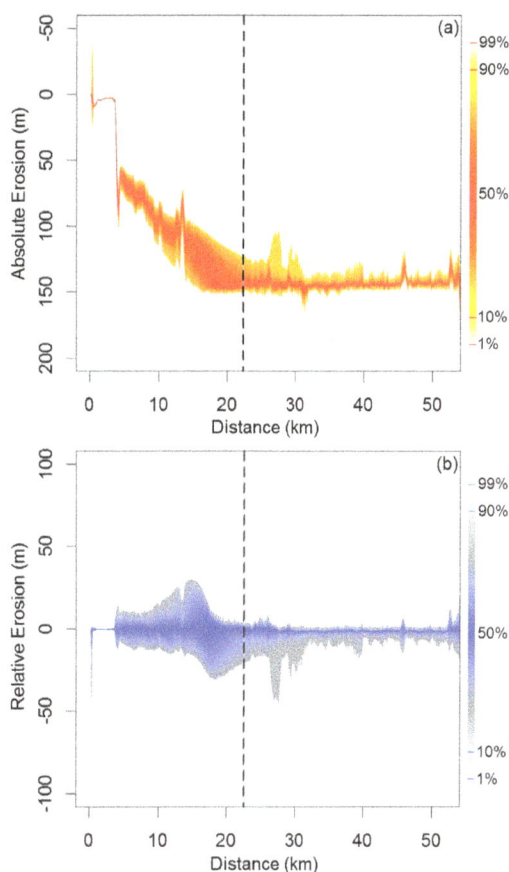

**Figure 5.** Simulated absolute change in coastline position (**a**) from 2010 to 2100 predicted using an ensemble of future wave climates. Relative change in coastline position (**b**), as referenced to the baseline for each member of the ensemble. The range of colours in each plot represents the ensemble percentiles as given on the right of the figures. The black dashed line represents the divide between the northern region (to the left) and the southern region (to the right) as defined in the text.

## 5.1   Spatial analysis

The simulated distribution of absolute erosion along the coast by 2100, for the ensemble of wave climate perturbations, is presented in Fig. 5a. Zero erosion represents the initial coastline position for 2010, and positive values represent a landward coastal retreat. Landward retreat is near zero at Flamborough Head and increases to a maximum of 150 m in the central sections of the coastline. Toward the south, absolute erosion reduces in a quasi-linear fashion to 145 m at Easington (far right in the plot). Within this southern section of the coast, there is little range in the absolute erosion produced by the ensemble. The largest range in absolute erosion occurs at between 10 and 30 km south of the northern domain boundary, where the difference between the 10th and 90th percentiles is around 60 m.

When compared with the baseline (Fig. 5b) the results reveal that future erosion rates could either accelerate or slow depending on the nature of the wave climate change. The negative skewing of the relative erosion, where future simulations are compared to the baseline, implies that a reduction in erosion rate for the coast as a whole is more likely, although relative erosion along the coast is highly heterogeneous (Fig. 5b). The southern region (defined as south of the sea wall at Hornsea, Fig. 1) shows little variation in relative erosion, and the 50th percentile is close to zero. As with the absolute erosion rates, the northern region (defined as north and including the sea wall at Hornsea, Fig. 1) exhibits the greatest range in relative ensemble erosion rates over the 90-year period. The first to third quartiles also show a wide range of values in this region, indicating a spread of retreat values throughout the ensemble. Depending on the wave climate attributes of the ensemble member, there is up to ±30 m (∼ 25 %) disparity in erosion relative to the baseline.

## 5.2 Impacts of changing offshore wave angle and height

Spatially averaging relative change in erosion for each ensemble member allows an assessment of the individual and combined influences of rotating offshore wave direction and perturbing wave height.

Figure 6a shows that the relationship between changing offshore wave direction and relative erosion is linear within the range of −20 to 0°, with anticlockwise rotation progressively reducing relative erosion. Clockwise rotation in the offshore wave climate of up to 18° increases erosion rates, although unlike the anticlockwise trend, this trend is nonlinear. There is further asymmetry between clockwise and anticlockwise offshore wave rotations where, under certain circumstances, clockwise rotation greater than 18° reduces recession in comparison to the baseline.

The influence of a changing wave height on relative erosion is presented in Fig. 6b. It reveals a weak relationship, where a reduction in relative erosion occurs with increasing wave height. Mean relative erosion ranges between −8 and 5 m under a wave height reduction of 0.4 m, and between −13 and 4 m under a wave height increase of 0.4 m.

The range in relative erosion for a particular offshore wave height or rotation is partially controlled by the corresponding perturbation. For example, the range of relative erosion determined for a fixed offshore wave rotation of −10° is a function of the range of wave heights. This suggests that both wave parameters influence relative erosion, although the strongest control remains offshore wave rotation, as this generates the smallest range in relative erosional for any fixed rotation value. To highlight the relationship between relative erosion and perturbations in offshore wave height and rotation, they are plotted together in Fig. 6c and d. As highlighted in previous plots, a clockwise rotation results in increased

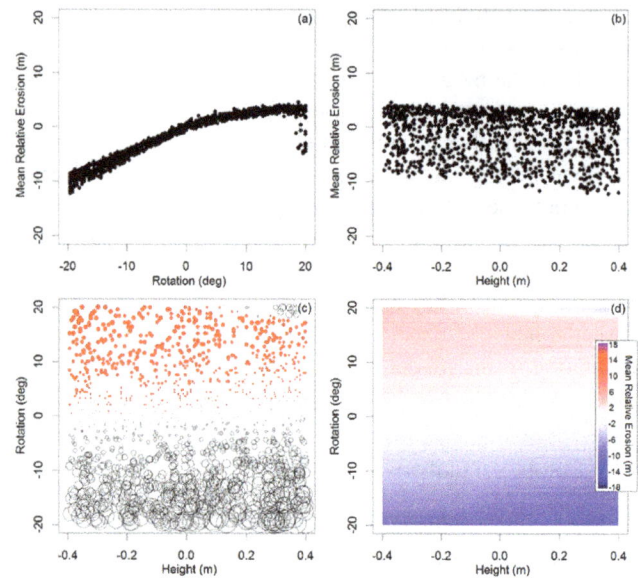

**Figure 6.** Wave rotation (**a**) and wave height (**b**) components of wave climate plotted against spatially averaged mean relative erosion over the 90-year period. Negative erosion values indicate a relative reduction in the erosion rate in comparison to the baseline. Wave height and rotation perturbation factors are plotted together (**c**). The size of each symbol is relative to the change in mean erosion rate imparted by that wave climate in comparison to the baseline scenario. Red dots represent increased erosion relative to the baseline and empty circles reduced erosion. The same data have been a contoured (**d**). The scale on this plot represents spatially averaged (mean value for the coast as a whole) relative erosion (m) after 90 years of simulation.

relative erosion and an anticlockwise rotation less relative erosion. The small, subtle effects of changing wave height are also highlighted: for any particular offshore wave rotation, the relative erosion rate decreases by a small amount as wave height increases. These relationships alter under the most extreme clockwise changes (under offshore wave rotations above 18°), where greater wave heights increase relative erosion.

## 5.3 North–south divide

The northern and southern regions of the coastline respond differently to changes in wave climate. To assess these differences, the relative change in erosion for each region is presented in spatially averaged forms (Figs. 7 and 8).

In the northern region, there is a highly linear coupling between offshore wave rotation and erosion, even under clockwise rotations. The reduction in relative erosion, apparent at the extreme of clockwise rotation for some ensemble members, where the whole coastline is considered (see Fig. 6), is not apparent in the northern region. For this region there is no definitive relationship between changing wave height and relative erosion.

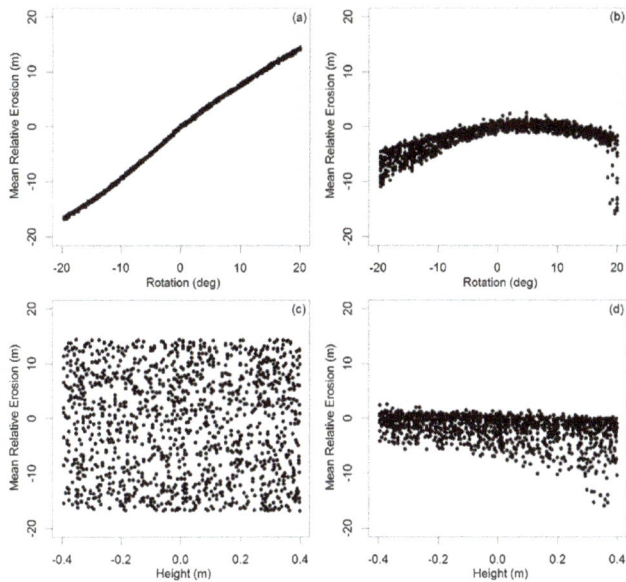

**Figure 7.** Wave rotation perturbation plotted against spatially averaged mean relative erosion for the north (**a**) and south (**b**). Negative values indicate a reduction in the erosion rate. Wave height perturbations are also plotted against spatially averaged mean relative erosion for the north (**c**) and south (**d**).

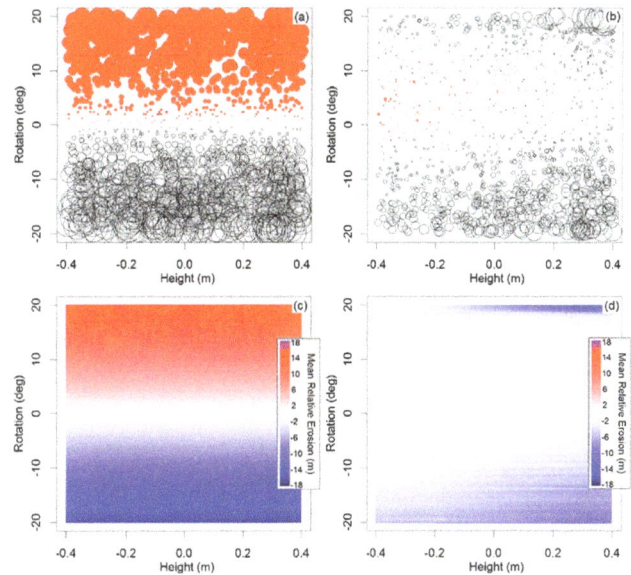

**Figure 8.** Wave height and rotation perturbation factors are plotted against each other for the north (**a**) and south (**b**) of the model domain. The size of each symbol is relative to the change in mean erosion rate imparted by that wave climate in comparison to the baseline scenario. Red dots represent increased erosion relative to the baseline and empty circles reduced erosion. Interpolated contour plots of the height change component of wave climate against the wave rotation component for the north (**c**) and south (**d**) of the model domain are also given.

In the south, maximum relative erosion occurs at clockwise rotations of around 8°. In comparison to the north, the range of relative erosion is lower in the south, suggesting a balance in the height and rotation perturbations. The relationship between increasing wave height and relative erosion reduction is broadly linear in the south. This relationship produces a weak gradient under small rotations in offshore wave climate, but with large clockwise rotations the gradient in increased. This trait is highlighted in Fig. 8, where, for the southern region, there is a relatively strong horizontal gradient in relative erosion at clockwise offshore wave rotations above 18°.

### 5.4 Temporal evolution

By plotting the average relative erosion against time, arising temporal divergences were elucidated (Fig. 9). Throughout the simulation, the average relative erosion rate remains near zero. Over the first 40 years of simulation, the range of possible erosion rates show little asymmetry, indicating a low tendency for either increased or decreased erosion rates. Modifications to the wave climate over this period are small, as the wave climate perturbations are applied linearly to the baseline climate for each scenario. As the wave factors begin to impart a larger influence, there is a non-linear response from the system. The range between both the outliers and the first and third quartiles get progressively larger. The data become negatively skewed, implying that a reduction in rel-

ative erosion is more likely given the input parameters of the ensemble.

### 6 Discussio

The following discussion highlights three overarching impacts of morphology on recession that may be extrapolated to similar coastlines. Detail is provided for the Holderness coast; however, separate analysis would be required to determine the same level of detail for a different coastline of similar morphology.

Overall, the response of the Holderness coastline to modified wave climates is a reduction in relative recession. This reduction is due to a tendency for the future wave climates to move the average offshore wave angle (with respect to coastline orientation) away from the sediment transport peak of 42° as defined in the underlying CERC equation. The coastline response is spatially heterogeneous, with different sections of the coastline exhibiting variable rates of erosion under differing wave climate scenarios. In the north, clockwise offshore wave climate rotations increase relative erosion and anticlockwise rotations reduce erosion. In the south, increased relative erosion peaks occur at a clockwise wave climate rotation of about 5°. The difference in recession rates for the northern and southern sections of the model is due to a combination of factors. Firstly, the angle between that of

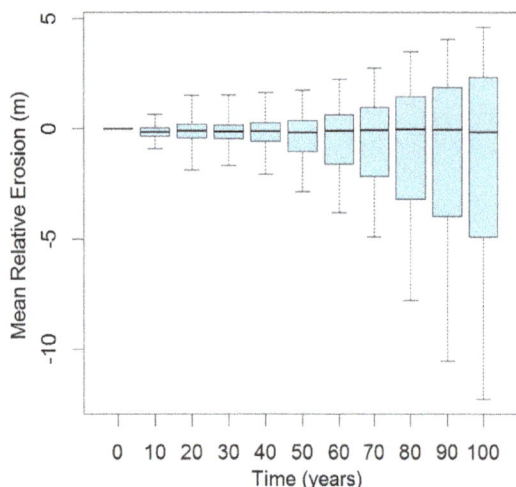

**Figure 9.** Box-and-whisker plot showing the evolution of spatially averaged mean relative coastal erosion, in comparison to the baseline, through time. The central bars represent the median value, the blue boxes show the interquartile range and the extended bars represent extreme values.

the wave direction and the coast increases towards the south as the coastline orientation changes. Secondly, there is a heterogeneous distribution of beach sediment through time and space, and hence variable protection from erosion, along the coast. Thirdly, the CEM accounts for shadowing of the coast by headlands, and thus Flamborough Head shadows the north and protects this region from erosion under certain wave climates. Here we discuss the relative importance of these three factors in driving the heterogeneous erosion rates.

### 6.1 Shoreline angle

Neglecting wave-shadowing effects (considered below), alongshore changes in coastline orientation tend to cause shoreline change. The dominant observed offshore wave direction in the historic data set is approximately 30°. This results in angles with the coastline of between 20 and 90° in the north region and an approximately constant 20° in the south. The variation in coastline orientation with respect to the incoming wave direction tends to create a gradient in transport, and associated shoreline change, that is strong in the north.

Changes in offshore wave-approach angles will tend to affect part of the northern section especially strongly. Modelled sediment transport is greatest when offshore waves impinge the coast at relative angles of between 30 and 50°, depending on the equation used for breaking-wave-driven, alongshore sediment transport (Ashton and Murray, 2006a). For the CERC equation, as used by the CEM, maximum sediment transport occurs at 42°. Parts of the coastline within the northern region are orientated close to this angle, with respect to the dominant offshore wave direction. Changes in relative wave-approach angles generally cause a shift in coast-

line diffusivity; the rate that shoreline curvature is smoothed out, or the rate that it is exaggerated in the case where waves approach from angles greater than the one that maximises alongshore sediment flux (Ashton et al., 2001). In the relationship between coastline diffusivity and relative wave angle, the gradient (slope of the curve) is greatest for the wave-approach angle that maximises alongshore sediment flux (Ashton and Murray, 2006a). For those parts of the coastline with orientations close to the one that maximises the net alongshore flux for a given distribution of wave-approach angles, changes in the dominant wave-approach direction will tend to have the greatest effect in changing coastline diffusivity, and therefore shoreline change rates. For some coastline segments, changing diffusivity could even involve a shift from a tendency toward diffusive coastline dynamics (accretion in concave-seaward segments) to anti-diffusion (erosion where the coastline is concave), and vice versa. These coastline-curvature effects partially explain the spatial variation in the relative sensitivity of the north and south coastal sections to changes in wave direction.

### 6.2 Sand protection and the impact of increasing wave height

The spatial variation in erosion contributes to the morphological sensitivity of the Holderness coastline, as the majority of transported sediment is derived from the cliffs within the coastal domain. In the simulation, beach volume is spatially heterogeneous at any particular time, tending to reflect pulsed fluxes of sediment migrating in the southerly longshore drift direction. These flux pulses are relatively small in the north, where sediment input is limited. They increase in size towards the centre of the model, where they reach a maximum, migrating at around 1 km yr$^{-1}$. Although differing in shape, these pulses are analogous to the "ord" sandbars described and discussed by Pringle (1985). The ords are formed in the northern part of the coast, where the shadowing effect of Flamborough Head begins to diminish. Their formation in the northern region of the model is controlled by the driving wave climate. The influence of high-angle waves (those approaching from angles greater than the alongshore-flux-maximising angle) in the northern section favours the development of simulated ords (Ashton and Murray, 2006a).

As described by Pringle (1985), ords and their migration affect coastal erosion. Where the ords cover and protect the shoreface and cliff base, erosion rates are reduced, while they are enhanced between the ords where the protection is reduced. The movement of these sand bodies along the shoreface, integrated over the decadal to centennial scale, reduces erosion rates along the entire coast, relative to a situation in which protective sand cover is absent. However, because of the concave-upward (exponential) relationship between beach width and cliff erosion rate in the model, the formation of ords in the model could also enhance long-term erosion rates relative to a smoother alongshore distribution of protective sediment. A thought experiment helps to explain this effect. Starting with an alongshore-uniform distribution of a given amount of sediment, rearranging that sediment into pulses tends to reduce cliff erosion rates where beach width is increased. This tends to reduce alongshore- or time-averaged cliff erosion rates. However, because of the concavity of the beach-width cliff-erosion-rate curve, compared to the decreased cliff erosion, where beach widths are increased, cliff erosion is increased more where widths are decreased. Therefore, the effect of increasing the heterogeneity of beach widths (whilst fixing the amount of sediment along a coastline segment) is to enhance averaged cliff erosion rates. For a coastline section in which part of the coast is bare of sediment, so that cliff-erosion rates locally (and temporarily) dictate shoreline change rates, the migration of ords could increase long-term erosion.

The relationship between protective beach cover and cliff erosion rates could explain the effect that changing wave height has on erosion rates. One possible explanation for this wave-height effect involves cliff erosion rates in the northern-most section of the coastline. As there is no influx of sediment around Flamborough Head, beach widths in the most northerly regions cannot grow sufficiently to protect the soft-till cliffs from erosion. Increasing wave height laterally extends this zone of reduced beach protection by increasing the gradient in sediment transport, generating more sediment input from cliff retreat and increasing the availability of sand for beach protection further south. The reduction in erosion created by the added beach protection outweighs the increased erosion along the cliffs in the north, reducing overall erosion along the coast when wave height is increased.

Another possibility (not mutually exclusive with the first one), involves the relationship between wave heights and the diffusivity of the wave climate. Holding the distribution of wave-approach angles fixed, increasing offshore wave height tends to increase coastline diffusivity (Ashton and Murray, 2006a). In the southerly parts of the domain, where the ords migrate along the coastline, increasing coastline diffusivity would tend to smooth out the ords, making them less pronounced (Ashton and Murray, 2006a). This tendency to reduce alongshore variability in beach width reduces the long-term shoreline erosion rate if the coastline is intermittently bare; as ords migrate through the domain making erosion rates intermittently limited by detachment rate (as opposed to transport limited). Determining the dominant cause for the wave-height effect on coastline erosion rates will require additional analysis of model results beyond this study.

## 6.3 Shadow zone

The final effect on variations in simulated erosion involves the shadow zone caused by Flamborough Head. As the shoreline extends seawards to a headland (Flamborough Head), portions of the shoreline are shadowed to waves of particular orientations. At each time step in the model, the wave direction is determined and the shoreline scanned for shore segments that are shadowed. Wave height, and concurrently sediment transport, within the shadow is set to zero. This method assumes the headland is a prominent coastal feature and therefore that wave energy retained for sediment transport following refraction is greatly reduced. Ashton and Murray (2006a) provide a detailed description of the shadow function with the CEM. The extent of the shadowed region changes with changing offshore wave direction at each time step. Within any stretch of coastline that is shadowed part of the time, moving down-drift from the headland, the degree of shadowing decreases, which tends to produce a divergence of alongshore sediment flux and associated erosion. Within a shadow zone, large-scale concave-seaward coastline curvature tends to develop from this shadow-related erosion, as the alongshore gradients in shoreline orientation tend to adjust, producing alongshore-uniform erosion rates (which are determined by the boundary conditions).

In our study area, due to the orientation of Flamborough Head, waves from the north create the largest shadow zone, while waves from the east and south produce no shadow zone. With an increasingly rotating offshore wave climate, the area shadowed by the headland is progressively modified and the sediment transport rates along the coastline changed accordingly. Anticlockwise rotations will increase shadowing in the northern section, which reduces net alongshore sediment fluxes, and therefore the potential for erosion. Conversely, clockwise rotation will reduce the shadowing and increase the potential for erosion. Perhaps most importantly, changing the portion of the waves from which the coastline is shadowed changes local wave climates along that coastline, and due to the shape of the coastline, the impacts of these changes are asymmetrical with respect to the direction of wave climate rotation. Anticlockwise rotation will remove more of the high-angle waves felt by coastline segments within the northern section. This loss of high-angle influence will tend to make the local wave climates more diffusive, favouring more sediment retention in the concave-seaward northern coastline section. Clockwise rotations, on the other hand, tend to make the local wave climate less diffusive, favouring an increase in the coastline curvature through erosion in the central northern section.

## 7 Caveats

Although the model produces a reasonable representation of erosion rates during calibration, it is important to highlight the caveats of using such a model for sensitivity analysis. The basis for many of these caveats surrounds the simplified representation of physical processes within the CEM.

The geology within each erodible rock type for the simulation is assumed to be homogeneous and free from any dominating, anisotropic features, such as rivers or major fractures. Smaller-scale features are integrated implicitly in the lithological and shoreface properties determined by the calibration phase. However, larger features, such as the stream mouth north of Withernsea, will not be properly reflected by the model. In addition to these natural heterogeneities, it is recognised that negation of coastal defence structures within the simulation will alter the evolving morphological characteristics of the represented coast. These defences act as non-erodible surfaces, arresting the landward retreat of the coastline. These features are intentionally ignored as we are assessing the natural morphological sensitivity of the coast.

The single division of rock susceptibility to erosion within the model allows Flamborough Head to remain a prominent headland. The complete geometric shadowing assumed by the model on the down-drift side of a headland forms a region in the model with no sediment transport processes. In reality, complex refraction, and diffraction, of the wave around the headland would occur, sending a part of the wave energy into the shadowed zone and modifying the approach angle of waves. However, the main effect of greatly reducing sediment transport within the shadow is captured.

The remaining caveats are concerned with the representation of the future state of the North Sea within the model. A future wave climate is unlikely to be similar to the simply perturbed historic wave climate used in this study. It assumes that weather patterns are essentially the same as they were in 2009–2010, and there has been no attempt to reflect possible changes in storminess. However, by using an ensemble approach, the range of likely effects on the morphological characters of the Holderness coastline is captured.

Finally, predicted sea-level rise for the Holderness coast should not have a significant influence on wave climate, and the direct influence on coastal recession is thought to be quasi-linear. Thus, the simulation undertaken here does not include sea-level rise and our results need to be considered in conjunction with estimates of recession rates from such rises. Bray and Hooke (1997) suggested an increase in recession rates of between 22 and 133 % by 2050 using a modified Bruun rule method, while Castedo et al. (2012) used their cliff recession model to derive a linear increase in recession rates ranging from 0.015 m yr$^{-1}$ for an annual sea-level increases in of 1 mm to 0.32 m yr$^{-1}$ for an annual sea-level increase of 10 mm.

## 8 Conclusions

An ensemble of 1350 simulations of coastal erosion is undertaken, each forced with a gradually perturbed version of a recorded, historic wave climate to represent the period from 2010 to 2100. A baseline run is undertaken using the historic wave climate from 2009 to 2010, cycled over 90-years. This provides a reference against which to compare the output from the simulations using the stochastically varied wave climate.

Considering the Holderness coast as a whole, anticlockwise rotation of the wave climate broadly reduces the rate of erosion, whilst clockwise rotations increase rates. Although the correlation is less strong, wave heights also have an impact on erosion; however, due to changes in sediment distribution, they unexpectedly lead to an average reduction in relative erosion with increasing height. The sensitivity of the coast to these changes in offshore wave climate is spatially variable, with broadly differing impacts in the northern and southern regions.

Fundamental changes to the system due to the changing offshore wave climate do not occur in the first 40 years of simulation. In the following years to 2100, landward retreat remains close to zero for the chalk outcrop in the north, which pins the system. The combination of incident wave angle, wave shadowing and variable beach protection results in northern regions of the coast exhibiting the greatest sensitivity to changes in wave climate. Difference in the relative influences of perturbations in offshore wave height and rotations of wave direction are found for the northern and southern regions of the Holderness coast. These differences suggest that erosion in the northern region is more sensitive to changes in offshore wave direction, while erosion in the southern region is sensitive to combined changes in wave height and wave direction.

**Acknowledgements.** The authors are grateful to Martin Hurst of the British Geological Survey for reviewing this paper and returning feedback that led to its improvement. This paper is published with the permission of the Executive Director of the British Geological Survey (NERC), and was supported by the Climate and Landscape research programme at the BGS.

Edited by: S. Mudd

## References

Adams, P. N., Anderson, R. S., and Revenaugh J: Microseismic measurement of wave energy delivery to a rocky coast, Geology, 30, 895–898, 2002.

Ashton, A. and Murray, A. B.: High-angle wave instability and emergent shoreline shapes: 1. Modeling of sand waves, flying spits, and capes, J. Geophys. Res. Earth Surf., 111, F04011, doi:10.1029/2005JF000422, 2006a.

Ashton, A. and Murray, A. B.: High-angle wave instability and emergent shoreline shapes: 2. Wave climate analysis and comparisons to nature, J. Geophys. Res. Earth Surf., 111, F04012, doi:10.1029/2005JF000423, 2006b.

Ashton, A., Murray, A. B., and Arnoult O: Formation of coastline features by large-scale instabilities induced by high-angle waves, Nature, 414, 296–300, 2001.

Bladé, I., Liebmann, B., Fortuny, D., and van Oldenborgh, G. J.: Observed and simulated impacts of the summer NAO in Europe: implications for projected drying in the Mediterranean region, Clim. Dynam., 39, 709–727, 2012.

Bray, M. J. and Hooke, J. M.: Prediction of soft-cliff retreat with accelerating sea-level Rise, J.Coastal Res., 13, 453–467, 1997.

Brown, S., Barton, M. E., and Nicholls, R. J.: The effect of coastal defences on cliff top retreat along the Holderness coastline, P. Yorks. Geol. Soc., 59, 1–13, 2012.

Castedo, R., Murphy, W., Lawrence, J., and Paredes, C.: A new process–response coastal recession model of soft rock cliffs, Geomorphology, 177/178, 128–143, 2012.

Catt, J. A.: The Pleistocene glaciation of eastern Yorkshire: a review, P. Yorks. Geol. Soc., 56, 177–207, 2007.

CCO: Channel Coastal Observatory, http://www.channelcoast.org/data_management/real_time_data/charts/?chart=72, las access: July 2013, 2013.

Ciavola, P.: Coastal dynamics and impact of coastal protection works on the Spurn Head spit (UK), Catena, 30, 369–389, 1997.

Clark, H. C. and Johnson M. E.: Coastal Geomorphology of Andesite from the Cretaceous Alisitos Formation in Baja California (Mexico), J. Coastal Res., 11, 401–414, 1995.

de Boer, G.: Spurn Head: its history and evolution, T. I. Brit. Geogr., 34, 71–89, 1964.

DEFRA: Charting Progress 2. Feeder Report: Ocean Processes, Department for Environment Food and Rural Affairs, on behalf of the UK Marine Monitoring and Assessment Strategy community, Department for Environment Food and Rural Affairs on behalf of the UK Marine Monitoring and Assessment Strategy community, Nobel House, London, UK, 2010.

Dickson, M. E., Kennedy, D. M., and Woodroffe, C. D.: The influence of rock resistance on coastal morphology around Lord Howe Island, southwest Pacific, Earth Surf. Proc. Land., 29, 629–643, 2004.

Gibbons, C. R.: A study of the different types of landslides and the associated rates of recession along the Holderness coast, East Yorkshire, MSc thesis, University of Leeds, 2004.

Grabemann, I. and Weisse, R.: Climate change impact on extreme wave conditions in the North Sea: an ensemble study, Ocean Dynam., 58, 199–212, 2008.

Hapke, C. and Reid, D.: National assessment of shoreline change, Part 4: Historical coastal cliff retreat along the California coast, US Geological Survey Open-file Report, 2007–1133, 1–51, 2007.

Hulme, M., Jenkins, G., Lu, X., Turnpenny, J., Mitchell, T., Jones, R., Lowe, J., Murphy, J., Hassell, D., Boorman, P., Macdonald, R., and Hill, S.: Climate-Change Scenarios for the United Kingdom: The UKCIP02 Scientific Report, Tyndall Centre for Climate Change Research, 2002.

Hurrell, J. W.: Decadal trends in the North-Atlantic Oscillation – Regional temperatures and precipitation, Science, 292, 676–679, 1995.

Hurrell, J. W. and van Loon, H.: Decadal variations in climate associated with the North Atlantic Oscillation, Clim. Change, 36, 301–326, 1997.

IECS: Humber Estuary and Coast, Institute of Estuarine and Coastal Studies, University of Hull, Hull, UK, 1994.

IPCC: Climate Change 2007: vol 4, Cambridge University Press, Cambridge, UK, 2007.

Komar, P. D.: The mechanics of sand transport on beaches, J. Geophys. Res., 76, 713–721, 1971.

Lee, E. M.: Coastal cliff behaviour: Observations on the relationship between beach levels and recession rates, Geomorphology, 101, 558–571, 2008.

Limber, P. W. and Murray, A. B.: Beach and sea cliff dynamics as a driver of rocky coastline evolution and stability, Geology, 39, 1149–1152, 2011.

Limber, P. W. and Murray, A. B.: Unraveling the dynamics that scale cross-shore headland relief on rocky coastlines, Part 2: Model predictions and initial tests, J. Geophys. Res. Earth Surf., doi:10.1002/2013JF002978, in press, 2014.

Limber, P. W., Patsch, K. B., and Griggs G. B.: Coastal sediment budgets and the littoral cut-off diameter: A grain-size threshold for quantifying active sediment inputs, J. Coastal Res., 24 (supplement 2), 122–133, 2008.

Limber, P. W., Murray, A. B., Adams, P. N., and Goldstein, E. B.: Unraveling the dynamics that scale cross-shore headland relief on rocky coastlines, Part 1: Model development, J. Geophys. Res. Earth Surf., doi:10.1002/2013JF002950, in press, 2014.

List, J., Farris, A. H., and Sullivan, C.: Reversing storm hotspots on sandy beaches: spatial and temporal characteristics, Mar. Geol., 226, 261–279, 2006.

List, J. H. and Ashton, A. D.: A circulation modeling approach for evaluating the conditions for shoreline instabilities, Coastal Sediments 2007, Am. Soc. of Civ. Eng., Reston, Va, 327–340, 2007.

May, V. J.: Flamborough Head, Volume 28: Coastal Geomorphology of Great Britain Chapter 4: Soft-rock cliffs - GCR site reports, J. S. Publications, Suffolk, UK, 1980.

Metropolis, N. and Ulam, S.: The Monte Carlo method, J. Am. Statist. Assoc., 44, 335–341, 1949.

Montreuil, A.-L. and Bullard, J. E.: A 150-year record of coastline dynamics within a sediment cell: Eastern England, Geomorphology, 179, 168–185, 2012.

Moore, L. J., McNamara, D. E., Brenner, O., and Murray, A.B., 2013. Observed changes in hurricane-driven waves explain the dynamics of modern cuspate shorelines, Geophys. Research Lett., 40, 5867–5871, doi:10.1002/2013GL057311, 2013.

Newsham, R., Balson, P. S., Tragheim, D. G., and Denniss, A. M.: Determination and prediction of sediment fields from recession of the Holderness Coast, NE England, J. Coastal Conserv., 8, 49–54, 2002.

Pendleton, L. H.: The economic and market value of coasts and estuaries: what's at stake?, Coastal Ocean Values Press, Washington, DC, USA, 2010.

Pringle, A. W.: Holderness coast erosion and the significance of ords, Earth Surf. Proc. Land., 10, 107–124, 1985.

Quinn, J. D., Philip, L. K., and Murphy, W.: Understanding the recession of the Holderness Coast east Yorkshire, UK: a new presentation of temporal and spatial patterns, Q. J. Eng. Geol. Hydroge., 42, 165–178, 2009.

Quinn, J. D., Rosser, N. J., Murphy, W., and Lawrence, J. A.: Identifying the behavioural characteristics of clay cliffs using intensive monitoring and geotechnical numerical modelling, Geomorphology, 120, 107–122, 2010.

Robert, C. P. and Casella, G.: Monte Carlo Statistical Methods. Secaucus, NJ, USA: Springer, New York, Inc., 2 Edn., 2004.

Robertson, I.: Erosion and Stability of Till Cliffs on the Holderness Coast, Ph.D. Thesis, University of Newcastle Upon Tyne, University of Newcastle Upon Tyne, Newcastle, UK, 1990.

Sallenger, A. H., Krabill, W., Brock, J., Swift, R., Manizade, S., and Stockdon, H.: Seacliff erosion as a function of beach changes and extreme wave runup during the 1997–1998 El Nino, Mar. Geol., 187, 279–297, 2002.

Scott, W.: Humber Estuary Coastal Authorities Group (HECAG), Flamborough Head to Gibraltar Point Shoreline Management Plan 2, Scott Wilson, Basingstoke, Hampshire, UK, 2009.

Slott, J. M., Murray, A. B., Ashton, A. D., and Crowley, T. J.: Coastline responces to changing storm patterns, Geophys. Res. Lett., 33, L18404, doi:10.1029/2006GL027445, 2016.

Sutherland, J. S. and Wolf, J.: Coastal defence vulnerability 2075, HR Wallingford Report SR590, H R Wallingford, Wallingford, UK, 2002.

Thornton, E. B., Sallenger, A., Conforto, S. J., Egley, L., McGee, T., and Parsons, R.: Sand mining impacts on long-term dune erosion in southern Monterey Bay, Mar. Geol., 229, 45–58, 2006.

Trenhaile, A. S., Pepper, D. A., Trenhaile, R. W., and Dalimonte, M.: Stacks and notches at Hopewell Rocks, New Brunswick, Canada, Earth Surf. Proc. Land., 23, 975–988, 1998.

Valentin, H.: Land Loss at Holderness, Applied Coastal Geomorphology, edited by: Steers, J. A., 1 Macmillan, 16–137, 1971.

Valvo, L., Murray, A. B., and Ashton, A.: How does underlying geology affect coastline change? An initial modeling investigation, J. Geophys Res. Earth Surf., 111, F02025, doi:10.1029/2005JF000340, 2006.

van den Berg, N., Falqués, A., and Ribas, F.: Modeling large scale shoreline sand waves under oblique wave incidence, J.Geophys Res. Earth Surf., 117, F03019, doi:10.1029/2011JF002177, 2012.

Walkden, M. J. A. and Hall, J. W.: A predictive mesoscale model of the erosion and profile development of soft rock shores, Coast. Eng., 52, 535–563, 2005.

Wang, X. L., Zwiers, F. W., and Swail, V. R.: North Atlantic Ocean Wave Climate Change Scenarios for the Twenty-First Century, J. Climate, 17, 2368–2383, 2004.

Woollings, T.: Dynamical influences on European climate: an uncertain future, Phil. Trans. R. Soc. A, 368, 3733–3756, 2010.

Woollings, T., Hannachi, A., Hoskins, B., and Turner, A.: A Regime View of the North Atlantic Oscillation and Its Response to Anthropogenic Forcing, J. Climate, 23, 1291–1307, 2010.

Woth, K., Weisse, R., and von Storch, H.: Climate change and North Sea storm surge extremes: an ensemble study of storm surge extremes expected in a changed climate projected by four different regional climate models, Climate Dynam., 26, 3–15, 2006.

Young, A. P. and Ashford, S. A.: Instability investigation of cantilevered seacliffs, Earth Surf. Proc. Land., 33, 1661–1677, 2008.

# Coastal vulnerability of a pinned, soft-cliff coastline – II: assessing the influence of sea walls on future morphology

A. Barkwith[1], M. D. Hurst[1], C. W. Thomas[1], M. A. Ellis[1], P. L. Limber[2], and A. B. Murray[3]

[1]British Geological Survey, Keyworth, Nottingham, UK
[2]Department of Geological Sciences, University of Florida, Gainesville, FL, USA
[3]Nicholas School of the Environment, Duke University, Durham, NC, USA

*Correspondence to:* A. Barkwith (andr3@bgs.ac.uk)

**Abstract.** Coastal defences have long been employed to halt or slow coastal erosion, and their impact on local sediment flux and ecology has been studied in detail through field research and numerical simulation. The non-local impact of a modified sediment flux regime on mesoscale erosion and accretion has received less attention. Morphological changes at this scale due to defending structures can be difficult to quantify or identify with field data. Engineering-scale numerical models, often applied to assess the design of modern defences on local coastal erosion, tend not to cover large stretches of coast and are rarely applied to assess the impact of older structures. We extend previous work to explore the influences of sea walls on the evolution and morphological sensitivity of a pinned, soft-cliff, sandy coastline under a changing wave climate. The Holderness coast of East Yorkshire, UK, is used as a case study to explore model scenarios where the coast is both defended with major sea walls and allowed to evolve naturally were there are no sea defences.

Using a mesoscale numerical coastal evolution model, observed wave-climate data are perturbed linearly to assess the sensitivity of the coastal morphology to changing wave climate for both the defended and undefended scenarios. Comparative analysis of the simulated output suggests that sea walls in the south of the region have a greater impact on sediment flux due to increased sediment availability along this part of the coast. Multiple defence structures, including those separated by several kilometres, were found to interact with each other, producing complex changes in coastal morphology under a changing wave climate. Although spatially and temporally heterogeneous, sea walls generally slowed coastal recession and accumulated sediment on their up-drift side.

## 1 Introduction

Soft sediment coastlines are highly dynamic environments, where the interaction of sea and land are constantly changing in response to natural and anthropogenic forcing with significant socioeconomic implications (Pendleton, 2010). In an attempt to reduce the loss of property under strongly erosional conditions, it has been general policy in the UK to build solid defences to halt land loss (Scott Wilson, 2009). This approach has been subsequently replaced with the adoption of managed retreat, however, around 44 % of the English and Welsh coastlines remain defended against erosion to some degree (DEFRA, 2010). Coastal defence strategies typically comprise "soft" engineering, usually beach nourishment, or "hard" engineering solutions. The latter consists of building structures designed to directly protect the coastline and fix its position (e.g. sea walls, riprap), encourage beach formation (e.g. groynes, jetties), or reduce the wave energy experienced at the shore (e.g. breakwaters) (Kamphuis, 2000). These structures, often placed on soft sediment coastlines, modify the sediment flux and hence the erosional and depositional processes. In the UK, the majority of sea walls were built during the Victorian era, with little knowledge of the impact on the environment (Brown et al., 2012; Bruun, 1995). More recently, advances in numerical simulation have allowed the impact of the placement of such structures to be assessed in detail with regard to the immediately surrounding area (e.g. Hanson, 1989). The difficulty with these approaches is that there are often non-local impacts to mesoscale morphology (defined as features up to tens of kilometres long, changing at the annual to decadal scale) that are difficult to quantify with field data or the commonly employed engineering-scale models (e.g. Ells and Murray, 2012; Slott et al., 2010). Furthermore, complex, non-linear interactions between multiple defence structures, wave forcing and sediment transport at the mesoscale are difficult to elucidate and quantify using local-scale models and field data.

The local effects of engineered coastal defences on coastline morphology and sediment transport are well known, and have been modelled using one-line modelling approaches (e.g. Hanson, 1989). Typically, whilst such defences may protect the local coast, interruption to longshore transport often causes down-drift increases in coastal erosion (Dean et al., 2013). Barrier structures (groynes and groyne fields) are favoured on coastlines subject to significant littoral drift. They act to reduce the local longshore sediment transport, trapping sediment to protect the beach. Barriers also set up gradients in longshore transport, which result in accretion of sediment on the up-drift side and erosion down-drift due to the loss of protective sediment influx (Kamphuis, 2000; Bruun, 1995; Bakker et al., 1970). Hence, groyne emplacement strategies are best coupled to complementary beach nourishment (Dean et al., 2013). Groynes generate an offshore current and may result in increased loss of sediment to the offshore (Kraus et al., 1994). Eventually natural bypassing will occur as the beach areas between groynes are filled and littoral transport occurs by transport around or over the groynes, or due to groyne permeability. Loss of sediment offshore during storm events may result in the areas between groynes needing to "refill", resulting in potentially significant down-drift erosion.

Sea walls are built in locations where it is desirable to stop coastal erosion and pin the coast. Potential increases in offshore sediment transport may result in a diminished beach fronting a sea wall. This sediment might otherwise contribute to beaches protecting the shoreline down-drift from the seawall structure. In addition, the prevention of erosion due to presence of the sea wall may also reduce the sediment supply to the coastline (Kamphuis, 2000; Kraus and McDougal, 1996).

Relatively few studies have investigated the mesoscale and far-field influence of coastal defences on coastline morphology and sediment transport. Bruun (1995) highlighted that barriers (e.g. groynes) influence local coastline development down-drift, enhancing erosion due to changes in local wave climate by refraction or diffraction. Barriers can also result in a wave of increased erosion propagating down the coastline,

potentially over several kilometres, due to the modification of the longshore sediment transport budget. These observations are supported by studies that have modelled mesoscale coastline evolution under conditions of beach nourishment (Ells and Murray, 2012; Slott et al., 2010). The authors found that nourishment at fixed locations not only mediated the coastline locally but can alter the evolution of the coastline tens of kilometres away. Ells and Murray (2012) extend this study to simulate the effects of sea walls on mesoscale coastline evolution. Their findings indicate that protection through either nourishment or hard-structure intervention results in accretion up-drift; that nourishment produces either accretion or erosion down-drift (depending on the surrounding coastline shape); and that hard-structure stabilisation generally causes increased divergent sediment flux down-drift, leading to increased erosion relative to an unprotected coast. Dickson et al. (2007) simulate the influence of climate change on an eroding coastal region at the decadal scale, focussing on the impacts of changing offshore wave height and direction, and the effect of sea level rise. Although not the main focus, the impact of engineered structures was also simulated through comparison of various coastal management and future wave-climate scenarios. The sensitivity of the erosional response to these scenarios, however, was not explored in detail.

In this paper, the influence of seawall structures on the mesoscale evolution of a soft-cliff, sandy coastline is studied through the use of the Coastline Evolution Model (CEM), developed by Ashton, Murray and others (Ashton and Murray, 2006a, b; Valvo et al., 2006; Ashton et al., 2001). Specifically, we focus on understanding the difference between the predicted behaviour of defended and undefended coastline in the face of wave-climate changes anticipated over the coming century. This paper extends the work of Barkwith et al. (2014), which assesses the sensitivity of erosion of an undefended pinned, soft-cliff, sandy coastline under a modified wave climate. The Holderness coastline of East Yorkshire is used as case study to develop a generalised understanding of the evolution of defended, pinned, soft-cliff, sandy coastal systems. The use of the term pinned in this instance refers to the low recession rate at the northern end of the coastline due to the influence of the chalk headland.

## 2 Holderness coastline

The Holderness coastline formed as the North Sea basin flooded during the Holocene Epoch (Shennan et al., 2000). The study domain is bounded by Flamborough Head in the north, where little sediment is thought to bypass into the littoral cell (Scott Wilson, 2009), and Easington in the south (Fig. 1). Flamborough Head is composed of slowly eroding Cretaceous chalk cliffs ca. 35 m high. The remaining 55 km of coast to the south of Flamborough Head is composed largely of Devensian glacial till and other deposits; these range between 2 and 35 m in thickness, thinning towards the

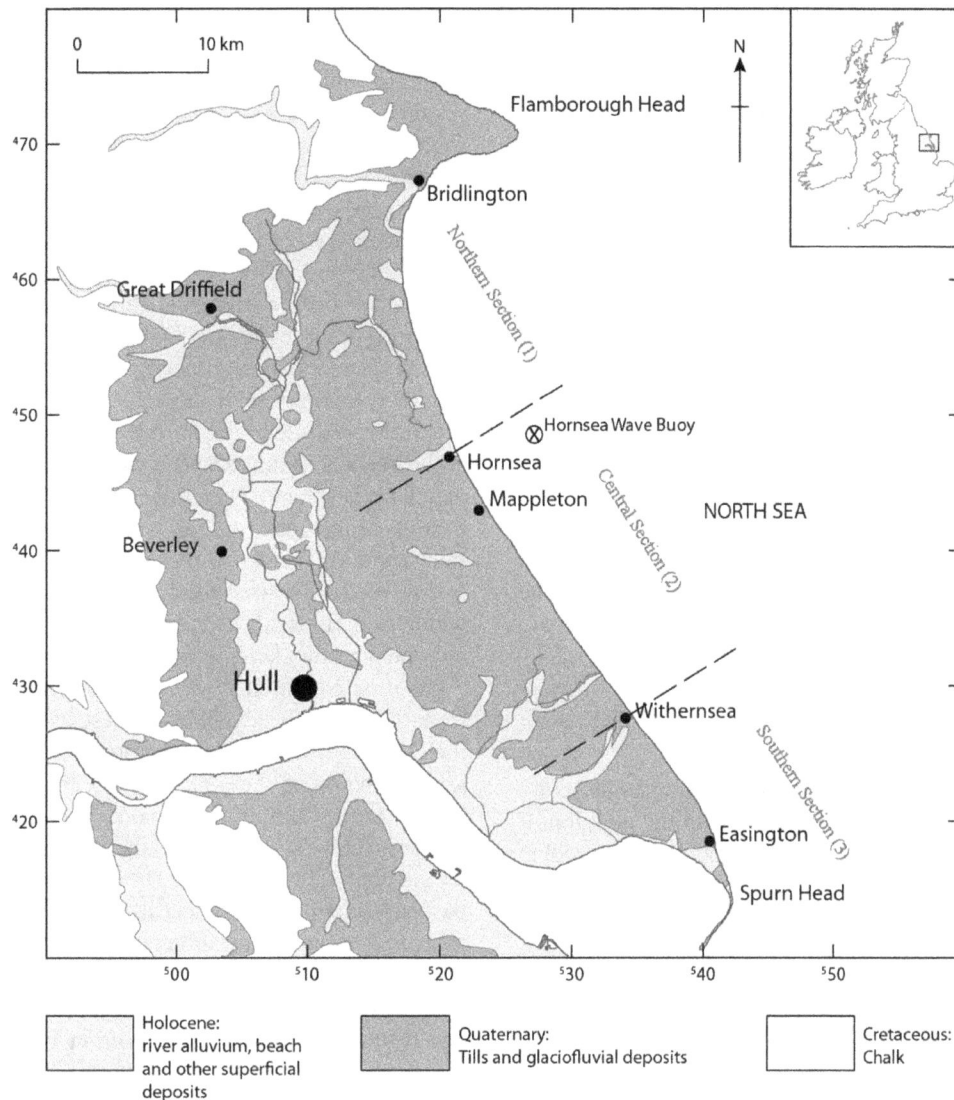

**Figure 1.** Geological composition of the Holderness coast (main) and the location of the region within the UK (insert). Also indicated are the positions of the Hornsea wave buoy, from which wave climate was recorded, and the division into northern, central and southern coastline regions, as referenced by the sea walls at Hornsea and Withernsea (dashed lines), to aid analysis. Modified from Barkwith et al. (2014).

south (Quinn et al., 2009; Catt, 2007). The glacial cliffs are easily eroded and are thought to be the dominant source of the littoral sand at the coast. Erosion occurs through wave action undercutting the cliff base, causing cliff collapse. The result is a rapidly eroding coastline. Recession rates for the Holderness coast have been documented in recent studies by Montreuil and Bullard (2012), Brown et al. (2012) and Quinn et al. (2009). Average recession rates are on the order of 1–2 m a$^{-1}$, but may be an order of magnitude greater during storm events, or local, large-scale collapses.

There has been a long history of defending the Holderness coastline from erosion (Brown et al., 2012). The earliest chronicled sea defences along this coastline were in place during the Abbacy of Burton between 1396 and 1399 (Burton, 2012). From the 19th century onwards there was

a policy of building large-scale sea walls at seaside towns, many popular as tourist destinations. Smaller-scale defence features, including groynes, revetments and rock armour have also been used at various locations along the coastline. Brown et al. (2012) document the changing position of the Holderness coastline cliff top since the mid-19th century, focusing particularly on areas adjacent to coastline defences. They found increased cliff retreat rates for up to several kilometres on the down-drift side of coastal defences, and attributed these to a negative gradient in longshore transport resulting in reduction of the natural beach defence. More recent, the repair of smaller-scale features has ceased, and in some cases, defences have been completely removed, allowing the coast to develop naturally (Brown et al., 2012). However, due to sociopolitical constraints, the removal of the larger sea defences protecting major towns and infrastructure is untenable and maintenance and repair will continue for the foreseeable future.

The offshore wave climate for the Holderness coastline is currently being recorded by the Hornsea Directional Waverider III Buoy (CCO, 2013). Deployed in June 2008, the buoy provides data on significant wave height, period and direction, amongst other parameters. Between 2009 and 2010, the wave input period used for this study, significant wave height varied between 0.2 and 3.5 m, with an annual mean of 0.9 m. The mean wave period for the same period was 7.8 s, ranging from 2.6 to 18.8 s. The dominant mode in wave direction was from the northeast, with a secondary mode from the southeast. There have been several studies that have focussed on the evolution of the North Sea wave climate, with respect to possible future climate change scenarios over the forthcoming century (for example, Sutherland and Wolf, 2002). The range of scenarios used and uncertainty in future storm and North Atlantic Oscillation (NAO) prediction make the North Sea wave climate difficult to predict (Bladé et al., 2012; DEFRA, 2010; Woollings, 2010).

## 3  Modelling

The model, calibration and setup are the same as those described by Barkwith et al. (2014), but with the addition of sea wall defences, represented by essentially non-eroding coastline. For clarity, model simulations that include sea wall structures in the future simulations are termed "defended" and those without such structures termed "undefended". A description of the modelling components, calibration procedure and the ensemble approach are contained in this section of the paper. For further details of the modelling procedure, the reader is directed to Barkwith et al. (2014).

### 3.1  CEM description

A modified version of the CEM (Ashton and Murray, 2006a, b; Valvo et al., 2006; Ashton et al., 2001) is implemented to represent numerically the processes within the coastal domain of interest. The model uses the Coastal Engineering Research Center (CERC) equation (Komar, 1971) to determine long-shore sediment flux. The CEM code has been modified to accept observed wave-climate data and include sediment input from cliff recession (Barkwith et al., 2014; Limber and Murray, 2011). Changes to the coastline position through time are functions of beach geometry and width (Ashton and Murray, 2006a); sea cliff height, lithology and cohesion (Limber et al., 2008); shoreface and offshore wave angles; wave shadowing by protruding coastline features; and wave energy delivered to the shore after attenuation through shoaling and refraction (Adams et al., 2002). Representing the long-term results of relatively short-term processes, the model implicitly averages over short-term events, such as cliff collapses, and over sub-grid scale, spatially random,

heterogeneous features. Such features, including heterogeneity in the geological substrate, the presence of fractures and grain size variability, are assumed to be evenly distributed within each cell (Dickson et al., 2004; Trenhaile et al., 1998; Clark and Johnson, 1995). Temporal processes active at frequencies below the scale of the time step in the model, such as tides, are also handled implicitly (List et al., 2006).

Different erosion rates for different lithologies can be specified within the version of the CEM used in this study. We use this facility to define the chalk cliffs at Flamborough Head as highly resistant to erosion. Conversely, the glacial till forming the remainder of the coast is defined as readily erodible, at rates consistent with those known from the Holderness coast. Sea wall defences have a near-zero erosion rate and, at the decadal scale, exhibit a similar erosional response as the chalk headland. Therefore, to avoid unnecessary modification of the code, the sea wall defences are assigned the same erosion potential as the chalk cliffs.

The model is discretised into uniform cells, 100 m in width, and run with a daily time step. Eastern and western domain boundaries consist of a no-flow condition, with a specified condition of zero sediment flux into the model from the north; this explicitly represents the absence of sediment transport around Flamborough Head. The Spurn Head spit, extending off the southern tip of the coast, and Humber estuary are simulated in the model as a sediment store and sink respectively. However, as their interactions and dynamics are complex (see Ciavola, 1997) they are not included in the analysis. Lithological and shoreface properties have been measured at specific locations along the coastline (Newsham et al., 2002). The data are spatially limited and are not representative of the coastline as a whole. Therefore, calibration was required to define these properties within the model before predictive simulations could be undertaken.

### 3.2  Calibration

Calibrating the model to observational recession data, by modifying the beach and rock properties, allows greater confidence to be placed in the initialisation of future simulation. Cliff erosion provides sediment which is subsequently transported along the coast via longshore drift. Not calibrating the model to observations of coastal retreat may significantly alter the amount of sediment in the system and therefore the system response to a changing wave climate.

Beach and rock properties (notably the erosional resistance and the fraction of fine-grade material in the eroding substrate and beach material) are initialised to be spatially homogenous within the modelling framework. To determine these values we apply a stochastic calibration approach using an ensemble of 2,000 models with varying rock and beach sediment properties. The wave climate for each member comprises 2 years (2009 and 2010) of observed daily significant wave height, angle and period, cycled for the duration of the simulation period.

Each ensemble member is initialised with a 10-year spin-up period, required to reach a dynamic steady state. Dynamic steady state is achieved when the amount of sand being transported along the coastline shows a repeatable response to a particular set of driving factors. Spin-up is undertaken using the 2-year, repeating, recorded wave-climate data and the response to the same events analysed to ascertain whether steady state has been achieved. Sediment transported along the coast is checked for the 8th and 10th year of spin-up for each ensemble member, to establish whether steady state has been attained. Following the spin-up phase, erosion is simulated for a period of 15 years for each ensemble member. This period matches that over which the observed recession rates were compiled by Montreuil and Bullard (2012). The simulated recession rates are compared to observed rates and the ensemble member with the lowest root mean square error (RMSE) is selected to provide the initial properties for the main modelling phase, which is run for the remainder of the current century.

### 3.3 Simulation setup

An ensemble of modified wave climates consisting of 1350 members drives the future simulations. The ensemble approach allows the sensitivity of coastal erosion to small changes in driving factors to be explored. The technique is suited to studying this stretch of coastline as it is a non-linear system (Barkwith et al., 2014) and the future driving wave climate is uncertain. The background wave climate for each ensemble member is formed from the 2 years (2009 and 2010) of observed daily significant wave height, angle and period, and is cycled for the 90 years of simulation. This wave climate is perturbed for each member by selecting changes at random from ranges of $\pm 20°$ rotation in offshore wave direction and $\pm 0.4$ m in significant wave height. These variations are applied linearly over the 90-year simulation. The defended and undefended coast scenarios use the same set of wave perturbations to allow comparison when assessing the impact of the sea wall defences on the evolution of the Holderness coastline with a changing wave climate. In order to elucidate the evolution of the coastline, baseline simulations are undertaken for the defended and undefended scenarios. Both baselines consist of a single 90-year simulation with no perturbations applied to the cycled, observed wave climate.

Barkwith et al. (2014) conclude that the sensitivity of erosion on the natural coast to changing wave climates is controlled by the current morphology of the Holderness coastline, via changing shoreline angle; the reduction in wave energy in the "shadowed" zone created by Flamborough Head; and the greater availability of beach sediment in the southern region of the model. To aid assessment of the impact of sea defences on erosion rates, the coastline was divided into three sections (Fig. 1) and cumulative erosion rates were averaged spatially for each section. Section 1 extends from Flambor-

ough Head southwards to Hornsea and includes the sea walls at Skipsea and Hornsea. Section 2 starts at the southern end of the Hornsea sea wall and continues to Withernsea, up to and including the sea wall along the town promenade. Section 3 extends from Withernsea south of the defences to Easington, where a long sea wall section protects the Easington Gas Terminal.

## 4 Results and analysis

Analysis of the results focuses on the patterns and rates of predicted coastline change evident from the inclusion of defences in the simulation. Results are presented and compared for the undefended and defended scenarios. Our analysis initially examines the spatial distribution of absolute and relative erosion along the Holderness coastline for the entire ensemble. By spatially averaging the relative erosion for each ensemble member, and plotting this value against the wave perturbation factors, the influences of rotating the wave climate and changing the wave height are examined. Finally the combined influences of a changing wave direction and height on erosion rates are explored, focussing on the difference between the undefended and defended scenarios.

### 4.1 Absolute and relative erosion

Absolute erosion over the 90-year simulation period is presented for the baseline (i.e. with no wave-climate modification) undefended and defended scenarios in Fig. 2. Total amounts of erosion appear very similar between these two scenarios away from the locations of sea defences. Reduced erosion is observed in the defended scenario on the up-drift flank of sea walls at Withernsea and Easington. The range of absolute erosion values under the same ensemble of wave perturbations is shown in Fig. 3a (undefended) and Fig. 3b (defended). Positive values represent a landward migration of the coast (i.e. erosion) and negative values land accretion. When compared to the undefended scenario, the sea wall at Skipsea (location included in Fig. 2) in the northern section (1) of the model does not appear to have a significant impact on surrounding recession rates under the differing wave climates. In the central section (2), maximum absolute erosion values are similar in both the defended and undefended scenarios, at ca. 150 m over most of this coastal section. Under clockwise rotation of wave direction and increased significant wave height, absolute erosion can be reduced in the stretches of coast between the sea wall structures, by as much as 100 m, when compared to the baseline simulation. Although the pattern of reduced erosion is spatially heterogeneous, the peaks correspond with the regions of lowest absolute erosion for the undefended scenario. Under the majority of simulated wave climates, the sea walls in the south at Withernsea and Easington (section 3) have less absolute erosion on their up-drift sides. In the southernmost part of section 3 this leads to an overall reduction in erosion

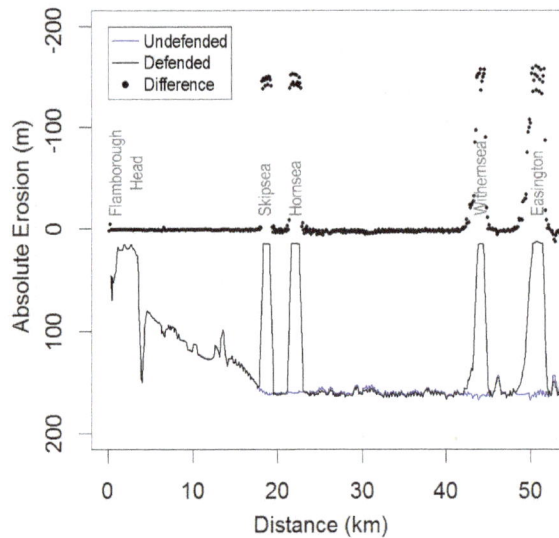

**Figure 2.** Absolute erosion after 90 years of simulation for the undefended (blue line) and defended (black line) coastlines under the baseline wave climate (2009–2010 repeated cycle). The difference between these scenarios is highlighted by the black points. Flamborough Head and the towns with sea wall defences that are included in the model are labelled in grey text at their respective location on the coastline.

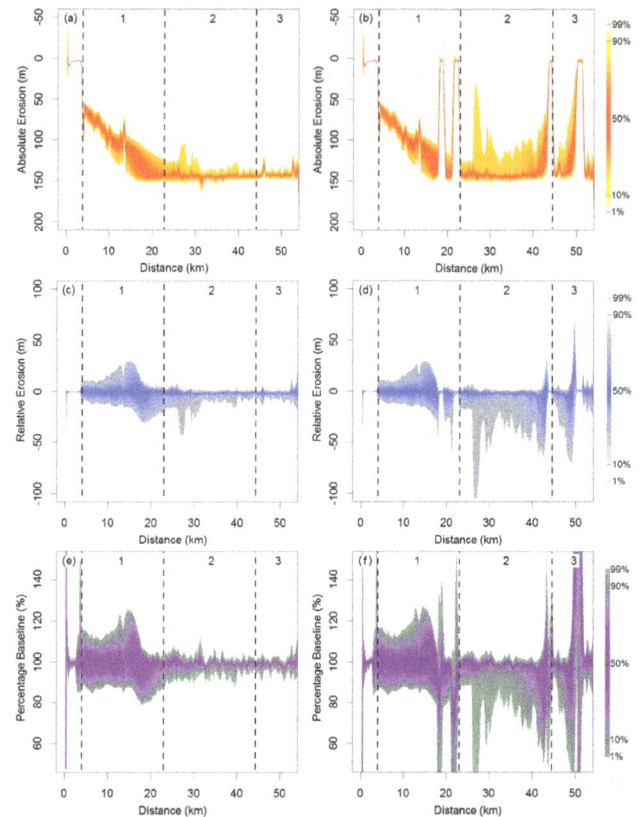

**Figure 3.** Simulated erosion for the Holderness coastline. Simulated absolute change in coastline position (2010–2100) predicted using an ensemble of future wave climates for undefended (**a**) and defended (**b**) coasts. Relative change in coastline position (relative to baseline simulation) for each member of the ensemble, for the undefended (**c**) and defended (**d**) simulations respectively. Percentage change in erosion relative to the baseline simulation for the undefended (**e**) and defended (**f**) simulations, respectively. The range of colours in each plot represents the ensemble percentiles as given on the right of the figures. The regions 1, 2 and 3 refer to the three coastal sections facilitating along-coast comparison, as defined in the text.

when sea defences are included in the simulation. Relative total (Fig. 3c, d) and percentage (Fig. 3e, f) erosion for the suite of ensemble members, as subtracted from the respective baseline, reflect the spatially heterogeneous recession pattern of the absolute erosion. Although there is a reduction in absolute erosion on the up-drift sides of the sea walls at Withernsea and Easington, the increased and decreased regions of relative erosion suggest that the recession rate is highly dependent on the perturbations of the wave climate. The low erosion rates assigned to sea wall structures during model initialisation cause the extreme values of percentage of baseline erosion at the location of sea wall structures (Fig. 3e, f), where a small change in absolute erosion may nevertheless equate to a large percentage change.

## 4.2 Wave direction

Spatially averaging the erosion for each ensemble member, relative to baseline erosion, allows the influence of wave-climate perturbations to be compared for both the undefended and defended scenarios. The data presented in Fig. 4 reveal the influence of wave-climate rotation on erosion rate, for the coast as a whole and each of coastal sections 1–3. When considering the coast as a whole, under counterclockwise rotations in wave climate (Fig. 4a), there tends to be a reduction in relative erosion for both the defended and un-

defended scenarios, with the coastal defences resulting in a lesser response at extremes in rotation.

Under clockwise rotations, there is a marked difference in the erosional response with and without sea defences. The undefended scenario suffers an increase in relative erosion with a clockwise rotation. However, due to the reduction in longshore transport of sediment, the response of the defended coast to the same wave-climate perturbations has an equal chance of also reducing the relative erosion. In section 1 (Fig. 4b) there is a well-defined relationship between the angle of rotation angle and the relative erosion for both scenarios. Differences in response appear at the extremes of wave rotation, where the overall change in erosion rate for the ensemble members is damped by the presence of defended structures. Under a clockwise rotation, erosion relative to the

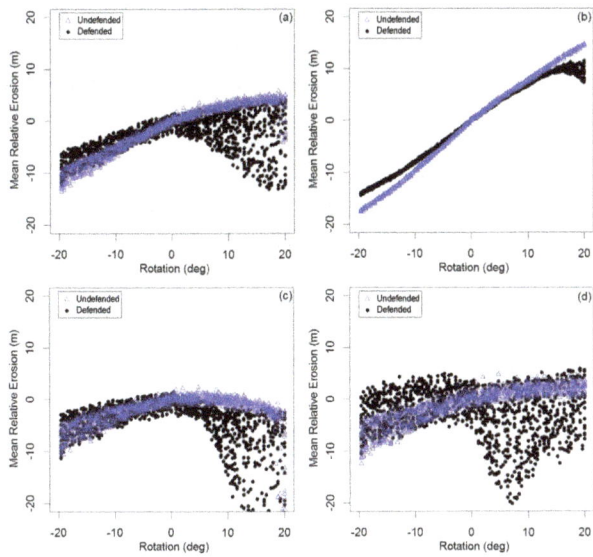

**Figure 4.** Perturbation in wave direction plotted against spatially averaged mean relative erosion for (**a**) the entire coast, (**b**) the northern section (1), (**c**) the central section (2), (**d**) and the southern section (3). Counterclockwise rotation of wave direction is negative, clockwise rotation is positive. Negative values in mean relative erosion indicate a reduction in coastal erosion in comparison to the baseline simulation.

**Figure 5.** Perturbation in significant wave height plotted against spatially averaged mean relative erosion for (**a**) the entire coast, (**b**) the northern section (1), (**c**) the central section (2), (**d**) and the southern section (3). Negative values in mean relative erosion indicate a reduction in coastal erosion compared to the baseline simulation.

baseline peaks at around 10° and reduces again with further rotation. In section 2, under the defended setup, the relative erosion peak is at a maximum where there is no rotation, reducing rapidly as clockwise rotation is applied (section 2; Fig. 4c). It is under these clockwise rotations where the response differs significantly between the defended and undefended coastlines. The undefended coast exhibits a relatively narrow band of erosional responses, while the defended coast shows considerable variability. In section 3 (Fig. 4d), the undefended response of erosion to rotations in wave direction is similar to the overall trend in erosion. When sea defences are introduced, complex patterns of erosional response merge, with large ranges of increased and decreased erosion rates at all rotations.

## 4.3  Wave height

For the undefended scenario, the relationship between perturbation in wave height and relative erosion for the whole coast (Fig. 5a) is less well defined than the relationship between rotation of wave direction and erosion (Fig. 4a). The reduction in mean erosion rates with an increase in wave height for the natural scenario was attributed by Barkwith et al. (2014) to increased protection in the southern sector of the coast provided by the increased availability of sediment. With sea defences included, this relationship is intensified, resulting in a stronger inverse relationship between erosion

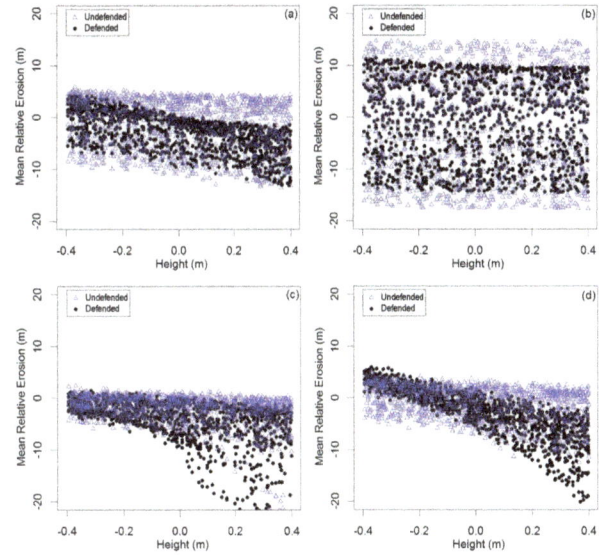

and wave height. This relationship is not so well defined in section 1 (Fig. 5b) and there is little correlation between perturbation in wave height and erosion rate for either the defended or undefended scenarios. In section 1, the lower range in erosion rate results from the influence of rotation of the wave direction. Both sections 2 and 3 (Fig. 5c, d, respectively) show similar patterns manifest in a greater range in erosion rate as significant wave height increases. In section 2, the undefended and defended coastlines respond similarly to change in wave height. In section 3, the relationship between wave height and erosion is increasingly inverse for the defended scenario when compared to the undefended coast. This suggests that the increase in sediment availability affords the coast greater protection from erosion.

## 4.4  Combined impact

Perturbations in significant wave height and wave direction for the coast as a whole and for each region are plotted for the defended scenario in Fig. 6. The size of the symbols in Fig. 6 is proportional to the relative erosion, compared to the baseline; red indicates increased erosion and open circles indicate reduced erosion. When the coast is considered as a whole, increased erosion occurs when wave height is decreased and the rotation in wave direction is clockwise. However, as with the plots assessing the individual influence of significant wave height and rotation of wave direction (Figs. 4, 5), behaviour averaged along the coast as a whole does not reflect the variations seen in detail for each of the

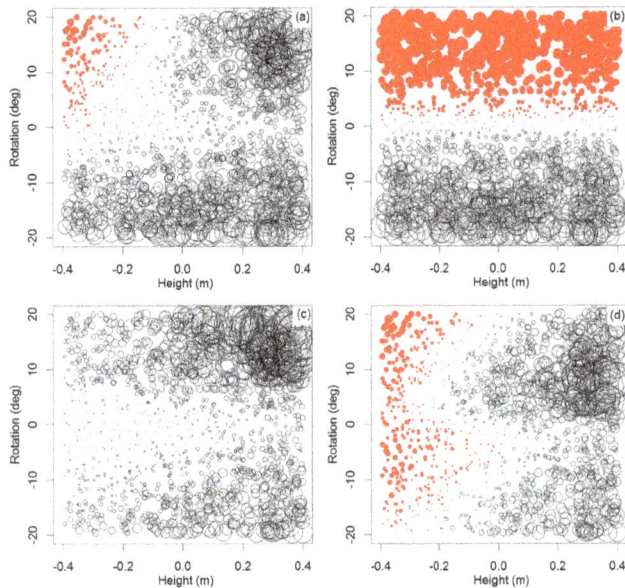

**Figure 6.** Plots of perturbation in wave direction and significant wave height for each member of the ensemble for **(a)** the entire coast, **(b)** the northern section (1), **(c)** the central section (2), **(d)** and the southern section (3). The size of each symbol is proportional to the change in mean relative erosion rate imparted by that wave climate in comparison to the baseline scenario. Red dots represent increased erosion relative to the baseline and empty circles reduced erosion.

three sections. In section 1, there is no correlation between wave height and erosion. Thus, increased erosion occurs at all significant wave heights under clockwise rotations of the wave direction (Fig. 6b). For section 2, in the centre of the coastline, peak erosion rates occur under the baseline wave climate, perturbations of the wave climate resulting in either similar or reduced erosion rates (Fig. 6c). The most complicated relationships occur in section 3 (Fig. 6d), where the divide between increased and reduced erosion is dependent on the combination of height and rotation perturbations. The near-vertical divide in the results suggests that perturbation of the significant wave height has a slightly greater influence on the erosion rate. There is also a strongly non-linear response to a clockwise rotation in wave direction; even small clockwise rotations cause a significant reduction in relative erosion. This is likely explained by a reduction in longshore transport of sediment, resulting from lower offshore wave angles just up-drift of a defended stretch of shoreline. The reduction of transport tends to decrease the sediment-flux divergence for some distance up-drift of the structure (Ashton and Murray, 2006a, b; 2001).

## 5 Conclusions

Defended structures have an impact on their immediate surroundings, on the adjacent mesoscale coastal morphology and consequently the vulnerability of the coast to changes in wave climate. Model simulations indicate that the impact of structures on erosion rates is minimal in the northernmost section of the coastline, where the sea defences at Skipsea and Hornsea do not heavily modify the available sediment load. This is manifest in the similarity of the absolute erosion rates for undefended and defended scenarios. In the central and southern sections and particularly on the up-drift side of the sea defences, differences in patterns of absolute erosion are more prominent.

Although coastal recession rates are similar for the majority of ensemble members under the defended and undefended scenarios, ensemble members with absolute erosion at the 20th percentile or below have increased beach thicknesses where sea defences are included. This increase in beach sediment is sufficient to protect the cliffs from erosion and reduce recession rates. These mesoscale effects extend over 15 km of coastline and are most prevalent when a +10° rotation is applied to the wave direction and significant wave heights are increased. Increased wave heights allow greater volumes of sediment to be transported from the north and the clockwise (positive) changes in wave climate lead to a "trapping" of sediment on the up-drift side of sea defences by reducing sediment flux around these structures.

The sensitivity of the coastline to changes in wave climate is also modified due to sea wall defences. In section 1, the effect of changing wave climate on erosion is damped with sea defences included. In sections 2 and 3, the interaction of defences and sediment transport create complex, non-linear responses, as revealed by the patterns of relative erosion. While the sensitivity to wave-climate changes is similar in central and southern regions for undefended coast, for the defended coast the behaviour in these sections differ markedly. The results suggest that multiple sea defences can have a coupled impact on erosional sensitivity. These specific impacts of coastal defence interactions are dependent on the sediment supply, the local recession rate of the coastline, the proximity of surrounding defences, wave climate, and the morphology of the coastline.

Future wave climates are unlikely to be similar to the simply perturbed current wave climate used in this study. We assume that weather patterns will be the same as they were in 2009–2010, and there has been no attempt to reflect possible changes in storminess. However, by using an ensemble approach, the range of likely effects on the morphological characters of the Holderness coastline is captured. When compared to the results of field studies of the impacts of defensive structures on coastal erosion rates (for example Brown et al., 2012), the simulated results do not represent well the increase in erosion rates often associated with the down-drift side of solid defences. This discrepancy could arise partly because the large-scale model, assuming shore-parallel contours, neglects localised complex wave refraction and shoaling patterns around the ends of structures, and the consequent effects on currents and sediment transport. However, in the model the large-scale reduction of alongshore sediment flux caused by a protruding defended coastline segment can cause enhanced down-drift erosion (Ells and Murray, 2012). The fact that the defended locations protrude seaward of the regional coastline trend, increasing wave-shadowing effects, could explain the unexpected lack of erosion down-drift of the defences in our results. The present simulation does, in any case, allow the impacts of individual and multiple coastal defences on recession rates to be assessed in this complicated environmental system, providing an important complement to field-based study.

**Acknowledgements.** The authors are grateful to Vanessa Banks of the British Geological Survey for reviewing this paper and returning feedback that lead to its improvement. This paper is published with the permission of the Executive Director of the British Geological Survey (NERC), and was supported by the Climate and Landscape Research programme at the BGS.

Edited by: G. Coco

# References

Adams P. N., Anderson R. S., and Revenaugh J: Microseismic measurement of wave energy delivery to a rocky coast, Geology, 30, 895–898, 2002.

Ashton A. and Murray A. B.: High-angle wave instability and emergent shoreline shapes: 1. Modeling of sand waves, flying spits, and capes, J. Geophys. Res. Earth Surf., 111, F04011, doi:10.1029/2005JF000422, 2006a.

Ashton A. and Murray A. B.: High-angle wave instability and emergent shoreline shapes: 2. Wave climate analysis and comparisons to nature, J. Geophys. Res. Earth Surf., 111, F04012, doi:10.1029/2005JF000423, 2006b.

Ashton A., Murray A. B., and Arnoult O: Formation of coastline features by large-scale instabilities induced by high-angle waves, Nature, 414, 296–300, 2001.

Bakker, W., Breteler, E. K., and Roos, A.: The dynamics of a coast with a groyne system, Coast. Engin. Proc., 1, 1001–1020, 1970.

Barkwith, A., Thomas, C. W., Limber, P., Ellis, M. A., and Murray, A. B.: 2014. Coastal vulnerability of a pinned, soft-cliff coastline, I: Assessing the natural sensitivity to wave climate, Earth Surf. Dynam., in press, 2014.

Bladé I., Liebmann, B., Fortuny, D., and van Oldenborgh, G. J.: Observed and simulated impacts of the summer NAO in Europe: implications for projected drying in the Mediterranean region, Clim. Dynam., 39, 709–727, 2012.

Brown, S., Barton, M. E., and Nicholls, R. J.: The effect of coastal defences on cliff top retreat along the Holderness coastline, P. Yorks. Geol. Soc., 59, 1–13, 2012.

Bruun, P.: The Development of Downdrift Erosion, J. Coastal Res., 11, 1242–1257, 1995.

Burton, E A.: Chronica Monasterii de Melsa, a Fundatione usque ad Annum 1396, Volume III, Editor Bond, E. A., Cambridge University Press, 2012.

Catt, J. A.: The Pleistocene glaciation of eastern Yorkshire: a review, P. Yorks. Geol. Soc., 56, 177–207, 2007.

CCO: Channel Coastal Observatory, http://www.channelcoast.org/data_management/real_time_data/charts/?chart=72 (last access: 15 July 2013), 2013.

Ciavola P.: Coastal dynamics and impact of coastal protection works on the Spurn Head spit (UK), Catena, 30, 369–389, 1997.

Clark H. C. and Johnson M. E.: Coastal Geomorphology of Andesite from the Cretaceous Alisitos Formation in Baja California (Mexico), J. Coastal Res., 11, 401–414, 1995.

Dean, R., Walton, T., Rosati, J., and Absalonsen, L.: Beach Erosion: Causes and Stabilization, Coastal Hazards, Springer, New York, USA, 319–365, 2013.

de Boer G.: Spurn Head: its history and evolution, T. I. Brit. Geogr., 34, 71–89, 1964.

DEFRA: Charting Progress 2. Feeder Report: Ocean Processes, Department for Environment Food and Rural Affairs, on behalf of the UK Marine Monitoring and Assessment Strategy community, Nobel House, London, UK, 290 pp., 2010.

Dickson, M. E., Kennedy D. M., and Woodroffe C. D.: The influence of rock resistance on coastal morphology around Lord Howe Island, southwest Pacific, Earth Surf. Proc. Land., 29, 629–643, 2004.

Dickson, M. E., Walkden, M. J. A., and Hall, J. W.: Systemic impacts of climate change on an eroding coastal region over the twenty-first century, Clim. Change, 84, 141–166, 2007.

Ells, K. and Murray, A. B.: Long-term, non-local coastline responses to local shoreline stabilization, Geophys. Res. Lett., 39, L19401, doi:10.1029/2012GL052627, 2012.

Hanson, H.: GENESIS: a generalized shoreline change numerical model. J. Coastal Res., 5, 1–27, 1989.

Kamphuis, J. W.: Introduction to Coastal Engineering and Management, Advanced series on ocean engineering, 16, Word Scientific, New Jersey, 2000.

Komar P. D.: The mechanics of sand transport on beaches, J. Geophys. Res., 76, 713–721, 1971.

Kraus, N. C. and McDougal, W. G.: The effects of seawalls on the beach: Part I, an updated literature review, J. Coastal Res., 12, 691–701, 1996.

Kraus, N. C., Hanson, H., and Blomgren, S.H.: Modern functional design of groyne systems, Coast. Engin. Proc., 1, 1327–1340, 1994.

Limber, P. W. and Murray, A. B.: Beach and sea cliff dynamics as a driver of rocky coastline evolution and stability, Geology, 39, 1149–1152, 2011.

Limber P. W., Patsch K. B., and Griggs G. B.: Coastal sediment budgets and the littoral cut-off diameter: A grain-size threshold for quantifying active sediment inputs, J. Coastal Res., 24 (supplement 2), 122–133, 2008.

List, J., Farris, A. H., and Sullivan, C.: Reversing storm hotspots on sandy beaches: spatial and temporal characteristics, Mar. Geol., 226, 261–279, 2006.

Montreuil, A.-L. and Bullard, J. E.: A 150-year record of coastline dynamics within a sediment cell: Eastern England, Geomorphology, 179, 168–185, 2012.

Newsham, R., Balson, P. S., Tragheim, D. G., and Denniss, A. M.: Determination and prediction of sediment fields from recession of the Holderness Coast, NE England, J. Coastal Conserv., 8, 49–54, 2002.

Pendleton, L. H.: The economic and market value of coasts and estuaries: what's at stake?, Coastal Ocean Values Press, Washington, DC, USA, 2010

Quinn, J. D., Philip, L. K., and Murphy, W.: Understanding the recession of the Holderness Coast east Yorkshire, UK: a new presentation of temporal and spatial patterns, Q. J. Eng. Geol. and Hydroge., 42, 165–178, 2009.

Scott Wilson: Humber Estuary Coastal Authorities Group (HECAG), Flamborough Head to Gibraltar Point Shoreline Management Plan 2, Scott Wilson, Basingstoke, Hampshire, UK, 2009.

Shennan, I., Lambeck, K., Flather, R., Horton, B., McArthur, J., Innes, J., Lloyd, J., Rutherford, M., and Wingfield, R.: Modelling western North Sea palaeogeographies and tidal changes during the Holocene, Geol. Soc., London, Special Publications, 166, 299–319, 2000.

Slott, J. M., Murray, A. B., and Ashton, A. D.: Large-scale responses of complex-shaped coastlines to local shoreline stabilization and climate change. J. Geophys Res. Earth Surf., 115, F03033, doi:10.1029/2009JF001486, 2010.

Sutherland, J. S. and Wolf, J.: Coastal defence vulnerability 2075, HR Wallingford Report SR590, Wallingford, UK, 2002.

Trenhaile A. S., Pepper D. A., Trenhaile R. W., and Dalimonte M.: Stacks and notches at Hopewell Rocks, New Brunswick, Canada, Earth Surf. Proc. Land., 23, 975–988, 1998.

Valvo L., Murray A. B., and Ashton A.: How does underlying geology affect coastline change? An initial modeling investigation, J.Geophys Res. Earth Surf., 111, F02025, doi:10.1029/2005JF000340, 2006.

Woollings, T.: Dynamical influences on European climate: an uncertain future, Phil. Trans. R. Soc. A, 368, 3733–3756, 2010.

# Morphological and sedimentological response of a mixed-energy barrier island tidal inlet to storm and fair-weather conditions

**G. Herrling and C. Winter**

MARUM – Center for Marine Environmental Sciences, University of Bremen, Bremen, Germany

*Correspondence to:* G. Herrling (gherrling@marum.de)

**Abstract.** The environment of ebb-tidal deltas between barrier island systems is characterized by a complex morphology with ebb- and flood-dominated channels, shoals and swash bars connecting the ebb-tidal delta platform to the adjacent island. These morphological features reveal characteristic surface sediment grain-size distributions and are subject to a continuous adaptation to the prevailing hydrodynamic forces. The mixed-energy tidal inlet Otzumer Balje between the East Frisian barrier islands of Langeoog and Spiekeroog in the southern North Sea has been chosen here as a model study area for the identification of relevant hydrodynamic drivers of morphology and sedimentology. We compare the effect of high-energy, wave-dominated storm conditions to mid-term, tide-dominated fair-weather conditions on tidal inlet morphology and sedimentology with a process-based numerical model. A multi-fractional approach with five grain-size fractions between 150 and 450 $\mu$m allows for the simulation of corresponding surface sediment grain-size distributions. Net sediment fluxes for distinct conditions are identified: during storm conditions, bed load sediment transport is generally onshore directed on the shallower ebb-tidal delta shoals, whereas fine-grained suspended sediment bypasses the tidal inlet by wave-driven currents. During fair weather the sediment transport mainly focuses on the inlet throat and the marginal flood channels. We show how the observed sediment grain-size distribution and the morphological response at mixed-energy tidal inlets are the result of both wave-dominated less frequent storm conditions and mid-term, tide-dominant fair-weather conditions.

## 1 Introduction

Tidal inlets at barrier island systems connect the open sea with the back-barrier tidal basin. Typically, they feature an ebb-tidal delta seawards and a flood-tidal delta landwards of a deep inlet throat that is bordered by shallow sandy shoals and marginal flood channels (Hayes, 1979). Both tidal flow constriction through the narrow inlet and wave energy dissipation on depth-limited ebb-tidal delta shoals account for local enhanced sediment transport and rapid morphological evolution.

Morphodynamics at mixed-energy tidal inlets are driven by the combined action of waves and tides and the relative contribution of these interacting forces largely determines the morphological and sedimentological response. Ko-

mar (1996), De Swart and Zimmermann (2009), Davis and FitzGerald (2009) and FitzGerald et al. (2012) give recent and comprehensive reviews on morphodynamic processes at a large variety of tidal inlet systems. The early work of Hayes (1975, 1979) and a recent study applying process-based models (Nahon et al., 2012) classified mixed-energy inlet regimes in a range between tide-dominated and wave-dominated and suggested corresponding inlet geometries that are in equilibrium with the long-term energetic input from waves and/or tides. Sha and Van den Berg (1993) developed a descriptive model to explain ebb-tidal delta symmetry, i.e., the orientation of the seaward inlet channel with respect to shallow ebb-delta shoals, as a response to the relative direction of waves to the interplay of tidal currents alongshore and within the inlet. Very few studies at mixed-energy tidal inlets have investigated the complex interaction of tide- and wave-driven processes and distinguished the contribution of each agent to residual sediment fluxes and morphological changes (e.g., Bertin et al., 2009; Elias and Hansen, 2013; Elias et al., 2006; Sha, 1989). Even fewer studies have managed to relate observed distributions of surface sediment grain sizes at tidal inlet systems to distinct physical drivers (e.g., Sha, 1990; van Lancker et al., 2004).

Recent studies have shown the applicability of process-based numerical models for sedimentological studies, for example, to simulate surface sediment grain-size distributions in combination with morphological changes (Kwoll and Winter, 2011; Van der Wegen et al., 2011a, b). This suggests the application of multi-grain-size models to decipher the morphological and sedimentological effect of different hydrodynamic drivers, i.e., different model boundary conditions.

In this study we aim to investigate the effect of tide- and wave dominance on residual sediment pathways at a mixed-energy barrier island tidal inlet Otzumer Balje in the southern North Sea. It serves as an example of a mixed-energy, slightly tide-dominant inlet regime with similar characteristics as described, for example, by Hayes (1979). This is achieved by simulating a storm event that represents a period of wave dominance and fair-weather conditions with waves smaller than average representing tide-dominated conditions. Real-time data of tides, wind and waves are applied as forcing conditions for each model scenario, and are assumed to be sufficiently representative to study the morphological and sedimentological responses to low- and high-energetic conditions. The following characteristics of tidal inlet systems are investigated:

1. It is commonly understood that ebb-tidal delta erosion during episodic storm events counteracts the continuous replenishment of the ebb-tidal delta during tide-dominated fair-weather conditions (De Swart and Zimmerman, 2009). We aim to show how this dynamic equilibrium behavior of either wave- or tide-dominated forcing conditions determines the sedimentology and morphology at a typical mixed-energy tidal inlet and the adjacent foreshore. After a synthetic separation of tide- and wave-dominated forcing conditions, we will point out relevant morphodynamics and sediment pathways that are due to the interaction of the driving forces leading to, for example, elongated channel fill deposits at the margin of the tidal inlet throat.

2. Son et al. (2010) postulated a dominant circular sediment pathway at the eastern ebb-tidal delta platform of the tidal inlet Otzumer Balje investigated here. Sediments are thought to be recycled between the ebb-tidal delta and the inlet throat without any evidence of sediment bypass to the downdrift beach. For the Dutch Ameland tidal inlet, authors have mentioned recirculation cells at the downdrift shoals of the ebb-tidal delta

supporting reversed net sediment transports towards the inlet throat, but claim only minor significance with respect to the overall sediment dynamics (Cheung et al., 2007; Elias et al., 2006; Sha, 1989). We evaluate the relevance of this recirculation cell at mixed-energy tidal inlets and identify the hydrodynamic drivers and interrelated mechanisms that induce these net circular sediment fluxes.

## 2  Study area

The tidal inlet Otzumer Balje is located between the East Frisian barrier islands Langeoog and Spiekeroog in the southern North Sea (Fig. 1). The back-barrier tidal basin represents a drainage channel system typical for the Wadden Sea. According to the classification of Hayes (1975, 1979), the study area is mesotidal with a mixed-energy to slightly tide-dominated regime. The tide is semidiurnal with a mean range of 2.8 m at Spiekeroog. The gorge in the inlet throat reaches maximum depths of approximately 24 m below German datum (around mean sea level) and a width of approximately 1 km. The residual flow in the inlet throat is ebb-dominant with maximum current velocities for neap to spring tides ranging from 0.5 to 1.0 and 0.8 to 1.6 m s$^{-1}$ for flood and ebb tide, respectively (Bartholomä et al., 2009).

Mean wind directions are from the westerly sector with mean velocities of about 7 m s$^{-1}$ observed at the offshore platform FINO1 at approx. 40 km off the East Frisian barrier islands. Here, mean significant wave heights of 1.4 m and mean peak periods of 6.9 s have been measured (data from May 2004 to June 2006, Federal Ministry for Environment, Nature Conservation and Nuclear Safety (BMU) and the Project Management Jülich (PTJ)). Extreme storms from the northwesterly sector can generate surge water levels of up to 2.5 to 3.3 m above mean high water at the coast. During the extreme storm event on 9 November 2007, known as "Tilo", significant wave heights of 10 m, maximum wave heights of 17 m and peak periods of up to 15 s were measured offshore at water depths of 30 m at the research platform FINO1 (Outzen et al., 2008). The combination of a tidal wave that travels from west to east and the dominant westerly wind and wave directions generates an alongshore eastward-directed net sediment drift. FitzGerald et al. (1984) estimated the net transport rate to be about 270 000 m$^3$ yr$^{-1}$ of sand.

The inlet consists of a variety of morphological features such as ebb- and flood-tidal deltas, inlet throat and marginal flood channels bordered by shoals and swash bars. The latter are sand bars with dimensions on the order of a few hundred meters superimposed onto the ebb-tidal delta that migrate onshore and coalesce to larger intertidal bar systems in the vicinity of the shore (FitzGerald, 1982); it should be mentioned at this point that the nomenclature "swash bars" is commonly used but is rather misleading as the physical processes involved are miscellaneous and not primarily related

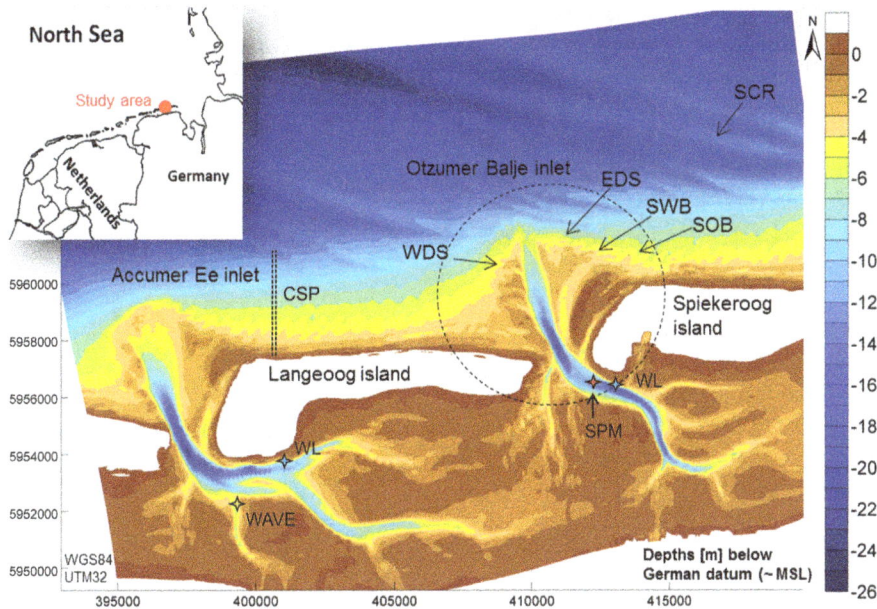

**Figure 1.** East Frisian barrier island system in the southern North Sea with the study area Otzumer Balje inlet between the islands Langeoog and Spiekeroog and nearshore morphological features such as the western/eastern ebb-tidal delta shoals (WDS/EDS), swash bars (SWB), shore-oblique sand bars (SOB) and shoreface-connected ridges (SCR). Measurement positions are indicated by wave (WAVE), water level (WL), suspended matter (SPM) and bathymetry at cross-shore profiles (CSP).

to swash. At the ebb-tidal delta of the Otzumer Balje inlet, their orientation is more shore-parallel compared to the shore-oblique sand bars that emerge downdrift of the ebb-tidal delta. The bed of the tidal inlet reveals different bed forms, from ripples to dunes. In the inlet throat, Noormets et al. (2006) measured three-dimensional sand dunes with mean lengths of 7.5 m and mean heights of 0.35 m. Medium to coarse, poorly sorted sands are found in the inlet channel; the ebb-tidal delta body mainly consists of fine sand but is superimposed by swash bars of medium-sized sand (Son et al., 2010).

## 3  Methodology

### 3.1  Modeling system

The modeling system Delft3D (Deltares, 2011) has been applied to set up and run high-resolution process-based morphodynamic models. The mathematical model solves the three-dimensional shallow-water equations and continuity equation on a staggered model grid by means of an implicit finite-difference scheme. The spectral wave model SWAN (Booij et al., 1999; Ris et al., 1999) is run in a stationary mode to simulate the wave propagation and deformation from the open sea to the shoreline. Wave measurements available at intervals of 30 min are applied as offshore boundary conditions. This coincides with the interval of the sequential two-way coupling between SWAN and the hydrodynamic module (Delft3D-FLOW) that allows the exchange of rele-

vant parameters on curvilinear model grids via a communication file. Wave parameters and the forcing terms associated with the wave radiation stresses computed by SWAN are read by the FLOW module. Once the assigned runtime of 30 min has been reached by FLOW, bottom elevation, water level and depth-integrated current fields are used as input to the computation in SWAN. The model will loop through these sequential module applications until the simulation is accomplished. The interaction of wave forces (radiation stresses), tidal currents and the changing bed- and water levels is thus realized by a fully coupled wave–current simulation.

Wave forces being computed in SWAN by radiation stress gradients are implemented as a shear stress in the flow module at the water surface layer. The ongoing debate about the vertical distribution of wave-induced radiation stresses that generally split up into a surface component, a bottom component and a body force and their implementation within 3-D momentum equations (discussed in, for example, Ardhuin and Roland, 2013; Ardhuin et al., 2008; Bennis et al., 2011) reflects on and indicates that important wave-induced processes interacting with the flow circulation may still be inadequately implemented in Delft3D. These model limitations are, however, accepted in the present study assuming minor effects on the sedimentology and morphology at the tidal inlet.

Important wave effects are incorporated in the 3-D simulations as wave-induced mass flux adjusted for the vertically nonuniform Stokes drift, additional turbulence and vertical mixing processes and streaming as an additional wave-induced shear stress in the wave boundary layer (Walstra et al., 2000). The effects of wave asymmetry on the suspended sediment transports are included based on the nonlinear wave approximation modified by Van Rijn et al. (2004) after the method of Isobe and Horikawa (1982). Mean and oscillatory bed shear stresses interact nonlinearly. Through the use of the parameterization of Soulsby et al. (1993), the wave–current interaction model of Fredsøe (1984) is applied to account for the wave-induced enhancement of the bed shear stress that affects the stirring of sediments and increases the overall bed friction.

The sediment transport formulation applied here differentiates bed- and suspended load mechanisms (Van Rijn et al., 2004). Suspended load is treated above a reference height, and bed load below (Van Rijn, 1993). For simulations including waves, the magnitude and direction of the bed load transport are calculated using an approximation method developed by Van Rijn (2003). The method computes the bed load transport accounting for the flow velocity in the bottom computational layer and the near-bed peak orbital velocity in the direction of wave propagation. Suspended sediment is entrained in the water column by imposing a reference concentration (Van Rijn, 2000) at the reference height. An advection–diffusion equation (Van Rijn et al., 2004) is solved for the current-related suspended transport. The settling velocity of sand is computed following the method of Van Rijn (1993), where different suspended grain-size diameters are accounted for by empirical formulations. The vertical sediment mixing coefficient follows directly from the vertical fluid mixing coefficient calculated by the $k - \epsilon$ turbulence closure model (Rodi, 1984).

The model is used to identify sediment transport patterns between consecutive morphological states and to differentiate between instantaneous and residual suspended load and bed load directions and quantities. For details on the equations and processes implemented in the modeling system Delft3D, the reader is referred to Lesser et al. (2004), Van Rijn et al. (2004) or the manual of Delft3D (Deltares, 2011).

## 3.2   Morphological acceleration factor

A morphological scale factor is applied to account for the acceleration of bed level changes during updates at each hydrodynamic time step (Roelvink, 2006). Through the use of this method, which aims to economize computational runtime, hydrodynamic timescales are adapted to much longer timescales of morphological evolution. Within this study, a morphological acceleration factor (MORFAC) of 20 is applied during a simulation of 17 tidal cycles between neap and spring tide (7 to 15 June 2007) in order to account for morphological changes that occur during approximately 5 months of fair-weather conditions. For the 5-day storm simulation (6 to 10 November 2007), no morphological acceleration is applied (MORFAC = 1).

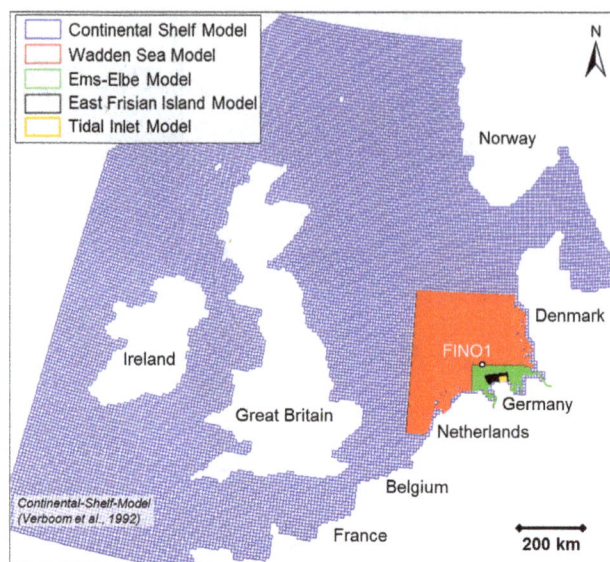

**Figure 2.** Cascade of five nested model grids and the position of wave measurements at FINO1 being applied as offshore boundary condition; wind- and atmospheric pressure fields computed by the German Weather Service cover all model grids.

## 3.3   Model nesting and boundary conditions

A hierarchical cascade of five model grids from the European continental shelf to the East Frisian barrier islands with decreasing spatial dimensions and increasing grid resolutions has been set up to derive water levels and wave climate at the study area (Fig. 2). Storm surge simulations in particular require large model domains as coastal surge is generated by wind drag effects and atmospherical pressure gradients acting over long distances on the open sea. The largest model with grid cell resolutions of 8000 m covers the continental shelf in the North Atlantic Ocean to the North Sea. Eight harmonic tidal constituents are applied to generate the astronomic tide at the sea boundaries of the continental shelf model (Verboom et al., 1992). It embeds the Wadden Sea model with average grid sizes of 1200 m covering the entire North Sea from the Dutch coast in the south to Denmark in the north. The Wadden Sea model, in turn, generates water level time series at the seaward boundary of the smaller Ems–Elbe model with grid resolutions of approx. 200 m. The latter is additionally forced at the seaward boundary by wave data observed at the research platform FINO1 located 45 km offshore in water depths of 30 m. The next smaller model covers the East Frisian barrier islands from Juist to Wangerooge with model grid resolutions of 60–120 m and supplies wave- and water level boundary conditions to the most detailed tidal inlet model covering only Langeoog and Spiekeroog. At the end of the model cascade, this three-dimensional model with 10 sigma layers over the vertical is dedicated to simulate the sediment dynamics at the tidal inlet Otzumer Balje and adjacent beaches (Fig. 1). It consists of 140 000 active grid cells with average grid resolutions of 60 m and up to 20 m in the breaker zones, assumed to be sufficiently resolved for proper generation of wave-induced alongshore currents during storm conditions. During fair-weather conditions, wavelengths are significantly shorter and the selected cross-shore grid resolution may not be ideally represented at the upper part of the beach, yet certain limitations are accepted in favor of reduced computational times.

## 3.4  Model bathymetry

Model bathymetries, i.e., depth schematizations for each particular model (Sect. 3.2), have been assembled by interpolating measured data of sea bottom elevations onto curvilinear model grids. Near coastal sub- and intertidal areas are covered by data of the years 2006, 2005 and 2001 based on conventional sounding methods (Federal Maritime and Hydrographic Agency, BSH). Elevations of inter- and supratidal barrier island beaches are partly covered by beach profiles of the year 2007 or high-resolution airborne lidar scans that are spatially limited and available for the years 2008, 2007 and 2005 (data with permission of the Coastal Research Station of Lower Saxony Water Management, Coastal Defense and Nature Conservation Agency – FSK-NLWKN).

## 3.5  Meteorological forcing

Storms in the central part of the North Sea are associated with low-pressure systems. During the investigated extreme storm event, Tilo, between 5 and 10 November 2007, with peak surge levels on 9 November 2007, maximum wind velocities of 33 m s$^{-1}$ and mean wind directions of north-northwest were recorded offshore (Outzen et al., 2008). Wind stress and horizontal atmospheric pressure gradients acted over a long fetch from the Arctic Sea across the entire North Sea inducing extreme surge setup superimposed by high astronomical tide. The storm surge simulations are forced by meteorological model data of the German Weather Service (DWD). Wind and atmospheric pressure fields are available at 1 h intervals and spatial resolutions of 7 and 2.8 km for the models COSMO-EU and COSMO-DE, respectively.

The simulation representing fair-weather hydrodynamic conditions is forced by time series of wind data measured at the research platform FINO1 (provided by BMU, PTJ). Real-time data between 7 and 15 June 2007 are imposed to the wave and hydrodynamic simulations to account for a meteorological forcing with nonstationary wind velocities and directions. The mentioned period was selected based on visual comparison of generated wind roses due to the selected data set and a 2-year data set. Thus the selected data do not fulfill long-term statistical correctness, but the overall distribution of wind directions and intensity are similar to the long-term trend. Wind directions of the selected data series are from the westerly sector with a short intermittent period of easterly winds. The selected data are assumed to be sufficiently representative to account for typical low-energy wind- and wave conditions.

## 3.6  Multiple-grain-size model

### 3.6.1  Bed layer model for multiple sediment fractions

A dynamic bed layer model is applied that permits the redistribution of multiple sand fractions in relation to imposed bed shear stresses. It thus enables the computation of spatial distributions of surface sediment grain-size fractions and to evaluate arithmetic mean grain sizes in response to different hydrodynamic conditions. Each sand fraction depletes or increases in the bed cell according to erosion or deposition processes in the sediment transport formulation. A coefficient according to each mass percent is applied in the transport equation to account for the availability of the mobilized sand fraction at a given bed cell. Thus, sediment transport occurs if the critical shear stress is exceeded for a certain grain-size fraction, whereas its load is additionally controlled by the relative availability of each sand fraction. The uppermost layer of the bed layer model, the so-called active layer, has a constant thickness and records the grain-size composition of the underlayers beneath. The underlayers account for the bed level change, while their thicknesses increase or decrease depending on the prevailing erosion or deposition of a certain grain-size fraction. In the present study, the selected active layer thickness is 0.25 m. At the start of the simulation, the total thickness of the underlayers is 10 m in order to guarantee enough sediment supply in case of locally strong erosion. Simulations presented in this study consider continuous bed level updating. This is clarified against the background that Delft3D allows simulations without bed level updating but redistribution of sediment fractions only. For details on the setup and functioning of the bed layer model, the reader is referred to Van der Wegen et al. (2011a).

In the present study, because of computational expenses, model simulations were restricted to a limited number of noncohesive sand fractions (five), with grain sizes of 150, 200, 250, 350 and 450 μm. As the focus is on the sediment dynamics at the tidal inlet, a characteristic gradation of rather coarse sediment fractions between 150 and 450 μm was selected. Areas exposed to a low-energy wave impact such as the back-barrier tidal flats or the lower shoreface where significantly finer grain sizes occur in nature are, according to the grain-size configuration selected here, not subject to significant morphological changes and thus grain-size sorting processes. Here, the initial mean surface sediment grain size does not change significantly during the simulations. This limitation is tolerated because back-barrier sediment dynamics and exchange processes between the back-barrier basin and the foreshore are not the focus of this study. Back-barrier tidal flats contain high amounts of fine sand and cohesive sediments and would require a different model setup and grain-size configuration.

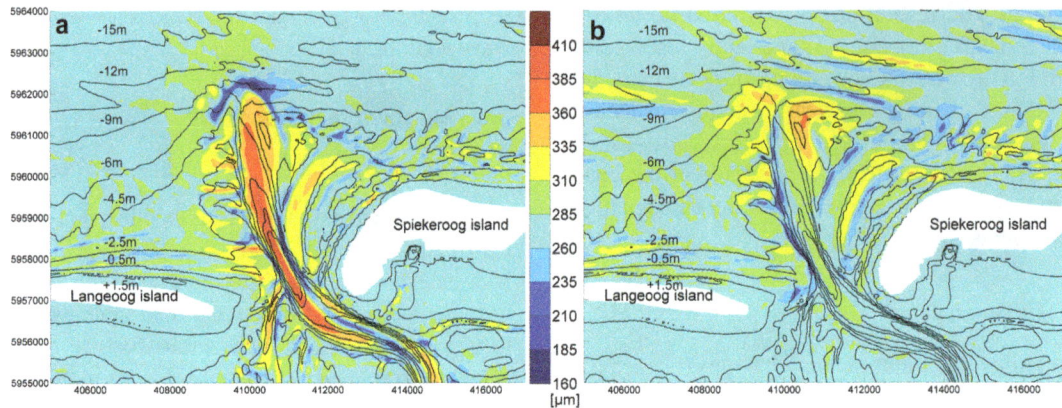

**Figure 3.** Distribution of arithmetic mean surface sediment grain sizes as a response to fair-weather (**a**) and storm (**b**) conditions; synthetic simulations were initiated with five uniformly distributed sand fractions between 150 and 450 $\mu$m.

### 3.6.2  Initial sediment grain-size distribution

Observations of surface sediments are restricted to the tidal inlet Otzumer Balje; samples of the ebb-tidal delta lobe are not available. The spatial inconsistency of measurements precludes the use of grain-size observations as initial conditions of simulations. The model is thus applied to allow for the redistribution of multiple sand fractions with the objective of generating a schematization of the observed surface sediments that can then be used to initiate more realistic simulations. According to this, two different model initializations are considered: on the one hand, the synthetic case of an initialization with uniform grain-size distribution; on the other hand, the more realistic "analysis simulations" taking into account a sediment distribution of nonuniform grain sizes that was generated by preceding model runs.

Synthetic simulations are initiated with uniform sediment type distribution. Five sediment fractions (150, 200, 250, 350 and 450 $\mu$m) are available at 20 mass percent each and thus represent an initial mean grain size of 280 $\mu$m. First, two synthetic simulations are run and forced by approximately 5 months of fair-weather conditions and 5 days of storm conditions to exemplify the sedimentological response to tide- and wave-dominant hydrodynamic conditions for simplified sedimentological settings. For fair-weather conditions, elevated shear stresses due to tide-induced currents in the inlet throat and marginal channels cause mean sediment grain sizes at the channel bottoms to increase, while fine sands are entrained, transported in the ebb direction and deposited at the ebb-tidal delta lobe (Fig. 3a). During storm conditions, however, sediment coarsening occurs due to wave-induced stirring and increased shear stresses at the ebb-tidal delta and adjacent shoals, whereas sediments at the tidal channels reveal only minor changes in mean grain size (Fig. 3b). Fine sediments being entrained by northwesterly waves on the elevated shoals are transported in the onshore direction and accumulate as elongated deposits along tidal channel margins. These simulations reveal synthetic sediment distribu-

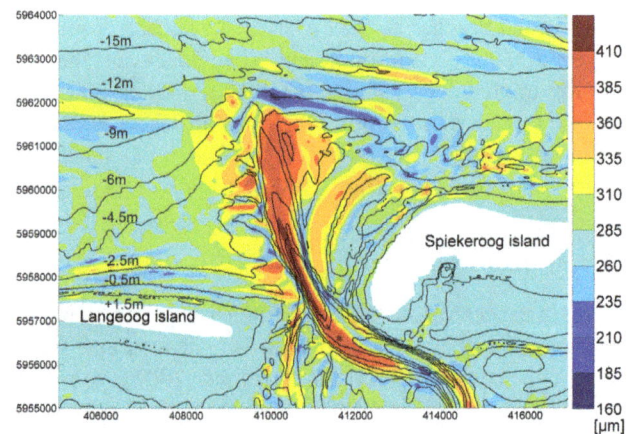

**Figure 4.** Arithmetic mean surface sediment grain size due to the distribution of five sand fractions between 150 and 450 $\mu$m generated by a series of three simulations with alternating hydrodynamic forcing due to 5 months of fair-weather conditions, an extreme storm event and again 5 months of fair-weather conditions.

tions since sediments in nature are nonuniformly distributed and sediment grain-size availability may be a crucial factor in the development of sediment fluxes; however, they allow for highlighting of the idealized response of surface sediments to distinct hydrodynamic forcing conditions.

As a second step, a more realistic sediment distribution is generated that is in response to a combination of fair-weather and storm conditions: a simulation of 5 months being forced by fair-weather boundary conditions is followed by a storm simulation of 5 days and another period of 5 months of fair-weather conditions. Sediment mass fractions at the end of each model run are turned over to the subsequent simulation. At the end of the sequence of simulations, the predicted sediment distribution schematizes the sedimentological response to a mixed-energy tidal inlet regime (Fig. 4). This sediment distribution is used for model validation purposes based on a qualitative comparison of predicted and observed grain-size distributions (Sect. 4.3). Furthermore, it is applied as an initial condition for simulations that aim to analyze morphodynamics and sedimentology in response to fair-weather and storm conditions (Sects. 5 and 6), because it allows for a more realistic schematization of surface sediments with consistency all over the model domain and one avoids having to rely on spatially limited sediment grain-size measurements.

# 4 Model validation

The model system Delft3D has been widely tested in morphodynamic modeling studies for various environments (e.g., Dissanayake et al., 2009; Lesser et al., 2004; Van der Wegen and Roelvink, 2012), yet it has been verified in comparably few morphological studies on nonidealized tidal inlets that take into account a real-world bathymetry (e.g., Elias and Hansen, 2013; Elias and van der Spek, 2006; Elias et al., 2012). The validation of simulated morphodynamics by field observations is generally difficult as in situ data are scarce and only available for very limited areas, if at all. This particularly applies to bathymetrical data measured just before and after a storm surge event. Data on storm-induced bed evolution are necessary for model calibration and verification purposes. Available observations and published data of the studied tidal inlet and adjacent barrier islands beaches are summarized and compared to modeled hydrodynamics, sediment dynamics and surface sediment grain-size distributions in order to determine the validity of the modeling approach below. Modeled data are from the two most detailed model domains of the cascade of nested model grids (Fig. 2).

## 4.1 Hydrodynamics

Time series of simulated water levels are compared to observations at available tidal gauges within the study area (data provided by the Federal Agency of Water and Navigation, WSV). Figure 5 shows modeled versus observed water level time series for the storm surge event at Spiekeroog tidal gauge. High water levels are generally well reproduced by the model, whereas low water levels show discrepancies. The phase lag between modeled and measured water level time series is in the range of 10–20 min. The root-mean-square errors (RMSE) for the water level amplitudes for the fair-weather and storm simulations are 12 and 19 cm at Spiekeroog and 14 and 22 cm at Langeoog.

Wave data measured during the storm event at the surf zone of the island of Norderney are available from a project report of the FSK-NLWKN (Kaiser et al., 2008). The island of Norderney is located 25 km to the west of the studied tidal inlet. The downward-looking ultrasound device mounted on a pole produced corrupted data once the distance between the water surface and the sensor was too small, due to unexpectedly high surge levels. Approximately 1 h before, on 9 November 2007 at 07:00 UTC, a time-averaged significant

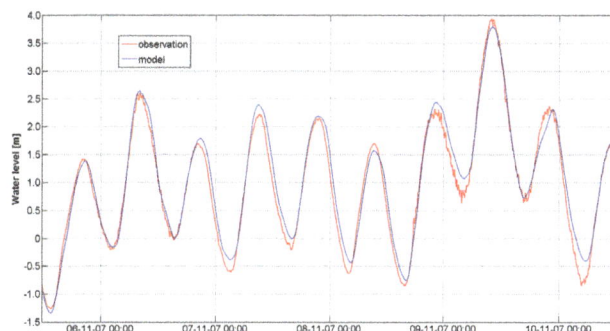

**Figure 5.** Comparison of modeled and observed water level time series at the water level gauge Spiekeroog for the storm event "Tilo" with peak surge levels on 9 November 2007.

wave height of 3.5 m is qualitatively compared to predictions of the storm simulation of the next larger model grid ("East Frisian Island Model", Fig. 2). The measured wave height is underestimated by 17 % in the simulation and thus confirms fair reliability of the predicted wave energy in the surf zone.

Time series of water levels and wave parameters were measured during the storm event at an observational pole located in the inner part of the Accumer Ee inlet separating the islands of Baltrum and Langeoog. The pole was operated between the years 2000 and 2007 by Helmholtz-Zentrum Geesthacht. Its configuration and functioning is described in Onken et al. (2007). The pole's location was not directly at the studied tidal inlet, but was still within the most detailed model domain (Fig. 1). The observation point is located at a hydrodynamic complex and morphologically dynamic location at the junction of the main tidal channel and a larger tributary. Here, bathymetrical information is only available for spring 2005 and has been incorporated in the model bathymetry, two years ahead of the chosen validation period of November 2007. Local differences between the real and the model bathymetry may influence the local wave regime, making a quantitative model validation based on the existing observations ambiguous. Observed wave parameters were calculated from water level elevations recorded at a frequency of 2 Hz taken from a floater guided along a rod with a magnetic readout. Spikes and stuck values were cleaned from the data. The effect of this data cleaning is shown in Lane et al. (2000) and the usage of floater-derived wave parameters for model validation in the North Frisian Sylt-Rømø Bight demonstrated in Schneggenburger et al. (2000).

Phases and amplitudes of the observed water levels and wave parameters are fairly well reproduced by the model in view of the complexity of the wave–current interactions at the measuring site (Fig. 6). For significant wave height (Hs), model data exhibit a bias of 0.24 m. Statistical analysis of model predictions with respect to the observations allows for evaluation of the RMSE of 0.19 m for the water level, 0.26 m for Hs, 0.69 s for the peak wave period (Tp) and 0.34 s for the mean wave period (Tm). Discrepancies

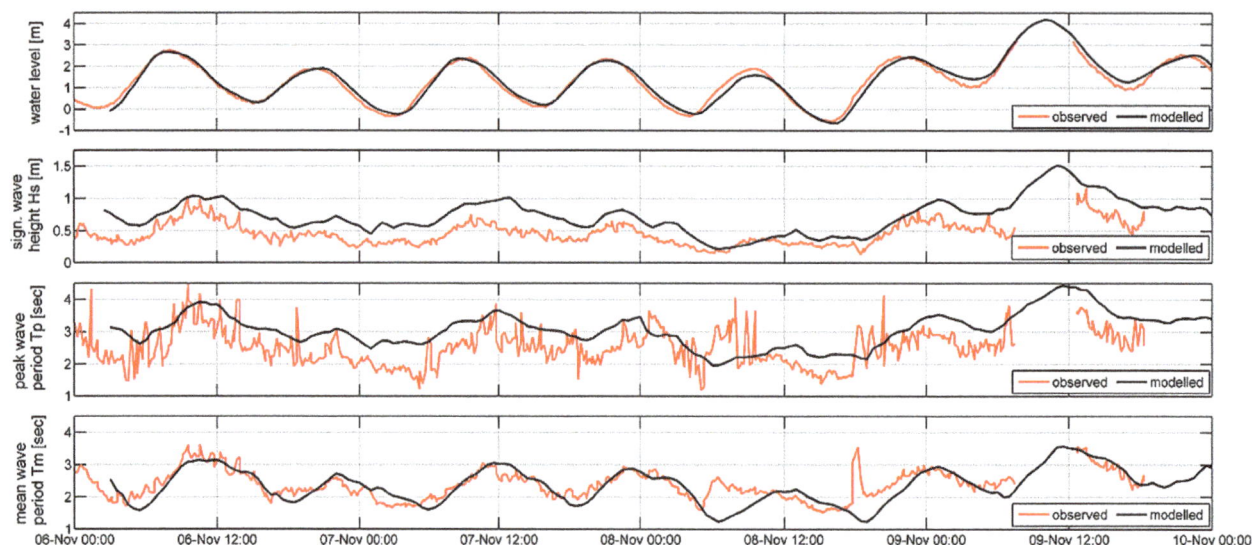

**Figure 6.** Observed and modeled water levels, significant wave heights (Hs), peak wave periods (Tp) and mean wave periods (Tm) at the back barrier of Langeoog during Tilo in November 2007 (observations provided by Helmholtz-Zentrum Geesthacht).

tend to be slightly larger during ebb and low water compared to flood. One explanation is that the measuring pole was exposed to the focused ebb currents in the tributary tidal channel that were opposite to wave propagation. The overestimated wave height during ebb may be a consequence of insufficient wave dissipation due to whitecapping incorporated by the saturation-based whitecapping formulation of Van der Westhuysen (2007). The enhanced dissipation of waves on negative current gradients after a recently published formulation (Van der Westhuysen, 2012) was successfully applied in Dodet et al. (2013). In the present study, however, uncertainty as to the bathymetry at the measuring site does not allow for the calibration of the model by the application of different whitecapping formulations.

It should be noted that no model calibration was performed by spatially varying bed roughness adaptation. Instead, the bed roughness was set to a uniform, constant value over the model domain (Manning parameter 0.024). Particularly against this background, this validation attests adequate model skill for the purpose of this study.

### 4.2 Sediment dynamics and morphology

Time series measurements of suspended matter (SPM) concentrations observed at the tidal inlet Otzumer Balje during the storm surge peak tide on 9 November 2007 show hourly mean (maximum) values on the order of 35 (65) mg L$^{-1}$ and 55 (95) mg L$^{-1}$ for maximum flood- and ebb-tide currents, respectively, at 0.5 m below mean low water level (Badewien et al., 2009). The three finest sediment fractions incorporated in the model simulation (150, 200 and 250 $\mu$m) reveal hourly mean (maximum) SPM concentrations of 45 (70) mg L$^{-1}$ during maximum flood-tide currents at 2 m below German

datum at the location of the measuring pole. During flood tide, SPM concentrations at the inlet are due to nearshore wave-induced sand resuspensions. The model reproduces suspended sediment dynamics fairly well for these conditions: hourly mean and maximum concentrations are overestimated by approximately 29 and 8 %, respectively. During ebb tide, however, predicted maximum SPM concentrations of 2 mg L$^{-1}$ are strongly underestimated with respect to measurements (95 mg L$^{-1}$). This can possibly be explained by the fact that fine sand ($< 150 \mu$m) and cohesive sediments that are typically flushed out of the back-barrier tidal flats during increased storm surge ebb-flows (Bartholomä et al., 2009; Cuneo and Flemming, 2000) are simply not incorporated in this model setup. However, discrepancies here are not relevant for this study, because the model is not applied to predict residual sediment rates between the foreshore and back-barrier basin.

Observations of morphological changes as a response to the storm event of November 2007 are available for two cross-shore profiles at the foreshore of Langeoog, both extending from the coastal dune up to a distance of 3900 m from the beach into water depths of 14 m below German datum. Data of profiles 37 and 38 measured at Langeoog on 15 October and on 12 and 22 November 2007 by the Coastal Research Station were processed with permission from NL-WKN (Kaiser et al., 2008). A direct comparison of observed and modeled morphological changes is not possible because bathymetrical data of spring 2006 being used to set up the model bathymetry do not coincide with the cross-shore profiles that were measured in October and November 2007. Morphological changes along these profiles are thus qualitatively compared to predicted patterns of erosion and sedimentation of the storm simulation (Appendix A).

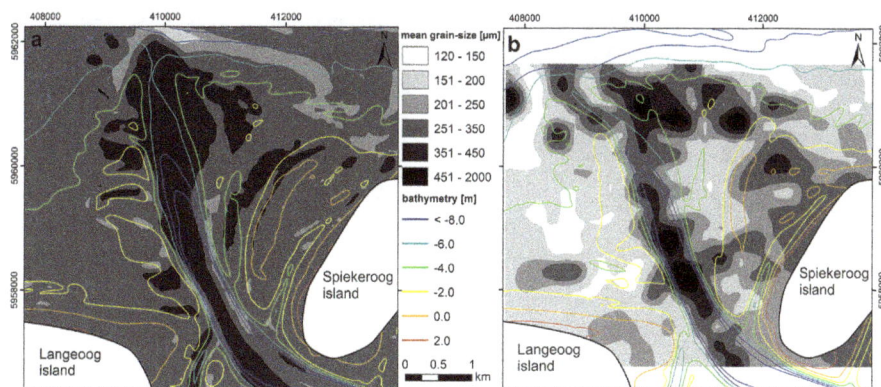

**Figure 7.** Modeled (**a**) and measured (**b**) arithmetic mean surface sediment grain-size distributions at Otzumer Balje inlet between Langeoog and Spiekeroog; depth isolines based on bathymetrical data measured in 2006/2007 (**a**) and 2004/2005 (**b**).

Morphological changes evaluated from observations are on the order of 0.5 m to 1.0 m at the surf zone within the first 500 m of the profiles (Figs. A1 and A2). This order of magnitude is reproduced by the model; in particular, the erosion of the upper beach and the filling of the trough of the first berm are generally captured by the model (Fig. A3). Between 500 and 2000 m from the beach, the downdrift migration of two shore-oblique sand bars through the transversal profiles generates alternations in erosion and deposition of approximately 0.5 m, which reveals good agreement with model predictions. At 2000–3500 m from the shoreline, mostly deposition on the order of 0.1 to 0.3 m is observed. The landward trough of the shoreface-connected ridge at the end of the profiles accumulates sand, whereas the adjacent slopes tend to suffer from erosion. Predicted sand depositions of 0.05 to 0.1 m in between depth isolines −6 and −9 m are underestimated by the model.

This qualitative analysis shows fair similarities between model predictions and observations in terms of magnitude and alteration from net sedimentation to net erosion at the described morphological compartments.

### 4.3 Sedimentology

Mapping of surface sedimentology of the whole domain of interest is not available. However, Son et al. (2010) compiled surface sediment grain-size distributions in the Otzumer Balje tidal inlet from a grid of Shipek sediment grab samples at distances of approximately 280 m for the year 2005. Their data are reinterpolated here to allow for comparison with modeled data. Modeled mean surface sediment grain sizes are due to redistributions of five sand fractions between 150 and 450 μm; and are generated by a series of three model runs with alternating hydrodynamic forcings due to fair-weather conditions, storm conditions and then fair-weather conditions once more (Sect. 3.5, Figs. 4 and 7a).

The initial bathymetry of the detailed tidal inlet model is based on bathymetrical data of the years 2006 and 2007 and thus different from the inlet morphology of the sediment

sampling campaign of 2005, here indicated by isolines based on available bathymetrical data of the years 2004 and 2005 (Fig. 7). The different morphological background explains the westerly bend of the channel through the ebb-tidal delta for the sampling state compared to a more straightened orientation in the model bathymetry.

Modeled and measured arithmetic mean surface sediment grain-size distributions show distinct similarities (Fig. 7). Surface sediments are coarsest at the inlet channel, the ebb-tidal delta and the eastern ebb-tidal delta shoal where swash bars migrate onshore. The central part of the ebb-tidal delta with medium to coarse sands is divided by a characteristic south–north-oriented pattern of finer mean grain sizes shown by both modeled and measured distributions. At the foreshore, modeled mean grain sizes are generally coarser with respect to measurements. This is explained by the selection of the initial mean surface sediment grain size of 280 μm composed of five uniformly distributed fractions that tend to be too coarse for specific areas, e.g., the foreshore or the back-barrier tidal flats, particularly as the performance of the model to predict surface sediment grain sizes decreases as morphological changes are small and thus sorting of sand fractions cannot take place.

At the western ebb-delta shoals, on the other hand, distinct grain-size patterns of medium sand that are predicted by the model cannot be validated by field data as the distance between sample positions (approx. 280 m) is too large for these spatial patterns in surface sediment grain sizes to properly be resolved.

## 5 Results

We compare the effect of an extreme storm surge event in the North Sea to a medium-term period (approx. 5 months) of representative fair-weather conditions on morphodynamics and sedimentology at the tidal inlet Otzumer Balje between the barrier islands Langeoog and Spiekeroog based on two model simulations.

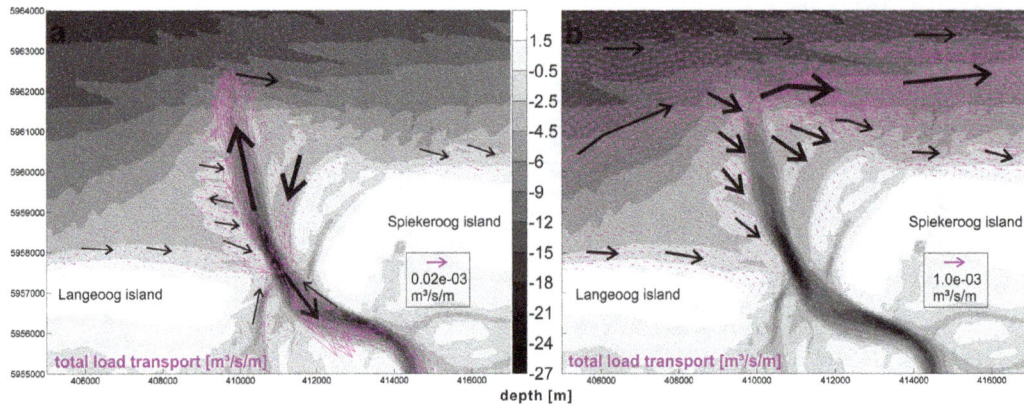

**Figure 8.** Residual total load transport for fair-weather conditions (**a**) and storm conditions (**b**); schematic main residual pathways indicated by black arrows.

Both simulations reveal sediment pathways for five distinct sediment fractions of 150, 200, 250, 350 and 450 $\mu$m. Residual total load (suspended and bed load) sediment transport of small sand fractions, i.e., 150 and 200 $\mu$m, shows pathways comparable to the residual suspended load sediment transport of all five fractions. On the other hand, residual total load transports of coarser sand fractions, i.e., 350 and 450 $\mu$m, resemble the pathways of the overall residual sediment transport due to bed load quantities. This is not unexpected and means that pathways of individual grain-size fractions do not give significant additional information compared to a presentation that only differentiates between pathways of bed load and suspended load transport. Figures of residual sediment transport pathways integrated over all five sand fractions are thus presented hereafter as total load, bed load or suspended load quantities.

It may be noted here that the morphological acceleration factor, i.e., MORFAC = 20 (Sect. 3.2), being applied during fair-weather simulations only accounts for an accelerated development of the morphology and the sediment grain-size distribution; however, it does not apply in the computation of sediment transport loads. The residual sediment transport load [m$^3$ s$^{-1}$ m$^{-1}$] is thus the time-averaged transport load over a runtime of 17 tidal cycles during fair-weather conditions and of 9 tidal cycles during the storm event.

### 5.1  Tide-dominated fair-weather conditions

Residual total sediment transport fluxes during fair-weather conditions are largest in the vicinity of the tidal inlet, particularly in the inlet throat and the eastern marginal flood channel (Fig. 8a). The residual total load sediment fluxes are differentiated into residual bed load transports (Fig. 9a) and residual suspended load transports (Fig. 9b). Residual suspended load quantities are approximately 4 times larger than the residual bed load quantities, but their residual directions are similar. North of the deepest location in the inlet throat, residual transport is ebb-dominant and directed towards the

ebb-tidal delta, whereas southwards it follows the inlet channel towards the flood delta and the back-barrier basin.

At the easterly end of Langeoog, alongshore net sediment drift supplies bed- and suspended load towards the inlet throat of the tidal inlet. At the western ebb-tidal delta shoal, a residual sediment import to the inlet throat takes place over the shallow shoals, whereas predominantly suspended sediment load is exported via ebb channels located in between these shoals.

At the northern part of the eastern ebb-tidal delta shoal, minor residual bed- and suspended load quantities are transported in a sharp bend from the center of the ebb-tidal delta to the eastern ebb-tidal delta shoal in a south-southeasterly direction. With increasing water depths landwards of the shoal, the sand is directed into a deeper, transverse tidal channel. Through this flood-dominant, marginal tidal channel significant residual suspended and bed load quantities are transported in a south-southwesterly direction back to the tidal inlet throat.

At the inlet widening towards the back-barrier tidal basin, the inlet throat is flood-dominant. Residual fluxes of predominantly suspended sediment point along the main inlet channel towards the flood-tidal delta and adjacent tidal flats. At the northern margin of the main channel and alongside the western head of Spiekeroog, minor residual bed- and suspended load fluxes are opposite and thus ebb-directed via a bordering transport pathway. Between the easterly end of Langeoog and the flood delta, a marginal tidal channel is also ebb-dominated and leads residual suspended and bed load fluxes out of the basin.

The mid-term fair-weather simulation reveals morphological and sedimentological changes at the tidal inlet and adjacent channels, at shore-parallel bars in the surf zone and shore-oblique sand bars (Figs. 11a and 12a). Sediment dynamics at the foreshore are insignificant and net morphological changes are below 0.05 m (Fig. 11a). Sediments being eroded in the inlet throat and tributary channels

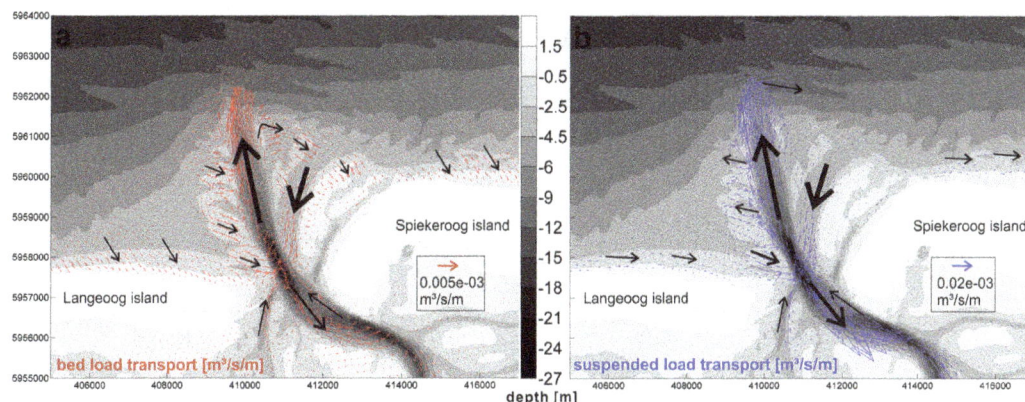

**Figure 9.** Residual bed load (**a**) and residual suspended load (**b**) transport for fair-weather conditions; schematic main residual pathways indicated by black arrows. Relative vector scaling indicates suspended load to be approx. 4 times larger than bed load transport.

are transported and deposited at the ebb-tidal delta and adjacent shoals. The most northern part of the ebb-tidal delta increases and protrudes offshore with net depositions exceeding 1.0 m at the ebb-delta lobe during the simulated period of 5 months.

The sediment distribution shows a relative coarsening of the mean surface sediment grain size on the order of 30–50 $\mu$m at the western ebb-tidal delta shoals and at the tidal channels, whereas the ebb-tidal delta lobe is fed by the entrained finer sand fractions and decreases the mean grain size by approx. 30–40 $\mu$m (Fig. 12a). Sedimentological changes are in relation to the initial, nonuniform sediment distribution (Fig. 4); thus the redistribution of surface sediments is small because the sedimentological response to fair-weather hydrodynamic conditions is largely included in the initial condition already. The absolute mean sediment grain-size distribution due to synthetic simulations (Fig. 3a), however, shows that the depositional area at the ebb-tidal delta experiences a grading of sediment grain sizes. The finest sand is deposited at the outermost ebb-tidal delta lobe where ebb-directed current velocities decrease due to increasing water depths.

Shore-oblique sand bars migrate eastwards in the same direction as the overall littoral sediment drift (Fig. 11a). Similar to fluvial low-energy bed forms, erosion takes place on the stoss side and sedimentation on the lee side. Once again, synthetic simulations reveal a gradient in mean surface sediment grain sizes with medium (fine to medium) sands at the upper stoss side and the crest (lee side and trough) (Fig. 3a); this indicates that tide-dominated alongshore current velocities during fair-weather conditions are strong enough to develop typical sediment gradients over morphological features.

## 5.2 Wave-dominated high-energy storm conditions

During the storm event, residual eastward-directed total sediment fluxes are predicted to be largest at the barrier island foreshore, particularly directly off the ebb-tidal delta,

whereas residual sediment load is very small in the tidal inlet throat (Fig. 8b). Across the tidal inlet, the residual total sediment transport direction is flood-dominant, but only a marginal amount of sand tends to be imported to the back-barrier basin. The residual total load transport is differentiated into residual bed- and suspended load transport vectors (Fig. 10a and b). Disregarding the residual transport directions, the relative scaling of the vectors indicates that the net suspended load quantity is overall approximately one order of magnitude higher than the net bed load quantity. The residual bed load transport is south-southeastward-directed, particularly at the eastern ebb-tidal delta shoal where it drives the migration of swash bars. The residual bed load transport direction agrees with the mean direction of wave propagation. Residual suspended sediment transport load is largest close to the ebb-tidal delta and in the extended surf zone from the islands' beaches to the transition of upper to lower shoreface. Here, residual directions are downdrift-oriented due to wave-induced alongshore currents that advect the entrained sand to the east.

During the storm event, significant morphological and sedimentological changes occur over large areas of the barrier island foreshore and upper shoreface, but particularly in the northern part of the ebb-tidal delta (Figs. 11b and 12b). High-energetic waves refract and break on the depth-limited ebb-tidal delta shoals, stirring large quantities of sediment. In the vicinity of the ebb-tidal delta, morphological changes along distinct linear patterns are predicted to be 1 m or more during this storm event (Fig. 11b). Fine sand fractions of 150, 200 and 250 $\mu$m are transported as suspended load by the combined flow of tide-, wind- and wave-induced currents downdrift to the east. Mostly medium-sized sands with sand fractions of 250, 350 and 450 $\mu$m remain, and as a result they increase the mean surface sediment grain size by up to 100 $\mu$m at the most seaward part of the ebb-tidal delta shoal (Fig. 12b).

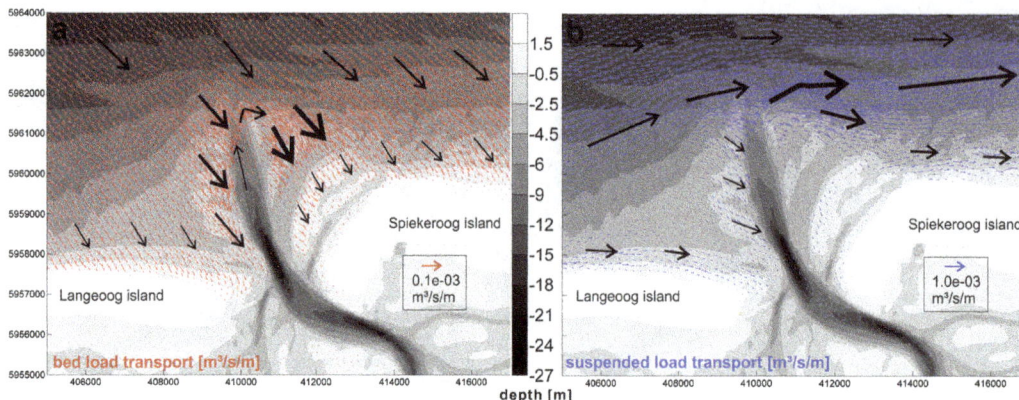

**Figure 10.** Residual bed load **(a)** and residual suspended load **(b)** transport for high-energy storm conditions; schematic main residual pathways indicated by black arrows. Relative vector scaling indicates suspended load to be approx. 10 times larger than bed load transport.

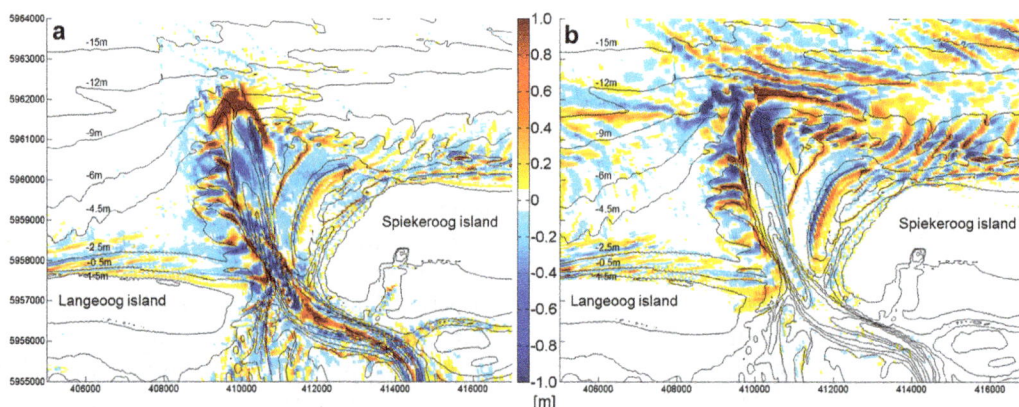

**Figure 11.** Morphological changes, i.e., sedimentation (red) and erosion (blue) as a response to fair-weather **(a)** and storm **(b)** conditions; analysis simulations were initiated with nonuniformly distributed sediment fractions (Fig. 4).

The morphology and sedimentology of the inlet throat and marginal flood channels are less affected, as the driving wave energy is dissipated at the shallow ebb-tidal delta shoals. In addition, bottom shear stress decreases as the tidal flow slows down through the relatively increased cross-sectional area of the inlet because of elevated surge levels.

Grain sizes increase insignificantly at the inlet gorge, whereas fine sands accumulate at the western margin of the inlet throat (Fig. 12b). Here, transport over the western ebb-tidal delta shoal directs finer sand south-southeastwards to the western margin of the inlet throat where settling starts, with increasing water depths and thus decreasing shear stresses causing a lateral shift of the inlet throat to the east (Fig. 11b).

At the eastern ebb-tidal delta shoal, alternating erosion and deposition patterns indicate a south-southeastward migration of large swash bars that are oriented almost parallel to the shore and thus deviate from shore-oblique sand bars (Fig. 11b). At the northeastern edge of the ebb-tidal delta shoal, shore-oblique sand bars connecting the eastern ebb-tidal delta with the downdrift surf zone migrate eastwards

during storm conditions. The sediment distribution due to the synthetic simulation being initiated with uniformly distributed sediment fractions predicts coarser grain sizes at the bed form crests with respect to the troughs of the shore-oblique sand bars (Fig. 3b). Analysis simulations of the storm event being initiated by a nonuniform sediment distribution (Fig. 4) reveal this gradient to be further enhanced (Fig. 12b).

At the lower shoreface, fine sand fractions are winnowed and eroded in the troughs between and at the landward slopes of shoreface-connected sand ridges being located in water depths of 15–20 m below German datum (Fig. 12b). Fine sand tends to accumulate on the crests and the seaward slopes of the shoreface-connected ridges. Thus the shoreface-connected ridges experience a positive morphological feedback and a downdrift migration (Fig. 11b).

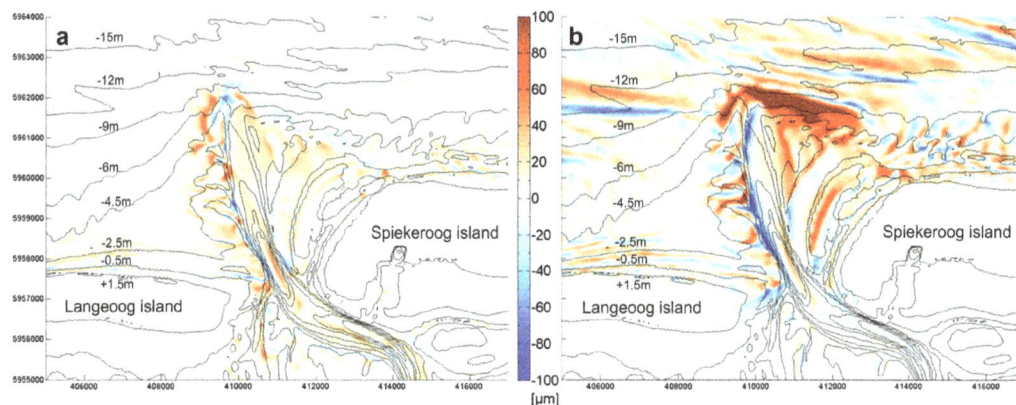

**Figure 12.** Sedimentological changes, i.e., relative increase (red) and decrease (blue) of mean surface sediment grain size as a response to fair-weather (**a**) and storm (**b**) conditions; analysis simulations were initiated with nonuniformly distributed sediment fractions (Fig. 4).

## 6    Discussion

### 6.1    Mixed-energy tidal inlet morphology and sedimentology in response to tide- and wave-dominated conditions

The main drivers determining the morphodynamic development of a mixed-energy tidal inlet system are commonly assumed to be waves that induce sediment stirring, transport and dispersal at the ebb-tidal delta and tidal currents in the inlet (e.g., FitzGerald et al., 2012; De Swart and Zimmerman, 2009). Mixed-energy barrier island tidal inlets are morphologically highly dynamic environments where both drivers continuously interact. Numerical model scenario experiments allow for the separation of processes and boundary conditions for greater understanding of the system. However, a potential model approach that reduces the forcing to either tides or waves alone would be misleading as the natural interaction at mixed-energy tidal inlets would be ignored. Instead, tide- and wave-dominated forcing conditions are represented here by realistic fair-weather and storm scenarios, respectively, which allow the evaluation of the morphological and sedimentological responses to distinct hydrodynamic drivers by preserving the mixed-energy regime of the system at the same time.

For typical mixed-energy tidal inlets, it is commonly assumed that ebb-tidal delta erosion during episodic storm events counteracts the continuous replenishment of the ebb-tidal delta lobe during tide-dominated fair-weather conditions (FitzGerald et al., 2012; Hayes, 1979). This study reproduces and thus confirms this hypothesis: model simulations of mid-term fair-weather conditions reveal that the morphological activity mainly focuses on the inlet throat. Eastward littoral drift along the foreshore beaches supplies fine sands to the inlet throat. In the deep inlet channel, bed shear stress due to tidal currents is strong enough to remove fine sands. As residual sediment fluxes in the seaward part of the inlet throat are ebb-directed, the entrained fine sands mainly feed the ebb-tidal delta terminal lobe. During storm conditions, wave refraction and shoaling over steep bottom gradients focus wave energy towards the ebb-tidal delta lobe and its shallow shoals where energy dissipates due to wave breaking. Here, the fine sand deposited during fair-weather periods is easily mobilized and transported eastwards by the ambient flow, dominated by alongshore velocity components induced by high-energy waves. These waves, approaching at an angle with respect to the shore, generate alongshore momentum flux that is greatest in the zone of breaking waves (Longuet-Higgins and Stewart, 1964). Furthermore, the model predicts transportation of coarse sand fractions as bed load in the landward direction and by migration of large swash bars superimposed onto the ebb-tidal delta shoals. As mentioned previously, the term "swash bars" is misleading as these sand bars are not primarily exposed to swash, and particularly not during storm surge conditions. Bertin et al. (2009) note a physical explanation based on a shore-directed component of forces due to wave breaking (i.e., radiation stress gradients) over the delta shoal that is not compensated for by pressure gradient terms in the momentum balance equation as the associated wave-induced water level setup is spread into the tidal inlet and the back-barrier basin; it is also referred to as the "bulldozer effect", i.e., the shoaling and prograding of the ebb-tidal delta (Hageman, 1969).

Sediment grain-size-sorting mechanisms and thus the spatial distribution of surface sediments are related to bed shear stress controlled by wave- and tide-induced flow: residual distributions of surface sediment grain sizes make it clear that both storm conditions with high-energy waves and fair-weather conditions where tidal currents dominate contribute to the sedimentology of barrier island tidal inlets and foreshore. At the tidal inlet, for instance, we can generalize that winnowing of fine sand at the inlet throat and marginal channels is attributed to tidal forcing, whereas high-energy waves are the driver for sorting mechanisms at shallow shoals of the ebb-tidal delta (Fig. 3). Simulations have shown that only the combined scenario forcing, i.e., alternating fair-weather and storm simulations, results in a surface sediment grain-size distribution that is in fair agreement with sedimentological field observations at Otzumer Balje inlet (Fig. 7; Son et al., 2010). This suggests that the combination of both hydro-dynamic forcing conditions is needed to determine the inlet sedimentology. Furthermore, in light of the analogy of modeled and observed sedimentological patterns, this confirms the model setup applied here to reliably simulate sediment dynamics in general and the evaluation of morphological and sedimentological features in response to representative boundary conditions in particular. The applied model resolution and necessarily reduced multi-fractional approach proves the ability to reproduce gradients in grain sizes on the spatial scale of morphological features equal and larger than swash bars and shore-oblique sand bars. Although smaller morphological features and bed forms such as ripples and dunes are not resolved in the model bathymetry, the modeling approach demonstrated here allows for the identification of distinct pathways of particular sediment grain-size fractions in response to wave–current interactions.

In the following, an example is given where simulated fluxes of particular sediment grain sizes combined with detailed information on three-dimensional hydrodynamics allow for the identification of larger scale sorting mechanisms at the ebb-tidal delta lobe and the upper shoreface.

Surface sediment grain-size composition reveals simulated mass fractions of up to 65 and 35 % for sand fractions of 150 and 200 $\mu$m, respectively, which accumulate at the ebb-tidal delta terminal lobe during tide-dominated fair-weather conditions. Here, predicted mean grain sizes are 170 $\mu$m and thus in fair agreement with observations of 120 to 150 $\mu$m at Otzumer Balje (Son et al., 2010) and 120 to 180 $\mu$m at "Harle" (Hanisch, 1981), the tidal inlet to the east of Spiekeroog. The two finest sand fractions of 150 and 200 $\mu$m are obviously stirred by wave action at the outer margin of the ebb-tidal delta but also bypass the inlet along the upper shoreface due to the storm-driven alongshore drift to the east. The finest fraction of 150 $\mu$m primarily settles at areas of reduced energy off the downdrift island of Spiekeroog within a shore-parallel band between the surf zone and the sloping faces of the shoreface-connected ridges. Here, after the storm simulation, 20–30 % of the surface sediment is made up of this finest grain-size fraction of 150 $\mu$m. Antia (1995) observed an almost shore-parallel elongated pattern of accumulated fine sands with mass fractions between 10 and 30 % for settling velocities of 1–1.5 cm s$^{-1}$, which translates to grain sizes of 115–150 $\mu$m after Gibbs et al. (1971). Antia (1995) also describes this pattern as being extended between two bands of medium sands, one at the surf zone and the other along the shoreface-connected ridges. The storm simulation reveals the physical process that explains this established deposit of fine sediments at the upper shoreface: wave-induced currents counteract the opposing westerly directed alongshore ebb-tidal currents in the expanded surf zone. The ebb-

tidal flow is restricted within this zone of wave-dominated alongshore currents and shifted to deeper waters outside the surf zone. This results in a band of reduced bottom shear at the interfacial boundary area of eastwards-directed wave-induced flow and westward-directed ebb-tidal flow. In this area, settling of fine-grained sand is possible. Inside the surf zone, alongshore wave-induced bottom currents are diverted slightly offshore at a shore-oblique angle due to the opposing ebb currents. In nature, enhanced offshore-directed currents due to undertow or downwelling (e.g., Niedoroda et al., 1984) may supply additional fine sand to the zone of reduced bottom shear. The latter process is most likely underestimated by the model. The onshore-directed wave streaming (Walstra et al., 2000) opposes and reduces the offshore-directed bottom shear stresses induced by undertow that is driven by wind surge and wave setup.

Besides these deposits of fine sand at the terminal lobe of the ebb-tidal delta and the shore-parallel band at the upper shoreface, additional characteristic spatial patterns are identified that stand out due to pronounced depositional processes within the surface sediment layer. Particularly for storm conditions, the simulation reveals elongated channel fill deposits of fine-grained sand at the northern fringe of the marginal eastern flood channel, which is even more pronounced at the westerly, sloping side of the inlet throat (Figs. 11b and 12b). The latter has been classified as channel margin linear bars (Hayes, 1979). Hubbard et al. (1979) called this a "zone of equilibrium", where landward wave-induced flow over the marginal shoal platform is opposed and dominated by the ebb-directed tidal jet in the inlet throat. As described earlier, we identified several similar zones of fine-grained deposits that evidently all have in common that tidal flow is partly or fully retarded and balanced by the opposing wave-induced momentum flux or vice versa. This results in a local reduction of bottom shear along the lateral interface of counteracting current fields and supports accumulation of fine-grained sediments. Wind- and wave-induced setup increases the water depth and the cross-sectional area within the inlet, supporting further reduction of bottom shear stresses.

All other patterns at the tidal inlet and the foreshore region can be explained by erosional processes where fine sands are winnowed from surface sediments and thus medium to coarse sediment grain sizes remain, e.g., the bottom of the tidal channels during tide-dominated fair-weather conditions and the ebb-tidal delta shoals during wave-dominated storm conditions.

## 6.2  Sediment recirculation patterns at the eastern ebb-tidal delta shoal

A simulation of tidal inlet morphology, sedimentology and sediment pathways calls for the identification of the communication and coupling of mesoscale hydro- and sediment dynamics between morphological units such as the ebb-tidal delta shoals, the inlet channels and the adjacent barrier coast.

Dissanayake et al. (2009) simulated the interaction between inlet tidal currents and alongshore tidal currents for an idealized tidal inlet by applying a process-based model. The residual flow pattern showed that a rotational current field is only developed to the east of the ebb-tidal delta. The physical description relates to the fact that strong directional velocity fields are developed to the west of the inlet when the alongshore current is eastward (westward) and the inlet current is landward (seaward), whereas the rotational current field supports the ebb-tidal delta growth in the east. These findings agree with the conceptual hypotheses of Sha (1989) and Sha and Van den Berg (1993). In a numerical model study for an idealized and a natural tidal inlet, Hench and Luettich (2003) give an additional explanation of how momentum balances contribute to circulation processes by tidal forcing alone. The inlet jet induces a "dynamical wall effect" with momentum imbalances due to tidal phase lags resulting in transient, cross-inlet elevation differences and thus secondary circulation for different stages of the tide. With respect to the symmetrical geometry of their idealized inlet, the authors were able to show that the morphology of the natural inlet, i.e., particularly marginal tidal channels, plays an additional role in focusing the identified fluxes. In contrast to these tide-controlled circulation cells, FitzGerald et al. (1976) and Smith and FitzGerald (1994) describe "sediment gyres" downdrift of the inlet due to wave refraction and swash over the ebb-tidal delta shoal platform that drive swash bars in a net landward direction, whereas wave-induced setup shoreward of the swash bars augments the inlet-directed currents in the marginal flood channel. Smith and FitzGerald (1994) conclude from sediment budgets due to assessed transport rates and morphological evolution analysis at the Essex River ebb-tidal delta system that the circulated sediment flux within the sediment gyres is estimated to be even larger than the amount that bypasses the inlet. Finley (1978) further adds that refraction of moderate waves around the inlet ebb-tidal jet is a process that contributes to ebb-tidal delta growth. The shoals are an efficient trap of reversed littoral sediment drift that would otherwise be carried alongshore.

These examples from the literature show the importance of recirculation cells for tidal inlet morphology and its budget in particular. Sediment dynamics involved are explained by physical processes that are controlled by either tides or waves. However, at mixed-energy tidal inlets, it is questionable which of the drivers contribute to the net circulation.

At the Otzumer Balje inlet, residual sediment fluxes reveal a pronounced recirculation cell at the eastern ebb-tidal delta shoal. The circular pathway of particular grain-size fractions is obviously of importance for the overall sediment dynamics. During storm conditions, individual swash bar migration and wave-induced bed load transport of medium sand point in the landward direction over the eastern ebb-delta shoal platform. During fair-weather conditions, however, residual transport is concentrated in the transverse, flood-dominated

tidal channel to the south of the eastern shoal platform and leads towards the southwest into the inlet throat. Once in the inlet throat, ebb-directed residual transport directs fine and medium sand to the ebb-tidal delta, where the cycle restarts. This suggests that solely the combination of wave-dominant storm and tide-dominant fair-weather conditions leads to net sediment fluxes describing a circular pathway to the east of the tidal inlet redirecting predominantly medium sand to the inlet throat.

The simulated sediment pathways confirm the conceptual model of Son et al. (2010), who assumed a recirculation cell over the eastern ebb-tidal delta shoal of Otzumer Balje tidal inlet in which sediment is recycled towards the inlet throat. Their hypothesis was primarily derived from the orientation of sedimentary structures found in box- and vibrocores. Sediment beds showed parallel lamination, which, according to the authors, originated from storm events that are better preserved in the long term than cross-laminated features generated during moderate conditions and that indicate dominant sediment pathways of medium-grained sand in a shoreward direction over the eastern ebb-delta shoal.

Separation into wave- and tide-dominated conditions allows for the differentiation of residual sediment fluxes that contribute to the recirculation cell. First, closed sediment circulation cells are not recognized for storm conditions. Here, wave-induced bed load transport is onshore-directed over the shoal platform, but no direct reversal to the inlet throat is evident. During fair-weather conditions, a complete circulation cell is weakly identifiable with prominent residual sediment transport through the ebb-dominated inlet throat and the flood-dominated eastern marginal channel but only minor residual transport in the shoreward direction over the eastern shoal platform. Hence, we conclude that – at least for the tidal inlet studied here – significant recirculation of sand to the inlet is only possible from a combination of both fair-weather and storm conditions.

In this context, we would like to address the aforementioned sediment bypass at the Otzumer Balje inlet. Son et al. (2010) suggest that there is no evidence for fine sand bypassing the tidal inlet. If at all, bypassing would take place along the subtidal margin of the terminal lobe and be independent of processes acting on the ebb-tidal delta. However, no evidence was given to support this hypothesis, as no data were collected from regions seaward of the ebb-tidal delta. Our simulations reveal sediment bypass to the downdrift beach and foreshore for both moderate and extreme conditions, in disagreement with the hypothesis of Son et al. (2010). The magnitudes of the bypass, seaward extent and the dominant grain size are primarily controlled by wave energy, i.e., wave-induced alongshore currents, and are consequently increased for storm compared to fair-weather conditions.

The question of whether the net volume of sand that is recirculated to the inlet throat is dominant over the bypassed quantity must be answered by future studies, as the simulated scenarios are representative of either tide- or wave-dominated conditions but nonrepresentative of the long-term regime of this mixed-energy tidal inlet. Ongoing research aims to elucidate the sediment budget at the tidal inlet.

## 7  Conclusions

This study identifies residual sediment fluxes of particular grain-size fractions and related morphological and sedimentological responses of a mixed-energy tidal inlet system. We use a process-based numerical modeling system to differentiate the effects of either tide- or wave-dominant forcing. During storm conditions, the ebb-tidal delta loses sand through wave impact. For fair-weather conditions, the ebb-tidal delta is replenished by ebb-directed residual sediment transports. The model simulations satisfactorily reproduce this well-known dynamic behavior. Sediment grain-size sorting mechanisms are also affected by the interacting tide- and wave-driven flow. We have shown that only a combined scenario forcing, i.e., alternating fair-weather and storm simulations, can result in a surface sediment grain-size distribution that is in agreement with measured grain-size distributions (Son et al., 2010). Medium-sized sand is found at either tidal inlet channels exposed to tidal-flow-induced bottom shear or at the ebb-tidal delta shoals where winnowing of fine sand is a result of wave stirring. Furthermore, it is shown that surface sediments at the barrier island foreshore and the inlet system in this setting can be explained by erosional, and not depositional, processes. Morphological patterns that are prone to depositional processes and accumulation of fine sand are identified to occur in zones of reduced bottom shear as a result of opposing tidal currents and waves.

The model study confirms the significance of the recirculation of sand at this tidal inlet. Mainly medium-sized sands are redirected to the inlet throat via a semicircular pattern across the eastern ebb-tidal delta and through the easterly marginal flood tidal channel. The combination of residual sediment fluxes of both scenarios, wave-dominated storm and tide-dominated fair-weather, is able to achieve this net sediment recirculation. The model shows additional sediment bypass, mainly by suspended sediment load, to the downdrift foreshore and beach, in disagreement with earlier findings of Son et al. (2010). The magnitude of the bypass, its seaward extend and the dominant grain-size fraction are primarily controlled by wave energy, i.e., wave-induced alongshore currents, and are consequently greater for storm compared to fair-weather conditions.

The overall shape of the tidal inlet of study in the German Wadden Sea appears to be similar to typical textbook tidal inlets described, for example, by Hayes (1979). Its geometry is characterized by a single ebb-dominated tidal inlet channel through the ebb-tidal delta with a slightly asymmetric outline of the shoals to the downdrift. This allows for the assumption that the processes and sediment pathway schemes discussed here are also applicable for many other mixed-energy tidal inlets within barrier island systems. This study thus reveals residual sediment transport pathways for tide- and wave-dominated conditions. It improves our understanding of complex sediment dynamics at mixed-energy tidal inlets as it identifies and qualitatively evaluates how the morphology and sedimentology respond to the contribution of distinct drivers that in nature are obscured by continuous interaction.

**Appendix A:**

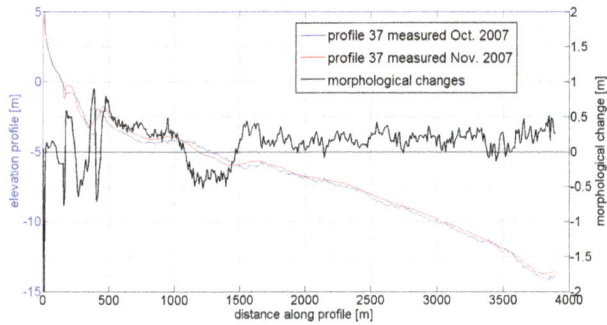

**Figure A1.** Elevation of cross-shore profiles at Langeoog foreshore measured in October and November 2007 and morphological changes as the response to the storm event "Tilo" (data of profile 37 with permission of the Coastal Research Station of Lower Saxony Water Management, Coastal Defense and Nature Conservation Agency, FSK-NLWKN).

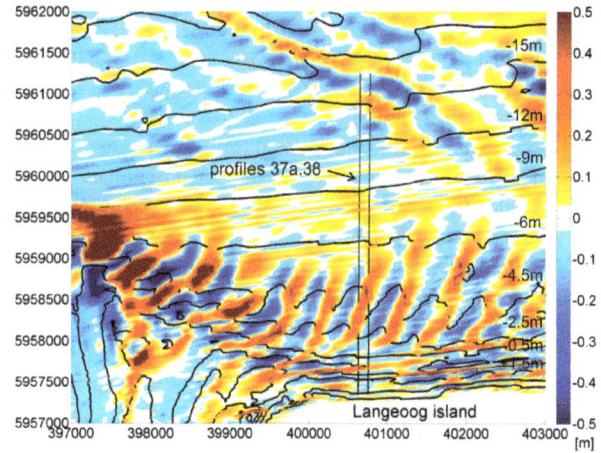

**Figure A3.** Morphological changes, i.e., sedimentation (red) and erosion (blue) patterns, predicted by the storm simulation at the foreshore of Langeoog with position of cross-shore profiles 37 and 38; the initial model bathymetry is based on observational data of 2006 and thus inconsistent with pre-storm cross-shore profiles of October 2007 that are shown in Figs. A1 and A2.

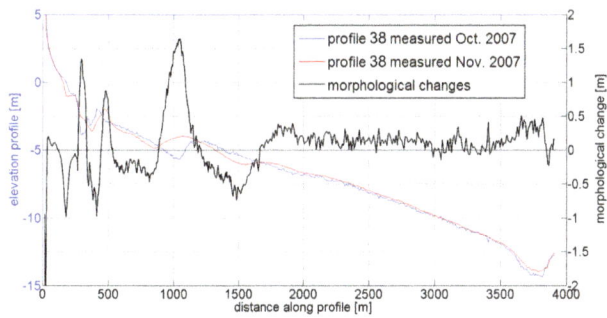

**Figure A2.** Elevation of cross-shore profiles at Langeoog foreshore measured in October and November 2007 and morphological changes as the response to the storm event "Tilo" (data of profile 38 with permission of the Coastal Research Station of Lower Saxony Water Management, Coastal Defense and Nature Conservation Agency, FSK-NLWKN).

**Acknowledgements.** This study is associated with and funded by the research project WIMO (www.wimo-nordsee.de) and is financed in equal parts by two ministries in Lower Saxony: the Ministry of Environment, Energy and Climate Protection, and the Ministry of Science and Culture.

We gratefully acknowledge the authorities and research institutes, namely the Federal Maritime and Hydrographic Agency (BSH) and the Coastal Research Station of Lower Saxony Water Management, Coastal Defense and Nature Conservation Agency (FSK-NLWKN) for furnishing bathymetrical data; the Federal Ministry for Environment, Nature Conservation and Nuclear Safety (BMU) and Project Management Jülich (PTJ) for providing measured data of waves and wind related to the research platform FINO1; the German Weather Service (DWD) for making meteorological model data available via the PAMORE database; and Senckenberg at Sea Wilhelmshaven (SaM) for furnishing in situ data of surface sediment grain size. We acknowledge the Helmholtz-Zentrum Geesthacht Zentrum für Material- und Küstenforschung GmbH, notably Rolf Riethmüller and Götz Flöser, for providing wave data recorded at the back barrier of Langeoog within the framework of the Coastal Observing System for Northern and Arctic Seas (COSYNA) and for text passages in Sect. 4.1 describing their data. Many thanks go to OpenEarth.nl (DELTARES), which is at the European forefront in disseminating high-quality data, tools and models, particularly for making the Dutch Continental Shelf Model available (Verboom et al., 1992).

The authors would like to thank Mick Van der Wegen, the anonymous reviewer and the editor, Giovanni Coco, for their valuable comments and suggestions, which helped to improve the quality of this paper.

Edited by: G. Coco

# References

Antia, E. E.: Sedimentary Deposits Related to Inlet-Shoreface Storm Flow Interaction in the German Bight, Estuar. Coast. Shelf Sci., 40, 699–712, doi:10.1006/ecss.1995.0047, 1995.

Ardhuin, F. and Roland, A.: The development of spectral wave models: coastal and coupled aspects, in: Proceedings of the 7th International Conference on Coastal Dynamics, 24–28 June 2013, University of Bordeaux, France, 25–38, 2013.

Ardhuin, F., Rascle, N. and Belibassakis, K. A.: Explicit wave-averaged primitive equations using a generalized Lagrangian mean, Ocean Model., 20, 35–60, doi:10.1016/j.ocemod.2007.07.001, 2008.

Badewien, T. H., Zimmer, E., Bartholomä, A., and Reuter, R.: Towards continuous long-term measurements of suspended particulate matter (SPM) in turbid coastal waters, Ocean Dynam., 59, 227–238, doi:10.1007/s10236-009-0183-8, 2009.

Bartholomä, A., Kubicki, A., Badewien, T. H., and Flemming, B. W.: Suspended sediment transport in the German Wadden Sea-seasonal variations and extreme events, Ocean Dynam., 59, 213–225, doi:10.1007/s10236-009-0193-6, 2009.

Bennis, A.-C., Ardhuin, F., and Dumas, F.: On the coupling of wave and three-dimensional circulation models: Choice of theoretical framework, practical implementation and adiabatic tests, Ocean Model., 40, 260–272, doi:10.1016/j.ocemod.2011.09.003, 2011.

Bertin, X., Fortunato, A. B., and Oliveira, A.: A modeling-based analysis of processes driving wave-dominated inlets, Cont. Shelf Res., 29, 819–834, doi:10.1016/j.csr.2008.12.019, 2009.

Booij, N., Ris, R. C., and Holthuijsen, L. H.: A third-generation wave model for coastal regions 1. Model description and validation, J. Geophys. Res., 104, 7649–7666, doi:10.1029/98JC02622, 1999.

Cheung, K. F., Gerritsen, F., and Cleveringa, J.: Morphodynamics and Sand Bypassing at Ameland Inlet, The Netherlands, J. Coast. Res., 231, 106–118, doi:10.2112/04-0403.1, 2007.

Cuneo, P. S. and Flemming, B. W.: Quantifying concentration and flux of suspended particulate matter through a tidal inlet of the East Frisian Wadden sea by acoustic doppler current profiling, in: Proceedings in Marine Science, Muddy Coast Dynamics and Resource Management, Elsevier, 2, 39–52, doi:10.1016/S1568-2692(00)80005-4, 2000.

Davis, R. Jr. and FitzGerald, D. (Eds.): Beaches and Coasts, 4th Edn., John Wiley & Sons, Oxford, UK, 2009.

Deltares: User manual Delft-3D FLOW, online available from: www.deltares.nl (last access: December 2011), 2011.

De Swart, H. E. and Zimmerman, J. T. F.: Morphodynamics of Tidal Inlet Systems, Annu. Rev. Fluid Mech., 41, 203–229, doi:10.1146/annurev.fluid.010908.165159, 2009.

Dissanayake, D. M. P. K., Roelvink, J. A., and Van der Wegen, M.: Modelled channel patterns in a schematized tidal inlet, Coast. Eng., 56, 1069–1083, doi:10.1016/j.coastaleng.2009.08.008, 2009.

Dodet, G., Bertin, X., Bruneau, N., Fortunato, A. B., Nahon, A., and Roland, A.: Wave-current interactions in a wave-dominated tidal inlet, J. Geophys. Res.-Oceans, 118, 1587–1605, doi:10.1002/jgrc.20146, 2013.

Elias, E. P. L., Cleveringa, J., Buijsman, M. C., Roelvink, J. A., and Stive, M. J. F.: Field and model data analysis of sand transport patterns in Texel Tidal inlet (The Netherlands), Coast. Eng., 53, 505–529, doi:10.1016/j.coastaleng.2005.11.006, 2006.

Elias, E. P. L. and Hansen, J. E.: Understanding processes controlling sediment transports at the mouth of a highly energetic inlet system (San Francisco Bay, CA), Mar. Geol., 345, 207–220, doi:10.1016/j.margeo.2012.07.003, 2013.

Elias, E. P. L. and Van der Spek, A. J. F.: Long-term morphodynamic evolution of Texel Inlet and its ebb-tidal delta (The Netherlands), Mar. Geol., 225, 5–21, doi:10.1016/j.margeo.2005.09.008, 2006.

Elias, E. P. L., Gelfenbaum, G., and Van der Westhuysen, A. J.: Validation of a coupled wave-flow model in a high-energy setting: The mouth of the Columbia River, J. Geophys. Res.-Oceans, 117, C09011, doi:10.1029/2012JC008105, 2012.

Finley, R. J.: Ebb-tidal delta morphology and sediment supply in relation to seasonal wave energy flux, North Inlet, South Carolina, J. Sediment. Res., 48, 227–238, doi:10.1306/212F743C-2B24-11D7-8648000102C1865D, 1978.

FitzGerald, D. M.: Sediment bypassing at mixed energy tidal inlets, in: Proceedings of the 18th International Conference on Coastal Engineering, 14–19 November 1982, Cape Town, South Africa, 1094–1118, 1982.

FitzGerald, D. M., Nummedal, D., and Kana, T. W.: Sand circulation pattern at Price Inlet, South Carolina, in: 15th Conference on Coastal Engineering, Honolulu, Hawaii, 1976.

FitzGerald, D. M., Penland, S., and Nummedal, D. A. G.: Control of barrier island shape by inlet sediment bypassing: East Frisian Islands, West Germany, Develop. Sediment., 39, 355–376, doi:10.1016/S0070-4571(08)70154-7, 1984.

FitzGerald, D. M., Buynevich, I., and Hein, C.: Morphodynamics and Facies Architecture of Tidal Inlets and Tidal Deltas, in: Principles of Tidal Sedimentology SE 12, edited by: Davis Jr., R. A. and Dalrymple, R. W., Springer, Netherlands, 301–333, 2012.

Fredsøe, J.: Turbulent boundary layer in wave-current motion, J. Hydraul. Eng., 110, 1103–1120, 1984.

Gibbs, R. J., Matthews, M. D., and Link, D. A.: The relationship between sphere size and settling velocity, J. Sediment. Res., 41, 7–18, doi:10.1306/74D721D0-2B21-11D7-8648000102C1865D, 1971.

Hageman, B. P.: Development of the western part of the Netherlands during the Holocene, Geol. en Mijnb., 48, 373–388, 1969.

Hanisch, J.: Sand transport in the tidal inlet between Wangerooge and Spiekeroog (W. Germany), Holocene Mar. Sediment. North Sea Basin Spec. Publ. 5, IAS, 35, 175–185, 1981.

Hayes, M. O.: Morphology of sand accumulation in estuaries: an introduction to the symposium, edited by: Cronin, L. E., Estuar. Res., 2, 3–22, 1975.

Hayes, M. O.: Barrier island morphology as a function of tidal and wave regime, in: Barrier islands, from the Gulf of St. Lawrence to the Gulf of Mexico, edited by: Leatherman, S., Academic Press, New York, 1–27, 1979.

Hench, J. L. and Luettich, R. A.: Transient Tidal Circulation and Momentum Balances at a Shallow Inlet, J. Phys. Oceanogr., 33, 913–932, 2003.

Hubbard, D. K., Oertel, G., and Nummedal, D.: The role of waves and tidal currents in the development of tidal-inlet sedimentary structures and sand body geometry; examples from North Carolina, South Carolina, and Georgia, J. Sediment. Res., 49, 1073–1091, doi:10.1306/212F78B5-2B24-11D7-8648000102C1865D, 1979.

Isobe, M. and Horikawa, K.: Study on water particle velocities of shoaling and breaking waves, Coastal Engineering in Japan, 25, 109–123, 1982.

Kaiser, R., Niemeyer, H. D., Dirks, H., and Witting, M.: KFKI-Projekt DÜNEROS, Schlussbericht 03KIS063, Norderney, 2008.

Komar, P. D.: Tidal-Inlet Processes and Morphology Related to the Transport of Sediments, J. Coast. Res., 23, 23–45, 1996.

Kwoll, E. and Winter, C.: Determination of the initial grain size distribution in a tidal inlet by means of numerical modelling, J. Coast. Res., 64, 1081–1085, 2011.

Lane, A., Riethmuller, R., Herbers, D., Rybaczok, P., Gunther, H., Baumert, H., and Riethmüller, R.: Observational data sets for model development, Coast. Eng., 41, 125–153, doi:10.1016/S0378-3839(00)00029-6, 2000.

Lesser, G. R., Roelvink, J. A., van Kester, J. A. T. M., and Stelling, G. S.: Development and validation of a three-dimensional morphological model, Coast. Eng., 51, 883–915, doi:10.1016/j.coastaleng.2004.07.014, 2004.

Longuet-Higgins, M. S. and Stewart, R. W.: Radiation stresses in water waves; a physical discussion, with applications, Deep-Sea Res., 11, 529–562, 1964.

Nahon, A., Bertin, X., Fortunato, A. B., and Oliveira, A.: Process-based 2DH morphodynamic modeling of tidal inlets: A comparison with empirical classifications and theories, Mar. Geol., 291, 1–11, doi:10.1016/j.margeo.2011.10.001, 2012.

Niedoroda, A. W., Swift, D. J. P., Hopkins, T. S., and Ma, C.-M.: Shoreface morphodynamics on wave-dominated coasts, Mar. Geol., 60, 331–354, doi:10.1016/0025-3227(84)90156-7, 1984.

Noormets, R., Ernstsen, V. B., Bartholomä, A., Flemming, B. W., and Hebbeln, D.: Implications of bedform dimensions for the prediction of local scour in tidal inlets: a case study from the southern North Sea, Geo-Mar. Lett., 26, 165–176, doi:10.1007/s00367-006-0029-z, 2006.

Onken, R., Callies, U., Vaessen, B., and Riethmüller, R.: Indirect determination of the heat budget of tidal flats, Cont. Shelf Res., 27, 1656–1676, doi:10.1016/j.csr.2007.01.029, 2007.

Outzen, O., Herklotz, K., Heinrich, H., and Lefebvre, C: Extreme waves at FINO 1 research platform caused by storm 'Tilo' on 9 November 2007, DEWI Mag., 33, 17–23, 2008.

Ris, R. C., Holthuijsen, L. H., and Booij, N.: A third-generation wave model for coastal regions 2. Verification, J. Geophys. Res., 104, 7667–7681, doi:10.1029/1998JC900123, 1999.

Rodi, W.: Turbulence models and their application in hydraulics, a state of the art review, International Association of Hydraulics Research, Delft, the Netherlands, 1984.

Roelvink, J. A.: Coastal morphodynamic evolution techniques, Coast. Eng., 53, 277–287, doi:10.1016/j.coastaleng.2005.10.015, 2006.

Schneggenburger, C., Günther, H., and Rosenthal, W.: Spectral wave modelling with non-linear dissipation: Validation and applications in a coastal tidal environment, Coast. Eng., 41, 201–235, doi:10.1016/S0378-3839(00)00033-8, 2000.

Sha, L. P.: Variation in ebb-delta morphologies along the West and East Frisian Islands, The Netherlands and Germany, Mar. Geol., 89, 11–28, doi:10.1016/0025-3227(89)90025-X, 1989.

Sha, L. P.: Surface sediments and sequence models in the ebb-tidal delta of Texel Inlet, Wadden Sea, The Netherlands, Sediment. Geol., 68, 125–141, doi:10.1016/0037-0738(90)90123-B, 1990.

Sha, L. P. and Van Den Berg, J. H.: Variation in Ebb-Tidal Delta Geometry along the Coast of the Netherlands and the German Bight, J. Coast. Res., 9, 730–746, 1993.

Smith, J. B. and FitzGerald, D. M.: Sediment Transport Patterns at the Essex River Inlet Ebb-Tidal Delta, Massachusetts, USA, J. Coast. Res., 10, 752–774, 1994.

Son, C. S., Flemming, B. W., and Bartholomä, A.: Evidence for sediment recirculation on an ebb-tidal delta of the East Frisian barrier-island system, southern North Sea, Geo-Mar. Lett., 31, 87–100, doi:10.1007/s00367-010-0217-8, 2010.

Soulsby, R. L., Hamm, L., Klopman, G., Myrhaug, D., Simons, R. R., and Thomas, G. P.: Wave-current interaction within and outside the bottom boundary layer, Coast. Eng., 21, 41–69, 1993.

Van der Wegen, M. and Roelvink, J. A.: Reproduction of estuarine bathymetry by means of a process-based model: Western Scheldt case study, the Netherlands, Geomorphology, 179, 152–167, doi:10.1016/j.geomorph.2012.08.007, 2012.

Van der Wegen, M., Dastgheib, A., Jaffe, B. E., and Roelvink, D.: Bed composition generation for morphodynamic modeling: case study of San Pablo Bay in California, USA, Ocean Dynam., 61, 173–186, doi:10.1007/s10236-010-0314-2, 2011a.

Van der Wegen, M., Jaffe, B. E., and Roelvink, J. A.: Process-based, morphodynamic hindcast of decadal deposition patterns in San

Pablo Bay, California, 1856–1887, J. Geophys. Res.-Earth, 116, F02008, doi:10.1029/2009JF001614, 2011b.

Van der Westhuysen, A. J.: Advances in the spectral modelling of wind waves in the nearshore, Delft University of Technology, Delft, the Netherlands, 207 pp., 2007.

Van der Westhuysen, A. J.: Spectral modeling of wave dissipation on negative current gradients, Coast. Eng., 68, 17–30, doi:10.1016/j.coastaleng.2012.05.001, 2012.

Van Lancker, V., Lanckneus, J., Hearn, S., Hoekstra, P., Levoy, F., Miles, J., Moerkerke, G., Monfort, O., and Whitehouse, R.: Coastal and nearshore morphology, bedforms and sediment transport pathways at Teignmouth (UK), Cont. Shelf Res., 24, 1171–1202, doi:10.1016/j.csr.2004.03.003, 2004.

Van Rijn, L. C.: Principles of sediment transport in rivers, estuaries and coastal seas, Aqua publications, Amsterdam, the Netherlands, 1993.

Van Rijn, L. C.: General view on sand transport by currents and waves: data analysis and engineering modelling for uniform and graded sand, Report Z28, Deltares (WL), Delft, the Netherlands, 2000.

Van Rijn, L. C.: Sand transport by currents and waves; general approximation formulae, in: Proceedings of the International Conference on Coastal Sediments, 18–23 May 2003, Clearwater Beach, Florida, USA, vol. 3, 2003.

Van Rijn, L. C., Walstra, D. J. R., and Van Ormondt, M.: Description of TRANSPOR2004 and implementation in Delft3D-ONLINE, Interim Rep. Prep. DG Rijkswaterstaat, Rijksinst, voor Kust en Zee, Delft Hydraul. Institute, Delft, the Netherlands, 2004.

Verboom, G. K., de Ronde, J. G., and van Dijk, R. P.: A fine grid tidal flow and storm surge model of the North Sea, Cont. Shelf Res., 12, 213–233, doi:10.1016/0278-4343(92)90030-N, 1992.

Walstra, D. J. R., Roelvink, J. A., and Groeneweg, J.: Calculation of Wave-Driven Currents in a 3D Mean Flow Model, in: Coastal Engineering, American Society of Civil Engineers, Reston, VA, 1050–1063, 2000.

# Permissions

All chapters in this book were first published in ESD, by Copernicus Publications; hereby published with permission under the Creative Commons Attribution License or equivalent. Every chapter published in this book has been scrutinized by our experts. Their significance has been extensively debated. The topics covered herein carry significant findings which will fuel the growth of the discipline. They may even be implemented as practical applications or may be referred to as a beginning point for another development.

The contributors of this book come from diverse backgrounds, making this book a truly international effort. This book will bring forth new frontiers with its revolutionizing research information and detailed analysis of the nascent developments around the world.

We would like to thank all the contributing authors for lending their expertise to make the book truly unique. They have played a crucial role in the development of this book. Without their invaluable contributions this book wouldn't have been possible. They have made vital efforts to compile up to date information on the varied aspects of this subject to make this book a valuable addition to the collection of many professionals and students.

This book was conceptualized with the vision of imparting up-to-date information and advanced data in this field. To ensure the same, a matchless editorial board was set up. Every individual on the board went through rigorous rounds of assessment to prove their worth. After which they invested a large part of their time researching and compiling the most relevant data for our readers.

The editorial board has been involved in producing this book since its inception. They have spent rigorous hours researching and exploring the diverse topics which have resulted in the successful publishing of this book. They have passed on their knowledge of decades through this book. To expedite this challenging task, the publisher supported the team at every step. A small team of assistant editors was also appointed to further simplify the editing procedure and attain best results for the readers.

Apart from the editorial board, the designing team has also invested a significant amount of their time in understanding the subject and creating the most relevant covers. They scrutinized every image to scout for the most suitable representation of the subject and create an appropriate cover for the book.

The publishing team has been an ardent support to the editorial, designing and production team. Their endless efforts to recruit the best for this project, has resulted in the accomplishment of this book. They are a veteran in the field of academics and their pool of knowledge is as vast as their experience in printing. Their expertise and guidance has proved useful at every step. Their uncompromising quality standards have made this book an exceptional effort. Their encouragement from time to time has been an inspiration for everyone.

The publisher and the editorial board hope that this book will prove to be a valuable piece of knowledge for researchers, students, practitioners and scholars across the globe.

# List of Contributors

**J. M. Turowski**
Swiss Federal Research Institute WSL, Zürcherstrasse 111, 8903 Birmensdorf, Switzerland
Helmholtz Centre Potsdam, GFZ German Research Centre for Geosciences, Telegrafenberg, 14473 Potsdam, Germany

**A. Badoux**
Swiss Federal Research Institute WSL, Zürcherstrasse 111, 8903 Birmensdorf, Switzerland

**K. Bunte**
Engineering Research Center, Colorado State University, Fort Collins, CO 80523, USA

**C. Rickli**
Swiss Federal Research Institute WSL, Zürcherstrasse 111, 8903 Birmensdorf, Switzerland

**N. Federspiel**
Swiss Federal Research Institute WSL, Zürcherstrasse 111, 8903 Birmensdorf, Switzerland
CSD Engineers SA, Hessstrasse 27d, 3097 Liebefeld (Berne), Switzerland

**M. Jochner**
Swiss Federal Research Institute WSL, Zürcherstrasse 111, 8903 Birmensdorf, Switzerland
Institute of Geography of the University of Berne (GIUB), Hallerstrasse 12, 3012 Berne, Switzerland

**J. Braun**
ISTerre, Université Grenoble Alpes and CNRS BP 53, 38041 Grenoble CEDEX 9, France

**C. Voisin**
ISTerre, Université Grenoble Alpes and CNRS BP 53, 38041 Grenoble CEDEX 9, France

**A. T. Gourlan**
ISTerre, Université Grenoble Alpes and CNRS BP 53, 38041 Grenoble CEDEX 9, France

**C. Chauvel**
ISTerre, Université Grenoble Alpes and CNRS BP 53, 38041 Grenoble CEDEX 9, France

**T. J. Coulthard**
Department of Geography, Environment and Earth Sciences, University of Hull, UK

**M. J. Van de Wiel**
Department of Geography, University of Western Ontario, London, Ontario, Canada

**T. Hoffmann**
Department of Geography, University of Bonn, Meckenheimer Allee 166, 53115 Bonn Germany

**S. M. Mudd**
School of Geosciences, University of Edinburgh, Drummond Street, Edinburgh EH8 9XP, UK

**K. van Oost**
KU Leuven – University of Leuven, Department of Earth and Environmental Sciences, Celestijnenlaan 200e, 3001 Leuven, Belgium

**G. Verstraeten**
Department of Earth and Environmental Sciences, University of Leuven, Celestijnenlaan 200e, 3001 Heverlee, Belgium

**G. Erkens**
Department of Physical Geography, University of Utrecht, Heidelberglaan 2, 3584 CS Utrecht, the Netherlands

**A. Lang**
School of Environmental Sciences, University of Liverpool, Liverpool L69 3GP, UK

**H. Middelkoop**
Department of Physical Geography, University of Utrecht, Heidelberglaan 2, 3584 CS Utrecht, the Netherlands

**J. Boyle**
School of Environmental Sciences, University of Liverpool, Liverpool L69 3GP, UK

**J. O. Kaplan**
Institute of Environmental Engineering, Ecole Polytechnique Fédérale de Lausanne, Station 2, 1015 Lausanne, Switzerland

**J. Willenbring**
University of Pennsylvania, Department of Earth and Environmental Science, 240 S. 33rd Street, Philadelphia, PA 19104-6313, USA

**R. Aalto**
College of Life and Environmental Sciences, University of Exeter, Rennes Drive, Exeter EX4 4RJ, UK

**E. D. Lazarus**
Environmental Dynamics Laboratory, Earth Surface Processes Research Group, School of Earth & Ocean Sciences, Cardiff University, Main Building, Park Place, Cardiff CF10 3AT, UK

**J. D. Pelletier**
Department of Geosciences, University of Arizona, Gould-Simpson Building, 1040 East Fourth Street, Tucson, Arizona 85721-0077, USA

**K. C. Rose**
Bristol Glaciology Centre, School of Geographical Sciences, University of Bristol, Bristol BS8 1SS, UK

**N. Ross**
School of Geography, Politics&Sociology, Newcastle University, Newcastle upon Tyne NE1 7RU, UK

**T. A. Jordan**
British Antarctic Survey, High Cross, Madingley Road, Cambridge CB3 0ET, UK

**R. G. Bingham**
School of GeoSciences, University of Edinburgh, Edinburgh EH8 9XP, UK

**H. F. J. Corr**
British Antarctic Survey, High Cross, Madingley Road, Cambridge CB3 0ET, UK

**F. Ferraccioli**
British Antarctic Survey, High Cross, Madingley Road, Cambridge CB3 0ET, UK

**A. M. Le Brocq**
School of Geography, University of Exeter, Exeter EX4 4RJ, UK

**D. M. Rippin**
Environment Department, University of York, York YO10 5DD, UK

**M. J. Siegert**
Grantham Institute and Department of Earth Science and Engineering, Imperial College London, London SW7 2AZ, UK

**S.-J. Kao**
Research Center for Environmental Changes, Academia Sinica, Taipei, Taiwan
State Key Laboratory of Marine Environmental Science, Xiamen University, Xiamen, China

**R. G. Hilton**
Department of Geography, Durham University, Durham, UK

**K. Selvaraj**
Research Center for Environmental Changes, Academia Sinica, Taipei, Taiwan
State Key Laboratory of Marine Environmental Science, Xiamen University, Xiamen, China

**M. Dai**
State Key Laboratory of Marine Environmental Science, Xiamen University, Xiamen, China

**F. Zehetner**
Institute of Soil Research, University of Natural Resources and Life Sciences, Vienna, Austria

**J.-C. Huang**
Department of Geography, National Taiwan University, Taipei, Taiwan

**S.-C. Hsu**
Research Center for Environmental Changes, Academia Sinica, Taipei, Taiwan

**R. Sparkes**
Department of Earth Sciences, University of Cambridge, Cambridge, UK

**J.T. Liu**
Institute of Marine Geology and Chemistry, National Sun Yat-sen University, Kaohsiung, Taiwan

**T.-Y. Lee**
Research Center for Environmental Changes, Academia Sinica, Taipei, Taiwan

**J.-Y. T. Yang**
State Key Laboratory of Marine Environmental Science, Xiamen University, Xiamen, China

**A. Galy**
Department of Earth Sciences, University of Cambridge, Cambridge, UK

**X. Xu**
School of Physical Sciences, University of California, Irvine, CA, USA

**N. Hovius**
Geomorphology, GFZ German Research Centre, Telegrafenberg, Potsdam, Germany

**P. D. Thorne**
National Oceanography Centre, Joseph Proudman building, 6 Brownlow Street, Liverpool L3 5DA, UK

**A. Barkwith**
British Geological Survey, Keyworth, Nottingham, NG12 5GG, UK

**C. W. Thomas**
British Geological Survey, Keyworth, Nottingham, NG12 5GG, UK

**P. W. Limber**
Nicholas School of the Environment, Duke University, Durham, NC, USA
Dept. of Geological Sciences, University of Florida, Gainesville, FL, USA

**M. A. Ellis**
Nicholas School of the Environment, Duke University, Durham, NC, USA

**A. B. Murray**
Nicholas School of the Environment, Duke University, Durham, NC, USA

**A. Barkwith**
British Geological Survey, Keyworth, Nottingham, UK

**M. D. Hurst**
British Geological Survey, Keyworth, Nottingham, UK

**C. W. Thomas**
British Geological Survey, Keyworth, Nottingham, UK

**M. A. Ellis**
British Geological Survey, Keyworth, Nottingham, UK

**P. L. Limber**
Department of Geological Sciences, University of Florida, Gainesville, FL, USA

**A. B. Murray**
Nicholas School of the Environment, Duke University, Durham, NC, USA

**G. Herrling**
MARUM – Center for Marine Environmental Sciences, University of Bremen, Bremen, Germany

**C. Winter**
MARUM – Center for Marine Environmental Sciences, University of Bremen, Bremen, Germany